Defining Media Studies

Reflections on
the Future of the Field

Edited by
Mark R. Levy
and Michael Gurevitch

Oxford University Press
New York, Oxford

P9-AGF-954

Oxford University Press

Oxford, New York, Toronto, Delhi, Bombay,
Calcutta, Madras, Karachi, Kuala Lumpur,
Singapore, Hong Kong, Tokyo, Nairobi,
Dar es Salaam, Cape Town, Melbourne,
Auckland, Madrid
and associated companies in Berlin and Ibadan

First published as Volume 43, Numbers 3 and
4, of *Journal of Communication*, by Oxford
University Press, 200 Madison Avenue, New
York, NY 10016.

Oxford is a registered trademark of Oxford
University Press.

Library of Congress Cataloging-in-
Publication Data
Defining media studies: reflections on the
    future of the field.
    edited by Mark R. Levy and Michael
    Gurevitch.
        p. cm.
    Includes bibliographical references (p.).
    ISBN 0-19-508787-9 (alk. paper)
    ISBN 0-19-508788-7 (pbk. alk. paper)
    1. Communication—Research.
    I. Levy, Mark R. II. Gurevitch, Michael.
P91.3.D44   1994
302.2'072—dc20   93-6066
                                          CIP

Contents

Defining Media Studies

Reflections on
the Future of the Field

7   **Preface**

14   **The Disciplinary Status of Communication Research**

14   *From Field to Frog Ponds* by Karl Erik Rosengren

26   *Communication—Embrace the Subject, not the Field* by James R. Beniger

34   *Why Are There So* Many *Communication Theories?* by Robert T. Craig

42   *The Past of Communication's Hoped-For Future* by Klaus Krippendorff

53   *Verbing Communication: Mandate for Disciplinary Invention* by Brenda Dervin

63   *Images of Media: Hidden Ferment—and Harmony—in the Field* by Joshua Meyrowitz

75   *The Consequences of Vocabularies* by Joli Jensen

83   *Against Theory* by Barbara O'Keefe

91   *Building a Discipline of Communication* by Gregory J. Shepherd

100   *Perspectives on Communication* by Kurt Lang and Gladys Engel Lang

108   *The Legitimacy Gap: A Problem of Mass Media Research in Europe and the United States* by Paolo Mancini

118   New Directions, New Agendas

118   *The Advent of Multiple-Process Theories of Communication* by Austin S. Babrow

127   *Communication and the New World of Relationships* by Mary Anne Fitzpatrick

135   *Target Practice: A Batesonian "Field" Guide for Communication Studies* by Horace Newcomb

141   *Harmonization of Systems: The Third Stage of the Information Society* by Sandra Braman

149   *Beyond the Culture Wars: An Agenda for Research on Communication and Culture* by Dennis K. Davis and James Jasinski

158   *The Hierarchy of Institutional Values in the Communication Discipline* by Jennifer L. Monahan and Lori Collins-Jarvis

166   *Argument for a Durkheimian Theory of the Communicative* by Eric W. Rothenbuhler

172   *Implications of Public Relations for Other Domains of Communication* by James E. Grunig

182   Connecting Communication Scholarship to Public Policy

182   *Making a Difference in the Real World* by Robert K. Avery and William F. Eadie

188   *A Policy Research Paradigm for the News Media and Democracy* by W. Lance Bennett

198   *The Centrality of Media Economics* by Douglas Gomery

207   *Reconnecting Communications Studies With Communications Policy* by Eli Noam

215   *The Traditions of Communication Research and Their Implications for Telecommunications Study* by Willard D. Rowland, Jr.

226   *Creating Imagined Communities: Development Communication and the Challenge of Feminism* by H. Leslie Steeves

238   *Scholarship as Silence* by David Docherty, David Morrison, and Michael Tracey

247   **Audiences and Institutions**

247   *The Rise and Fall of Audience Research: An Old Story With a New Ending* by Sonia M. Livingstone

255   *Active Audience Theory: Pendulums and Pitfalls* by David Morley

262   *The Past in the Future: Problems and Potentials of Historical Reception Studies* by Klaus Bruhn Jensen

271   *Reopening the Black Box: Toward a Limited Effects Theory* by Herbert J. Gans

278   *Realism and Romance: The Study of Media Effects* by Gaye Tuchman

284   *Revealing the Black Box: Information Processing and Media Effects* by Seth Geiger and John Newhagen

293   *Framing: Toward Clarification of a Fractured Paradigm* by Robert M. Entman

301   *Communication Research in the Design of Communication Interfaces and Systems* by Frank Biocca

311   *The Future of Political Communication Research: A Japanese Perspective* by Ito Youichi

322   *Has Communication Explained Journalism?* by Barbie Zelizer

331   **Rethinking the Critical Tradition**

331   *Can Cultural Studies Find True Happiness in Communication?* by Lawrence Grossberg

340   *Critical Communication Research at the Crossroads* by Robert W. McChesney

347   *Rethinking Political Economy: Change and Continuity* by Eileen R. Meehan, Vincent Mosco, and Janet Wasko

359   *Back to the Future: Prospects for Study of Communication as a Social Force* by Dan Schiller

367   The Search for a Usable History

367   *The Past and the Future of Communication Study: Convergence or Divergence?* An exchange by Everett M. Rogers and Steven H. Chaffee

374   *Genealogical Notes on "The Field"* by John Durham Peters

382   *History, Philosophy, and Public Opinion Research* by Susan Herbst

388   The Academic Wars

388   *Communication in Crisis: Theory, Curricula, and Power* by Pamela J. Shoemaker

396   *The Curriculum Is the Future* by Lana F. Rakow

405   *Fragmentation, the Field, and the Future* by David L. Swanson

415   *The Purebred and the Platypus: Disciplinarity and Site in Mass Communication Research* by Anandam P. Kavoori and Michael Gurevitch

424   *Communication Research: New Challenges of the Latin American School* by José Marques de Melo

433   Index

Preface

This book is devoted to a collective reconnaissance of communication scholarship and its future. It represents an attempt to define the field of media studies, not by presenting one unified voice, but rather by pooling the insights and concerns of fifty-nine communication scholars, representing a wide range of intellectual perspectives. By media studies, we mean scholarly work that is organized around an attempt to understand technologically mediated communication, its codes, processes, and consequences for the human experience.

We do not lay claim to the study of all human communication; thriving and important areas such as interpersonal communication, organizational communication, and rhetoric are largely absent from this volume. This is not by design, but rather the outcome of the crop of essays we received in response to our "Call for Papers." For better or worse, then, the essays collected here focus largely—but not exclusively—on what used to be called "mass communication" and the "mass media." We think, however, that those concepts have begun to outlive their usefulness. The landscape of the "old" mass media (newspapers, magazines, radio, television, and film) is changing. Challenged by the new communication technologies of cable, home video, computer networks, and the like, the "old" mass media and the "new" are converging, their hardware and software increasingly interlinked. The new forms of television, the Net of computers, and even virtual reality (just to name the most obvious) are rightly thought of as communication media. As media studies scholars, we have the opportunity and the obligation to study them. Hence the title of this book.

Since assuming the editorship of the *Journal of Communication* in 1991 we have become increasingly aware of the need to assess the state of media studies. Over the last decade, there have been significant paradigmatic shifts in the various "mother disciplines" from which communication scholarship arises. Intellectual sea-changes in both the social sciences and the humanities are now spreading to and re-forming media studies. The paradigmatic debate (or "dialogue") that dominated communication scholarship in the '70s and the early '80s has been replaced by new and different intellectual nudgings, by the injection into communication scholarship of recently emergent perspectives such as feminism, postmodernism, and neofunctionalism.

At the same time, the structure of media systems around the world has been dramatically remolded, either by politics or technology. As a result, new economic and regulatory arrangements are emerging. These changes in communication structures, technologies, and policy represent important new areas for research. Communication scholarship would have been sorely negligent if it did not respond to this challenge, and to a greater or lesser degree the research agenda of our field has done so.

The symbolic peg for the collective "navel gazing" that constitutes this book was the 10th anniversary of the publication of a special issue of the *Journal,* titled "Ferment in the Field." Whatever actual impact "Ferment" had on the field, it has become a touchstone to which scholars turn when they seek to familiarize or to reacquaint themselves with the state of communication research in 1983. Ten years later it was time to repeat that exercise.

All the essays in this book first appeared as the Summer and Autumn 1993 issues of the *Journal* under the shared title "The Future of the Field: Between Fragmentation and Cohesion." Because we wanted to cast the widest possible net for those special issues, we issued a "Call for Papers" that would throw the symposium's door wide open to all. In that call we attempted to foreground the intellectual and institutional changes mentioned above by posing five deliberately provocative propositions, asking potential authors to center their essays on one or more of the following:

(1) Past controversies in communication scholarship have been largely resolved, and no new controversies of that theoretical order have emerged. The "itch" to discover a universal paradigm of communication has been replaced by a comfortable acceptance of theoretical pluralism.

(2) Communication scholarship is unwilling and unable to influence either the practice of journalism and communication or the formulation of communication policy. In the future, it should focus more on socially relevant research.

(3) Communication scholarship lacks disciplinary status because it has no core of knowledge and thus institutional and scholarly legitimacy remains a chimera for the field.

(4) The political Cold War has ended, but ideological and methodological battles, such as those between psychological, cultural, economic, textual, and technological determinists, continue to fragment our field.

(5) The question of media "effects" remains the perennial black box of communication research and still poses the most unanswered questions.

In typical academic fashion, some scholars took up our challenge and structured their essays around one or more of these propositions. Others ignored them altogether, striking out instead in self-propelled but equally interesting directions. The voice of communication science can be heard in these essays, alongside that of cultural studies. It is, we believe, a feisty mix of argument, apologia, and advice. We hope that this reflects the diversity of voices in the field, although we have not tried for a strict statistical representation of all the ICA constituencies.

The Strength of the Field

Because much has changed since "Ferment," both in the landscape of communication scholarship and in the "communication world" outside, the field of media studies finds itself lacking neither intellectual nor disciplinary problems. But before we turn to some of these "problems" as they are identified in this collection, it would be appropriate to itemize some of the ways in which the health of the field manifests itself:

• *Defining Media Studies* spotlights the creativity and originality of communication scholars. Our contributors have provided a wide array of fresh and provocative ideas. Readers have a rich menu from which to select, and they will undoubtedly identify their own favorites.

• There is ample evidence of a search for broader and deeper intellectual foundations for the field. This can be seen in those essays that invoke and consider the ideas of classical philosophers, political theorists, sociologists, and the historical roots of media studies generally.

• There is growing awareness that we must relate our research to issues of communication performance. The application of humanistic and democratic norms to assessments of communication performance, from the interpersonal to the institutional, might well become a focus of future scholarly activity, as more and more scholars find it rewarding to engage with the communication needs of society.

• Attention to curricular reform in media studies is evident in a number of contributions. This is especially welcome, since it suggests that while our field appears to be "hot"—more students, more programs (some highly visible cuts notwithstanding), more journals than ever before—we have not been lulled into pedagogic complacency.

Not complacent, we live nevertheless in an age of diminished expectations and of diminished resources. That narrowing of dreams and opportunities is a subtext that runs through many of the essays in this volume. The contributors address a wide range of issues and concerns and to avoid redundancy we will refrain from rehashing them here. But we want to briefly note and comment on some of them.

The Changing of the Paradigmatic Guard

First, as we suggested above, the paradigm cleavage that once undergirded much controversy in the field no longer seems to anchor and structure the debate. Instead, we find a diversity of perspectives. We find both centripetal and centrifugal tendencies: paradigms at war with each other; paradigms whose epistemologies intersect, and the occasional search for an overarching, comprehensive conceptual framework for communication research. If we were forced to classify the essays in this book we would divide them by their academic orientation rather than by their paradigmatic bases. Thus, communication scholarship as it is represented here consists of roughly a two-and-a-half-modal distribution: one part hard-edged behavioral and social science; one part interpretative humanistic study; and a third, smaller dash of communication policy studies.

It is not that past controversies "have been largely resolved" (as we

playfully suggested in the "Call for Papers"). Rather, some cross-camp influence has taken place. A generally critical spirit seems to be diffused
now well beyond the ranks of those who once defined themselves as
critical scholars, and members of the self-styled critical school now pursue questions and deploy methods of which they were harshly critical in
the past. The erosion of the former intellectual divide may have also been
hastened by the loss of confidence in the underlying theories of media
and society, brought about by events outside the field: the collapse of
Marxist-based regimes, polities, and parties triggered theoretical self-
scrutiny on the Left, while the growing impoverishment of public communication processes in many western democracies ("led" by the United
States) has eroded liberal pluralism's confidence in the capacity of mar-
ket-driven communication to nourish public life and promote "good citi-
zenship."

Still, there is at least one fortunate result of this double-barreled crisis
of spirit: a welcome reduction in the polemical heat that permeated the
field a decade or two ago. In fact, hardly any of the essays in this collection take the polemical stance of arguing the superiority of one worldview over another. Of course, the basic approaches and their associated
issues have not entirely faded away, but the tone of intellectual arro-
gance—a feature of past debates—is by and large absent here.

At the same time, the proliferation of new approaches purporting to assume the mantle of "paradigms" raises a number of important issues.
First, we might indeed ask "what makes a paradigm a paradigm?" The
"old" paradigms possessed both theoretical and normative universality.
Their theoretical bases were broader than the field of media studies to
which we have applied them. They carried with them a theory of society
and they thus provided a link with other branches of academe. Previous
paradigms also had an ideological anchor, offering a vision of the "good
society." Perhaps the new approaches that increasingly seek to influence
our work should be put to some similar tests of their paradigmatic status.

Second, the multiplicity of approaches has not been an unalloyed boon
for the field. Some aspects of these approaches give us pause. One of the
costs of multiple approaches, for example, is the introduction into the literature of yet more layers of obscure jargon comprehensible to increasingly narrower groups of scholars. The issue is not merely the imposition
of barriers against comprehension, but perhaps more important, the exclusionary consequences of this trend. It has made too much of our writing inaccessible to ourselves, and it threatens to make much of media
studies equally inaccessible to discussions in the public forum. In addition, we now often confront a near-total relativism which unfortunately
has come to be a hallmark of much postmodern theorizing in our field.
Intellectual conversation becomes increasingly unwieldy and unproductive when all conceptual frameworks are regarded as equally valid. Besides, it is a nihilistic abnegation of responsibility to refuse to make choices even if we accept (as some do not) that all "social reality" is socially

constructed. We face no more pressing challenge than figuring out how media studies can participate in the work of renewing our professional (and personal) faith in the possibility and value of reasoned discourse.

Coherence, Identity, and Legitimacy

The essays collected here suggest an impressive variety of approaches amongst communication scholars, including a salubrious new diversity among researchers in the United States. Past stereotypes of a so-called American model of communication research no longer seem to apply. In general, research approaches and styles are much less coterminous with geographic location than was the case in the past—a clear sign of the global integration of communication scholarship. The down side of such diversity (as one contributor puts it) is that such wide-ranging inclusiveness conveys the impression of a field that is everywhere and nowhere. The result is a three-fold crisis of coherence, of identity, and of legitimacy.

Contributors who address this three-fold crisis seem broadly divisible into two camps—"redeemers" and "reformers." The redeemers believe that media studies can, indeed should, become a discipline in its own right, and be recognized as such by other established disciplines. Some, in fact, go even beyond "discipline" and "field" to argue that it is the all-embracing *subject* of communication upon which we should fix our gaze. By contrast, the "reformers" find that ambition grandiose and unrealizable, and instead offer many suggestions for improving the status and condition of the field by building on existing strengths.

Considering the redeemers first, one cannot escape the impression that their view is more of a plea, a hope, an expression of a felt need, rather than a statement concerning a currently emergent and advancing project. The case for disciplinarity is still unproven. Thus far there are few signs that media studies can be intellectually organized and convincingly presented as a discipline in its own right. In fact, one wonders whether those who struggle to "redeem"-the field as a discipline ignore its unique attribute: its reach into such diverse domains as political economy, technology, cultural studies, linguistic analysis, and audience behavior. Thus, the field remains multidisciplinary and multileveled. Indeed, it should be asked whether the search for disciplinarity is worth the efforts involved, and whether the costs of that quest, both in intrinsic intellectual terms and in terms of the relationship with the established disciplines, might be too high.

The reformers, on the other hand, address more immediate and perhaps somewhat more mundane concerns. Thus, for example, the recent constrictions of academic budgets everywhere have raised the uncomfortable question of how we persuade ourselves, our students, our colleagues in other fields, and the academic powers-that-be that despite our multidisciplinary, multifaceted character, we are not "all over the place" and that we deserve to thrive and prosper. Some reformers contend that the answer to those questions should turn on the theoretical excellence and conceptual creativity of our work. Others suggest pursuing a host of integrative strategies of research and analysis.

The quest for disciplinary status is linked to the issue of legitimacy, explored in a number of essays. There is a pervasive worry that our intellectual bona fides remains suspect, invisible to most practitioners in neighboring disciplines. That worry probably can be traced back to the point in the academic institutionalization of the field (especially in the United States) when communication scholarship gained institutional autonomy from other disciplines. It appears, however, that the gaining of departmental autonomy was insufficient to instill a sense of disciplinary autonomy. Even more painful, perhaps, was the sense that the autonomy gained simply implied disappearing from the "parents'" sight. Why that state of affairs persists to the present is an interesting question for sociologists of knowledge and of higher education.

Then there is the issue of relevance. There is a nagging, depressing sense of the marginality of our intellectual enterprise to the public discourse about the very subject we claim to study. What contribution, if any, ask some of our authors, have we made, or can we hope to make to the processes of policy-making at various levels that continuously shape the face of the means of communication in many countries? The silence of our field, argue some, is deafening.

But perhaps we should not be too hard on ourselves. The relevance of our research to the process of policy formation may not turn solely on our own efforts. Making a contribution to that process hinges in some significant measure on the receptivity of policymakers. We cannot, of course, convert all policymakers and media practitioners into readers of our academic journals and other publications. The gulf between these "two cultures"—the one academic, the other practitioner/policymaker—is not easily bridged. But it is certainly not unique to our field, and we might have to continue to live with it.

At the same time, the issue of relevance should not be judged strictly in terms of the linkages between research and public policy. Research produces not only findings entailing policy recommendations, but also terminologies and sensitivities that percolate into the public sphere in many subtle ways and that shape the ways in which policy issues are perceived and the intellectual climate that surrounds these issues. In the long run, that is probably our best hope.

Finally, a word of thanks to our colleagues Marjorie R. Ferguson, Edward L. Fink, and Vicki S. Freimuth for their help in creating "The Future of the Field" issues of the *Journal of Communication*; and to *Journal* editorial office staffers Jane Twomey, Archana Kumar, and Vicki Sneed for their tireless efforts to meet deadlines. We also want to extend a very special thanks to Jay Blumler, for sharing his insights about the essays that follow and about the shape and direction of communication research as it emerges from these essays. He is, indeed, a virtual coeditor. The views and arguments presented in the individual essays—and the responsibility for them—are the authors' own. We of course accept full responsibility for the overall structure and contents of this book.

M. R. L. and M. G.
College Park, Maryland

From Field to Frog Ponds

by Karl Erik Rosengren, University of Lund

In my contribution to *Ferment in the Field* 10 years ago, I drew upon a typology for schools of sociology originally presented by the Anglo-Ameri-
can dyad Gibson Burrell and Gareth Morgan (1979). Using their typology
to characterize the then state of affairs in media and communication studies, I made some predictions about future developments (Rosengren,
1983). Since then, some schools of thought and research in sociology and
communication have more or less disappeared, while others have seen
the light of day. But the basic dimensions of the typology are still valid.
Consequently, the typology as such may be used, first, to characterize the
present situation in communication research as compared to the situation
a decade ago; second, to compare previous predictions with the present
situation; and third, to make new predictions.

Burrell and Morgan's typology (Figure 1) is based on two dimensions
derived from a multidimensional property space for schools of sociological thought and research (originally eleven dimensions). The two dimensions cover assumptions about (a) the nature of social science (objectivis-
tic/subjectivistic), and (b) the nature of society (regulation/radical
change, consensus/conflict). Crossing the two dimensions yields a four-
fold typology: four main *paradigms,* which Burrell and Morgan called
radical humanism, radical structuralism, interpretive sociology, and
functionalist sociology, respectively.[1] Each paradigm encompasses a
number of different research traditions.

Today and Yesterday

Though the Burrell and Morgan typology mirrors the situation in sociology in the late 1970s, many of the same schools and groupings appear in
communication research, albeit sometimes under different names. There
is one basic difference, however, from the situation a decade ago. The

---

[1] Burrell and Morgan use the term *paradigm* much in the natural science sense originally in-
augurated by Kuhn (1970), a sense that is not at all applicable to most disciplines in the humanities and the social sciences. What they call paradigms, then, should rather be called
schools or *traditions* of research (cf. Rosengren, 1983, 1989, 1991a).

Karl Erik Rosengren is a professor of media and communication studies at the University of
Lund, Sweden.

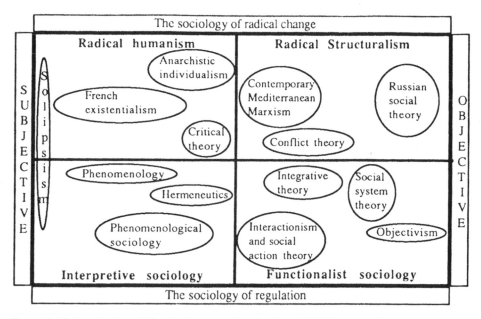

Figure 1. Schools of sociological research (after Burrell & Morgan, 1979).

relative importance of the two dimensions in the typology has shifted. In the late 1970s the regulation/radical change dimension was predominant, sometimes to an extent that called for well-founded warnings, good examples of which may be found in the *Ferment in the Field* issue (cf. Lang & Lang, 1983). Today, the subjectivistic/objectivistic dimension has the upper hand—in the humanities and the social sciences at large, as well as within communication. It does not take much cogitation to understand why this is so.

Because of long-term global political and intellectual change, the radical schools in the upper half of the typology find themselves in a somewhat awkward situation. From a *political* point of view, Marxist socialism has proved unviable; *intellectually,* Marx is on his way back to a niche among other niches in the mausoleum of the classics. During the last decade or so the horizontal divide of the typology has become less important, or at least less visible. Representatives of schools in the upper half of the typology are less anxious to discuss the very dimension that once defined their existence. Vague terms such as *critical, radical,* and *left* have recently become increasingly more popular, while more precise terms like *Marxist* or *Socialist* are on the wane.

Interest has focused on the vertical divide, differentiating between subjectivistically and objectivistically oriented sociology and communication research. Humanistically oriented research in sociology and communication, always an important tradition, has grown stronger, vitalizing the debate in a way that—although perceived by Katz (1983) and others—was only beginning 10 years ago. This general development has two articulations. First, the acting and willing subject, the human individual *qua* human individual, is in focus much more than before. Second, the historical perspective has grown ever stronger, a welcome complement to the sometimes rather one-sided, ahistorical perspective of the old behavioral and social science approaches.

The new situation admits—indeed, calls for—debates of a kind different from those dominating the last few decades. An imperative portal to such debates was offered by Curran in his reappraisal of what he called "The New Revisionism," a "reversion to certain discredited wisdoms of the past, a revivalism masquerading as new and innovatory thought," but also "a reformulation that could potentially strengthen the radical tradition of communications research," bringing about a "sea change" in media and cultural studies (Curran, 1990, *passim*).

Leaving aside for the time being attempts to rekindle the debate in terms of Burrell and Morgan's horizontal divide, to substitute scholarship for political activity (or vice versa), there is no denying that Curran makes it clear that there are some very basic problems to address. One way to do so is to look, not at what is happening or has happened, but at what has *not* happened—and why. Let us turn to yesterday's predictions.

Predictions Dated

Ten years ago I focused on potential developments different from the ones noted above. My attempts at prediction have proved only partly successful; it is interesting to compare them with actual developments during the last decade. Basically, I was hoping that the ferment we perceived would be replaced by vigorous growth, stemming from both mutual confrontation and mutual cooperation between the various schools and traditions of research. That is hardly what has happened, however. Instead, adherents of the various quasi-paradigms have increasingly avoided both confrontation and cooperation, preferring instead to isolate themselves in a number of self-contained enclaves—in spite of sometimes heroic efforts to the contrary.[2] A case in point is recent developments in audience research. Currently, half a dozen research traditions are developing, more or less independently of each other, with only weak signs of either cooperation or confrontation (Jensen & Rosengren, 1990; Rosengren, 1993; Vorderer & Groeben, 1993). Although I have no hard data, I believe that

---

[2] For instance, a whole International Communication Association conference, *Beyond Polemics: Paradigm Dialogues,* resulting in the two volumes edited by Brenda Dervin and her associates (Dervin, Grossberg, O'Keefe, & Wartella, 1989), or conferences with themes such as the one convened in 1990 by Larry Grossberg, Ellen Wartella, and their associates in Urbana–Champaign: *Towards a Comprehensive Theory of Audience Research.*

because of tendencies like these, the field today is characterized more by fragmentation than fermentation.

Communication across the vertical divide has become increasingly rare. At the same time, the horizontal divide has become more or less obsolete. The old front-lines are hardly visible any more. Negative bickering has replaced productive confrontation. It is as if the field of communication research were punctuated by a number of isolated frog ponds—with no friendly croaking between the ponds, very little productive intercourse at all, few cases of successful cross-fertilization.

Personally, I am convinced that a fundamental cause for the present fragmentation in communication studies is the fact that an increasing number of research traditions lack the basic preconditions for cumulativity. Cumulative growth in science and scholarship presupposes both confrontation and cooperation. Both of these activities presuppose very precise knowledge about what the whole thing is about. And this presupposes close relationships between three elements central to all scholarly activity: substantive theories, formal models, and empirical data. As we shall see, this basic precondition is not always fulfilled in contemporary communication research.

Theories, Models, and Data

In modern behavioral and social science, as well as in some humanistic disciplines (especially linguistics), there is a growing consensus that it is important, if not downright mandatory, to make a sharp distinction between substantive theories and the formal models used to explicate and analyze them (cf. Rosengren, 1992).

Substantive theory is often expressed in ordinary verbal language—using, of course, the special terminology necessary to express sophisticated thought. Formal models, however, are substantively empty, since they are expressed in terms of logical, mathematical, or statistical language (often visualized in terms of graphic models). It is very important to let substantive theory, formal models, and empirical data interact in a cumulative, spiraling process of knowledge building.

The world of phenomena will always be that "buzzing, blooming thing out there," just as theory is at best only a dim mirror image of that blooming richness. On the other hand, formal models, with their clarity and unambiguousness, call for maximum explicitness in the processes of translation between both theory and model, and model and data. They also create increased clarity with respect to the process of translation between theory and reality as represented by data. It is this demand for explicitness in the process of translation that initiates the spiraling process of knowledge building between substantive theory, formal models, and empirical data. It also makes confrontation and cooperation between different research traditions possible and meaningful.

Actually, it is precisely because of such an interplay between empirical data, substantive theory, and formal models—be they graphic, logical, mathematical, or statistical—that the modern behavioral, cognitive, and social sciences have been able to produce the fantastic development they have demonstrated during the last 50 years or so. The insistence on translation between theory and model, and model and data in order to make possible the theoretically relevant analysis of empirical data in terms of theoretically derived formal models has brought about an enormous growth in general knowledge. Translation between theory and model has thus made possible strong generalization.

We can find a good, relatively recent example of what strong theory combined with formal models and empirical data can lead to in Mary Douglas's grid/group typology as applied to empirical data and formal analysis (Douglas, 1970, 1982). It is a formalization now used in the study of substantively very different sociocultural phenomena (cf. Gross & Rayner, 1985; Wildawsky, 1989). The importance of formal models as such may be seen, for instance, in the vitalizing influence of advanced multivariate models such as those used in the LISREL approach on a number of social science research traditions (Joreskog & Sorbom, 1988). As a rule, however, such processes demand at least a decade or two in order to reach a stage of maturity.

The essential interplay between substantive theory, formal models, and empirical data often seems to be lacking in emerging traditions of communication research, both in the humanities and the social sciences. More specifically, substantive theories in verbal form are used directly to interpret and explain empirical reality and data derived from that reality, without relating the theory or data to explicit, formal models. However, formal models are beginning to be introduced to clarify things, and this is beginning to initiate a positive spiral of development in which substantive theory, formal models, and empirical data are combined to produce certified, precise knowledge. Typical examples of such a development may be found in traditions such as diffusion of news (Rosengren, 1987a), agenda setting (Brosius & Kepplinger, 1993), spiral of silence (Noelle-Neumann, 1991), and uses and gratifications research.

Unfortunately, not all emerging traditions follow this course. This is especially true of the schools of communication research in the two subjectivistic or humanistic cells of the Burrell and Morgan typology, which show a tendency not to draw on all three elements necessary for cumulative growth. They are often characterized by rich substantive theories, and sometimes also by quite impressive empirical data. *But they tend to avoid formal models.* The important links between substantive theory and formal model, between formal model and empirical data are thus weak or nonexistent. One reason for this sad state of affairs probably is a deep mistrust of quantification. At the same time, formalization is often mistaken for quantification. Logic tends to be forgotten for mathematics and statistics. As decades of dramatic developments in linguistics demonstrate, however, the strength of formalized qualitative models is formidable. And there have also been breakthroughs in other fields of the humanities and social sciences based on qualitative models. (For a good introduction to the area of formalized qualitative models, see Ragin, 1987.)

There is no reason at all, then, why qualitative studies should not be formalized. Since formal models are often absent from qualitative studies, true cumulativity cannot come about. Without the formal models, which alone make the confrontation between substantive theories and empirical data really fruitful, strict comparisons between different studies are not possible. Thus, key results cannot be carefully compared and, consequently, cannot be falsified. The wheat cannot be sifted from the chaff. Recently emerging traditions of reception analysis and lifestyle-oriented audience research offer two examples of this general tendency (see below).

The challenge facing any emerging tradition of communication research in the humanities and the social sciences is precisely this: to carefully heed the distinction between substantive theory and formal models, and to start a productive interplay between empirical data and those two vehicles of scientific analysis.

If such an interplay should not come about, I am convinced that even the most promising tradition of communication research will be stuck at the basically descriptive, narrative, and—in the technical sense of the word—anecdotal level, which seems to mark the beginning of most new traditions of communication research. The absence of translation between substantive theory and formal model, between formal model and empirical data will prevent strong generalization. Consequently, such a tradition will remain a fad among other fads, and just like all other fads, it will ultimately disappear. Without that indispensable interplay between substantive theory, formal models, and empirical data, there will be no productive processes of confrontation and cooperation between various schools of research. Consequently, there will be no cumulative growth.

Some concrete examples of differential development in three different traditions of mass communication research—uses and gratifications research, lifestyle-oriented research on media use, and reception analysis— illustrate and support my argument.

Models Make a Difference

Research is a very time-consuming business. The uses and gratifications tradition has been around for half a century. The roots of lifestyle-oriented mass media research may be traced back at least to the turn of the century. Studying developments within old traditions such as these can help us understand the situation in new traditions of research whose history covers only a decade or so—for instance, the emerging tradition of reception analysis.

The history of uses and gratifications (U&G) research has been traced more than once (cf. Blumler & Katz, 1974; Rosengren, Wenner, & Palmgreen, 1985; Swanson, 1993). From its beginning it has been characterized primarily by a very productive interplay between substantive theory, formal models, and empirical data. Empirical data have been collected in a number of ways, ranging from informal interviews to focused interviews, to survey schedules and questionnaires, to laboratory experiments. They also have been analyzed in terms of formal models ranging in complexity from simple cross tabulations to advanced multivariate statistical modeling. As a result, the tradition has developed from an initial stage characterized by descriptive studies, to a typological phase, to an explanatory phase characterized by a full-fledged interplay between substantive theory (e.g., expectancy-value theory), formal models (e.g., LISREL analyses), and sophisticated designs of data collection (e.g., combined longitudinal/cross-sectional designs). It has now reached a stage in which detailed and precise comparative studies, over both time and space, have become possible (Johnsson-Smaragdi, 1992; Rosengren, 1991b; cf. Rosengren, McLeod, & Blumler, 1992). This development took half a century.

Lifestyle-oriented mass communication research has its roots in classical social science, including the work of Veblen, Simmel, and Weber (cf. Johansson & Miegel, 1992, pp. 7ff.). Looking back on its long history, it is striking to observe how, after the first insightful, sometimes brilliant observations made by the classics, lifestyle research for a very long time never really left the stage of verbal description and verbal theories illuminated by striking examples. On the other hand, the two areas in which the notion of lifestyle has received most interest recently—market research and the sociology of medicine—sometimes use sophisticated models of analysis, but seem to be strangely uninterested in the role of theory as exemplified by the classics in the area (cf. Berardo, Greca, & Berard, 1985; Castro, Newcomb, & Cadish, 1987; Holman & Wiener, 1985; Veltri & Schiffman, 1984). They thus seem to combine theoretical indifference or even naïveté with some sophistication in the use of formal, statistical models.

By and large, when we look at the history of lifestyle research, we find either theoretically informed empirical studies lacking contact with formal models, or theoretically uninformed research with results couched in terms of formal models. Studies based on full-fledged combinations of substantive theories, formal models, and empirical data seem to be few and far between.

This paradoxical situation is probably one reason why lifestyle research, in spite of having a history more than twice as old as U&G research, needed so long to reach its take-off stage. Only in the last decade or so have any signs of such a development turned up. In France, of course, there is Bourdieu (1979/1984); in the U.S., Mitchell (1983). In Germany, one may point to Lüdtke (1989) and Opaschowsky (1991); in England, to Featherstone (1987). Some interest in lifestyle-oriented media research also has surfaced recently in Sweden, in a series of publications by Johansson and Miegel (1992), Loov and Miegel (1989), Reimer (1992), Reimer and Rosengren (1990), and Rosengren (1987b).

This new international wave of lifestyle studies has combined typological and other theoretical work with formal models and empirical data in terms of both quantitative and qualitative research. Consequently, it has been able to move relatively quickly from the original descriptive level to the explanative level. The stage thus seems set for strong, rapid, and cumulative development in lifestyle-oriented media research during the next decade or two. But what about that relatively new tradition within audience research, reception analysis?

Reception analysis has a number of roots in the humanities and the social sciences (Jensen & Rosengren, 1990). Two of these roots are located in disciplines that for decades have systematically applied a continuous interplay between substantive theory, formal models, and empirical data: linguistics and cognitive psychology. We can find attempts in this direction especially in reception analyses drawing upon cognitive psychology (cf. Hoijer, 1990, 1993). In the type of reception analysis that borrows terms and concepts from modern and not-so-modern linguistics, using them primarily metaphorically and/or by way of analogy, this type of systematic interplay between substantive theory, formal models, and empirical data has been used rather sparingly so far.

Prominent members of the linguistically oriented branch of reception analysis are by no means unaware of this state of affairs. Indeed, they even seem to make a point out of it. This specific tradition is dominated by so-called qualitative studies, based on anecdotal data and defined as unformalized, exegetic studies of the meaning of individual experience (cf. Swanson, 1993). In contrast to many qualitative studies in the psychologically oriented branch of reception analysis, these studies, as a rule, neglect otherwise generally accepted tests of reliability, validity, and representativeness. They are also said to have an explanatory value in their own right. Precisely what kind of explanatory value they actually have, however, is never made quite clear. What has been produced instead is an impressive number of ingenious and insightful case studies (cf. Jensen & Jankowski, 1991).

Looking Toward the Future

There is always that wonderful challenge facing any emerging tradition of research: to confront some interesting, so-called qualitative results with formal models, be they qualitative or quantitative, and to translate them into instruments that may be applied in studies of representative samples from carefully defined populations. Only such processes of translation admit generalization. This is indeed a challenge, and challenges tend to be accepted, sooner or later.

So, given what I have discussed above, it does not seem very farfetched to predict that in the near future some enterprising group of communications scholars will concentrate on some variables found to be highly strategic in the substantive theories of so-called qualitative reception analysis. Aiming at empirical and theoretical generalization, they will start the laborious work of relating the substantive theories to relevant formal models. They will translate the variables into empirical measurements applicable in studies of representative samples of carefully defined populations under carefully defined circumstances, and analyze the empirical data in terms of both the formal models and the substantive theories. A good example of what is already being done may be found in Ohler (1992).

As a matter of fact, that is precisely what happened half a century ago, with Herzog's and Arnheim's path-breaking qualitative studies from the early 1940s—as Curran so forcefully demonstrated in his thoughtful article about the new revisionism (Curran, 1990). Because of this, a great deal of new knowledge about media use and its causes and consequences has been produced within the U&G tradition, even if it took more than half a century. In that way, U&G was saved from the fate of being just a fad among so many other fads in communication studies. In principle, the same thing has happened to lifestyle-oriented media research during the last two decades—albeit on a much broader scene, and only after half a century of retardation. Personally, I'm convinced that the same development will soon take place with respect to the many interesting results right now being gained within so-called qualitative reception analysis. The only question is who will do it. The reception analysts themselves, of course, are first in line. Should they prefer not to take their chance, instead continuing to grind out more case studies of this or that "interpretive community," others will do it for them. Whoever does it, though, it will take time. We can only hope that it will not take half a century.

Combinations, Comparisons, and Confrontations

After a period of fermentation in the field (if ever there was any field in the strict sense of the word) we seem to have ended up with fragmentation and threatening stagnation. Those who hoped for positive confrontation and cooperation have reasons to be disappointed. Instead, a bland acceptance of or indifference toward research traditions other than one's own seems to dominate. Tendencies such as these may well be major causes behind the somewhat uncertain disciplinary status still plaguing our field.

There certainly is no viable mono-causal explanation of this sad state of affairs. For my part, however, I have pointed to the fact that one basic precondition for cumulative growth is sometimes lacking in contemporary communication research—especially in some humanistic traditions. There sometimes seems to be an aversion toward a vital element of all scholarship and research: the formal model, a necessary complement to substantive theory and empirical data. One reason for this may be the mistaken assumption that formalization means quantification. In the strict sense of the word, of course, qualitative studies can be formalized as much as quantitative studies. What are traditionally and loosely called qualitative studies of reception (usually a small number of informal talks with TV viewers, romance readers, or Madonna fans) may sometimes produce brilliant ideas. But brilliant ideas deserve, need, and should be able to stand up to hard tests. Such tests call for a combination of substantive theory, formal models, and empirical data. Unfortunately, some communication scholars are unwilling to draw on studies of that type.

The solution seems to be twofold. Humanistically oriented communication scholars must overcome their aversion to formal models. Social-science-oriented communication scholars must be willing to draw on the sometimes very productive insights gained by humanistically oriented scholars. There is a strong need for "combinations, comparisons, and confrontations" to be carried out in the precise terms offered by studies combining substantive theory, formal models, and empirical data (Rosengren, 1991a, 1992). A promising area in which to undertake a coordinated attempt in this direction may well be reception analysis. But there are a host of other options.

References

Berardo, F. M., Greca, A. J., & Berard, D. H. (1985). Individual lifestyles and survivorship: The role of habits, attitudes and nutrition. *Death Studies, 9*, 5–22.

Blumler, J. G., & Katz, E. (Eds.). (1974). *The uses of mass communications: Current perspectives on gratifications research*. Beverly Hills, CA: Sage.

Bourdieu, P. (1984). *Distinction*. Cambridge, MA: Harvard University Press. (Original work published 1979)

Brosius, H. B., & Kepplinger, H. M. (1993). Linear and nonlinear models of agenda setting in television. *Journal of Broadcasting & Electronic Media, 36*, 5–24.

Burrell, G., & Morgan, G. (1979). *Sociological paradigms and organisational analysis*. London: Heineman.

Castro, F. G., Newcomb, M. D., & Cadish, K. (1987). Lifestyle differences between young adult cocaine users and their nonuser peers. *Journal of Drug Education, 79*, 89–111.

Curran, J. (1990). The new revisionism in communication research: A reappraisal. *European Journal of Communication, 5*, 135–164.

Dervin, B., Grossberg, L., O'Keefe, B., & Wartella, E. (Eds.). (1989). *Rethinking communication, Vols. I-II*. Newbury Park, CA: Sage.

Douglas, M. (1970). *Natural symbols*. London: Barrie & Rockliff.

Douglas, M. (1982). *In the active voice*. London: Routledge & Kegan Paul.

Featherstone, M. (1987). Lifestyle and consumer culture. *Theory, Culture & Society, 4,* 55–70.

Gross, J. L., & Rayner, S. (1985). *Measuring culture*. New York: Columbia University Press.

Hoijer, B. (1990). Reliability, validity and generalizability. Three questions for qualitative reception analysis. *The Nordicom Review, n.v.*(1), 15–20.

Hoijer, B. (1993). Reception of television narration as a socio-cognitive process: A schema-theoretical outline. In K. E. Rosengren (Ed.), Special issue: Audience research, *Poetics, 21,* 283–304.

Holman, R. H., & Wiener, S. E. (1985). Fashionability in clothing: A values and lifestyles perspective. In M. R. Solomon (Ed.), *The psychology of fashion* (pp. 87–98). Toronto: Lexington Books.

Jensen, K. B., & Jankowski, N. W. (Eds.). (1991). *A handbook of qualitative methodologies for mass communication research*. London: Routledge.

Jensen, K. B., & Rosengren K. E. (1990). Five traditions in search of the audience. *European Journal of Communication, 5,* 207–239.

Johansson, T., & Miegel, F. (1992). *Do the right thing: Lifestyle and identity in contemporary youth culture*. Stockholm, Sweden: Almqvist & Wiksell International.

Johnsson-Smaragdi, U. (1992). Learning to watch television: Longitudinal LISREL models replicated. *Lund Research Papers in Media and Communication Studies, 4.*

Joreskog, K. G., & Sorbom, D. (1988). *LISREL 7: A guide to the program and applications* (2nd ed.). Chicago: SPSS Publications.

Katz, E. (1983). The return of the humanities and sociology. *Journal of Communication, 33*(3), 51–52.

Kuhn, T. S. (1970). *The structure of scientific revolutions* (2nd ed., enlarged). Chicago: University of Chicago Press.

Lang, K., & Lang, G. E. (1983). The "new" rhetoric of mass communication research: A longer view. *Journal of Communication, 33*(3), 128–140.

Loov, T., & Miegel, F. (1989). The notion of lifestyle. *Lund Research Papers in the Sociology of Communication, 24.*

Lüdtke, H. (1989). *Expressive Ungleichheit: Zur Soziologie des Lebensstiles* [Expressive difference: Notes on the sociology of lifestyles]. Opladen: Leske & Budrich.

Mitchell, A. (1983). *The nine American lifestyles*. New York: Warner Books.

Noelle-Neumann, E. (1991). The theory of public opinion: The concept of the spiral of silence. In J. A. Anderson (Ed.), *Communication yearbook 14* (pp. 256–287). Newbury Park, CA: Sage.

Ohler, P. (1992, October). The perspective of film psychology. In *The viewer as TV-director: Understanding individual patterns of exposure and interpretation*. International symposium, Hans-Bredow-Institut, Hamburg, Germany.

Opaschowsky, H. W. (1991). Freizeit, Konsum und Lebensstil [Leisure, consumption, and lifestyle]. In R. Szallies & G. Wiswede (Eds.), *Wertewandel und Konsum: Fakten, Perspektiven und Szenarien für Markt und Marketing, 2. überarbeitete und erweiterte Auflage* [Value change and consumption: Facts, perspectives, and scenarios for market and marketing. 2nd, revised and extended edition] (pp. 109–134). Landsberg/Lech, Germany: Verlag Moderne Industrie.

Ragin, C. C. (1987). *The comparative method. Moving beyond qualitative and quantitative strategies*. Berkeley: University of California Press.

Reimer, B. (1992). Inte som alla andra. Ungdom och livsstil i det moderna [Not like everybody else. Youth and lifestyle in modernity]. In J. Fornas, U. Boethius, H. Ganetz, & B. Reimer (Eds.), *Unga stilar och uttrycksformer* [Young styles and forms of expression] (pp. 163–201). Stockholm, Sweden: Symposion.

Reimer, B., & Rosengren, K. E. (1990). Cultivated viewers and readers: A life-style perspective. In N. Signorielli & M. Morgan (Eds.), *Cultivation analysis* (pp. 181–206). Newbury Park, CA: Sage.

Rosengren, K. E. (1983). Communication research: One paradigm, or four? *Journal of Communication, 33*(3), 185–207.

Rosengren, K. E. (1987a). Conclusion: The comparative study of news diffusion. *European Journal of Communication, 2,* 227–55.

Rosengren, K. E. (1987b). *Livsstil och massmediekultur. En projekt-beskrivning* [Lifestyle and mass media culture. A project description]. Lund, Sweden: Department of Sociology, University of Lund.

Rosengren, K. E. (1989). Paradigms lost and regained. In B. Dervin, L. Grossberg, B. O'Keefe, & E. Wartella (Eds.), *Rethinking communication, Vol I: Paradigm dialogues: Theories and issues* (pp. 21–39). Beverly Hills, CA: Sage.

Rosengren, K. E. (1991a). Combinations, comparisons and confrontations: Towards a comprehensive theory of audience research. *Lund Research Papers in Media and Communication Studies, 1.*

Rosengren, K. E. (1991b). Media use in childhood and adolescence: Invariant change? In J. A. Anderson (Ed.), *Communication yearbook 14* (pp. 48–90). Newbury Park, CA: Sage.

Rosengren, K. E. (1992). Substantive theories and formal models: Their role in research on individual media use. *Lund Research Papers in Media and Communication Studies, 4.*

Rosengren, K. E. (1993). Audience research: Back to square one—At a higher level of insight? In K. E. Rosengren (Ed.), Special issue: Audience research. *Poetics, 21,* 239–241.

Rosengren, K. E., McLeod, J., & Blumler, J. G. (1992). Comparative communication research: From exploration to consolidation. In J. G. Blumler, J. M. McLeod, & K. E. Rosengren (Eds.), *Comparatively speaking: Communication and culture across space and time* (pp. 271–298). Newbury Park, CA: Sage.

Rosengren, K. E., Wenner, L. A., & Palmgreen, P. (Eds.). (1985). *Media gratifications research: Current perspectives*. Beverly Hills, CA: Sage.

Swanson, D. L. (1993). Understanding audiences: Continuing contributions of gratifications research. In K. E. Rosengren (Ed.), Special issue: Audience research. *Poetics, 21,* 305–328.

Veltri, J. J., & Schiffman, L. G. (1984). Fifteen years of consumer lifestyle and values research at AT&T. In R. E. Pitts & A. G. Woodside (Eds.), *Personal values and consumer psychology* (pp. 271–285). Toronto: Lexington Books.

Vorderer, P., & Groeben, N. (1993). Audience research: What the humanistic and the social science approaches could learn from each other. In K. E. Rosengren (Ed.), Special issue: Audience research. *Poetics, 21,* 361–376.

Wildawsky, A. (1989). Choosing preferences by constructing institutions: A cultural theory of preference formation. In A. A. Berger (Ed.), *Political culture and public opinion* (pp. 21–46). New Brunswick, NJ: Transaction.

Communication—Embrace the Subject, not the Field

by James R. Beniger, University of Southern California

No Renaissance scholar could have asked for a more exciting or promising subject of study than contemporary communication. Never before have more disciplines or more minds converged more rapidly on a single phenomenon: information and its patterning, processing, and communication as central to culture, cognition, and social behavior. Throughout academia, ideas for interdisciplinary collaboration and synthesis in these areas are in the air, and often well under way, most notably in cognitive science and cultural studies.

Evidence to bolster such enthusiasm abounds. Bibliometric studies find increasing convergence on the concepts and literature of information and communication across the range of academic disciplines—from the humanities and social sciences to the cognitive, behavioral, and life sciences, computer science, and mathematics. Undergraduate enrollments continue to climb in courses on mass media and popular culture, literary and critical studies, and cognitive and computer science. Terms like *information, communication, computer,* and *media* are routinely used as adjectives to name the age in which we live.

Although no one discipline could possibly embrace the entire range of academic interest in information and communication, certainly any organized field that calls itself communication might be expected to play a central role. Alas, just the opposite has been true. The American field of communication, at least at its institutional core of research and training, associations and conferences, textbooks and journals, remains today not far advanced beyond its aims of nearly a half century ago. Even more striking than the field's insularity from developments outside its institutional boundaries, its belated and grudging acknowledgment of European social and literary theory notwithstanding, has been the utter lack of interest in communication—the field, not the subject—by the leading scholars of other disciplines.

Once again evidence abounds. Not only are the most cited books on topics relevant to communication rarely written by those formally in the field, but these works cite few, if any, who are. Most of today's leading

James R. Beniger is an associate professor at the Annenberg School for Communication, University of Southern California.

scholars of communication, in the United States no less than in the rest of the world, are neither formally educated in the American field nor consider it exclusively—or even primarily—their academic or intellectual home. Bibliometric studies over the past 15 years depict the field of communication as an intellectual ghetto, one that rarely cites outside itself and is even more rarely cited by other disciplines. Even the "Call for Papers" of this special issue raises the proposition that "communication scholarship lacks disciplinary status," so that "institutional and scholarly legitimacy remains a chimera for the field." Not surprisingly, communication departments continue to hire from outside the field—a largesse rarely reciprocated by disciplines of greater legitimacy.

What Might Be Done

What American communication might do to make itself more central to studies in its own field should by now be obvious: It ought to embrace its traditional albeit now greatly expanded subject matter, communication, while at the same time abandoning all institutional vestiges of its narrow and long outmoded approaches to the field. A particular set of changes that would well position communication relative to other evolving fields, among the subjects of their greatest mutual interest, can be seen as a long overdue shift from what might be called the *Three Rs* to the *Four Cs*.

By the Three Rs, I mean the traditional fields of education: readin' (input and decoding), writin' (encoding and output), and 'rithmetic (computation or processing). Larger implications of the Three Rs might be found both in the impoverished model of education that they imply and in similarly objective linear models like, for example, Claude Shannon's 45-year-old mathematical model of communication.

As the essence of objective linear models of information processing and communication, however, the Three Rs do indeed well represent the American field's other outmoded baggage of the late 1940s: simple linear causation, cybernetics, vulgar computer metaphors and computer modeling, and the persistently embarrassing media effects controversy, what the "Call for Papers" describes as "the perennial black box of communication research" that "still poses the most unanswered questions."

Much as such models capture the intellectual flavor of the post-World War II period in which the American field of communication was born, the Three Rs also evoke other salient features of that time: Systematic study of mass media like motion pictures and radio was still relatively new then, as was the medium of television. Also new were even the most basic systematic techniques for the conduct and analysis of survey research. Because the behaviorism of Watson and Skinner continued to dominate American psychology, the human mind itself remained—for the nation's behavioral scientists—a black box. As McCarthyism came to dominate American society in the early 1950s, critical studies of media and politics disappeared from public view. High culture, then a central focus of American universities, remained distinctly separate from its imagined other half, so-called popular or low culture, then virtually untouched as a subject of scholarly interest.

Even so brief a summary of the intellectual origins of the field helps to explain the central and still dominant features of the distinctively American approach to communication. Abandon its core of 50-year-old social science (garnished though it might be with a pinch of LISREL) and what remains of this field?

What remains are virtually all the interesting new developments, many imported from abroad, that have invigorated other disciplinary approaches to information and communication in recent decades. Most of these have only marginal if not fugitive status in American communication, however, and even these footholds exist only thanks mostly to younger scholars and to refugees from other disciplines and countries. It is in this residue of approaches, most not dreamed of a half century ago and all still new to American communication, that we might find the focus of a revitalized field, a focus I shall call the Four Cs after its four key elements: cognition, culture, control, and communication.

The Four Cs

By *cognition,* I mean not only the subject itself, long neglected as an integral part of communication as process, but also the work of the new cognitive science, a burgeoning field spanning disciplines from philosophy, anthropology, and linguistics to psychology, artificial intelligence, and neuroscience.

By *culture,* I mean the traditional subject, including its cognitive, external informational (symbolic), and purely material realms, so that cultural artifacts like an automobile, for example, might be simultaneously good to get around with, good to show off one's status with, and also good to think with (in the sense of Lévi-Strauss). Attention to culture necessarily links the humanities (especially literary studies and philosophy) to areas of cognitive science (especially in anthropology and linguistics) and to other social science approaches to mass media and popular culture.

By *control,* I mean not cybernetics (an important early contribution largely spent by the 1960s) but rather the much more modest insight that all social behavior—including all communication—is by definition goal-directed. The intellectual history of this approach subsumes those of pragmatism, phenomenology, natural language philosophy, social action theory, symbolic interactionism, and the new sociology of science, among other intellectual developments of the past century. Goal-directed behavior and the resultant conflicts among actors competing for resources find no place in the purely objective models manifest in most American research on communication.

Finally, by *communication* I mean the goal-directed movement and processing of information, whether in flows small (like pointing, alluding, and evoking) or large (like teaching, convincing, and simulating), as well as the residues of such flows in cultural programming (socialization) and media content. Here communication serves to link the material and symbolic levels of both culture and cognition, and these to each other, as well as to link cognitive culture to social behavior. In so doing, communication grounds behavior in cognition, and thus the individual—through culture—in the dyad, the group, the organization, and the society.

As one of the Four Cs, communication does not represent a *subject* of study, or an end in itself, but rather a means to another end—a *method* for integrating the concepts, models, and data of many disciplines. All human behavior is instigated, shaped, and constrained by information and communication, after all, both from within—by socialization, perception, and cognition—and from without—through human interaction, social structure, and technologies. Cognition, in turn, is shaped through communication by factors ranging from ideology to advertising, popular culture to journalism—factors that themselves reflect structural constraints like those of technology and political economy. Such external constraints are in turn influenced from within, both through organizational and other interpersonal communication and—ultimately—by individual cognition.

Empiricism, Pluralism, and Paradigm

Reconstituted in terms of the model and method implied by the Four Cs, the field would no longer concentrate nearly as much on particular manifestations of communication. The field would instead devote itself to a more systematic and integrative understanding of a much wider set of phenomena that are simultaneously cognitive, cultural, behavioral, and social.

Communication might in this way transcend what the "Call for Papers" describes as "the 'itch' to discover a universal paradigm," certainly a debilitating affliction in a postmodern age when—in a growing number of disciplines—text reigns supreme at the expense of theory. At the same time, the field might also avoid what the "Call for Papers" terms "a comfortable acceptance of theoretical pluralism," which in practice has meant an increasing proliferation of ever more "theories" (often known by the names of their perpetrators) whose proclamations—aided by opportunistic publishers—daily clog our university mailboxes. To the extent that we can develop the Four Cs *qua* method while forswearing the grail of paradigm, we might stem the flood of marginal theories while focusing our widest possible efforts—however modestly—on a common integrative model.

As just one of many obvious examples of phenomena that might be integrated in this way, consider the current study of popular culture. As even the most diehard hopes for distinguishing high from low culture have been abandoned, over the past decade, the study of popular culture has flourished. Too often in the United States, however, popular content has been treated as an end in itself, by unabashed fans, rather than as a means to some more general understanding. So integral is popular culture to the understanding of communication, interpersonal no less than mass, that the field can hardly afford to leave the subject entirely to specialists. What are we to do?

Historical evidence suggests that we increasingly think with—and are thus influenced by—mass-produced sound bites, images, and icons, among other constituents of cognitive structure. Popular culture might therefore be studied as a component not only of material artifacts and media content, but also of cognition—a possibility largely overlooked by cognitive scientists. Both fields might profit from the understanding of, for example, the particular cognitive role of icons of popular culture— along with those of ancient myth, the Bible, Shakespeare, and other literature, art, and history—as constituents of everyday thought. By studying popular culture in this way, communication scholars could hope to make significant contributions both to cognitive science, as most broadly defined, and to work on artificial intelligence and expert systems in particular, even though cognitive anthropology has already staked out many of these opportunities.

Toward a Theoretical Synthesis

Despite the integrative potential of the Four Cs, the suggested synthesis cannot yet be considered a theoretical one. Communication still awaits a unifying theory, and indeed much of any theory to call its own. Even to achieve the relatively lofty status of an academic discipline, however, would not require an overarching theory like biology's modern synthesis of evolution, genetics, and ecology or the classical theory and neoclassical synthesis that informs modern economics. Except for economics and linguistics, after all, none of the social sciences or humanities has achieved anything approaching even theoretical consensus, let alone a grand synthesis of theory.

What might communication do to foster development of its own integrative theory and the potential for theoretical synthesis? Many academic fields, including most in the social sciences and humanities, have steadily developed while unified by only a common subject matter or model like those implied by the Four Cs. Much like the concepts of power in political science and the allocation of scarce resources in economics, the concepts of information and communication—as specified by both the Three Rs and the Four Cs—afford a third, potentially powerful, distinct yet complementary, and inherently integrative means by which to approach *all* social and behavioral phenomena.

Thus might the field of communication hope to join economics and political science in a tripartite division of social science, defined as the study of information and communication, of the power emergent from but dependent on them, and on the role of such power and control in the allocation of resources—including those of both information and power. Is it difficult to imagine a future in which no educated person could encounter the term *political economy* without reflexively adding a third concept like information, communication, or media? By such means might communication subsume much of the subject matter of sociology and anthropology, both currently in eclipse, and at the same time appropriate many of the social and behavioral aspects of psychology.

Under the four Cs, communication might find a more distinct place among the social sciences, by virtue of its several theoretical and methodological subfields that would necessarily center on the *exchange and flow* of information quite apart from considerations of cognition and culture per se. Modeled as one essential aspect of *all* social exchanges and flows, much as Cooley once modeled transportation, information provides the field of communication with an identifiable and measurable entity and potential commodity, distinct from those based on matter or energy, with the commoditization of information as the obvious focus of yet another important subfield.

In this way communication could complement a number of other fields that also study behavior and society as exchanges and flows: ecology, the study of energy flows; kinship analysis, the study of *ideal* flows of people through marriage, residence, and inheritance; demography, the study of *real* flows of people from birth through menarche, marriage, child bearing, emigration and immigration, to death; and input–output economics, the study of flows of money or value added. Unlike more narrowly defined studies of exchange and flow, however, communication would be able to contribute findings concerning the creation, marketing, exchange, and flow of information to a wider synthesis involving *both* the subjective (culture, cognition, and purpose in the form of *intended* control) *and* the objective (the Three Rs and *actual* control).

Because it would be predicated on the strictest attention to information, communication, and media in all aspects of human life, such a synthesis of subject matter and models, if not of theory, would force the study of any one behavioral or social phenomenon to confront all others. This would hardly end the "ideological and methodological battles" that the "Call for Papers" says "continue to fragment our field," but it would achieve an even more productive end: It would assure that such battles, the lifeblood of any discipline, take place on common ground in compatible if not identical language. In this way, the field of communication might aspire to become what Comte had hoped to make of sociology— queen of the social and behavioral sciences. Certainly the field meets one necessary condition: Its essential subject matter, information and communication, plays a fundamental role in the central theories and models of all relevant disciplines.

Communication—The Very Idea

If communication is to achieve anything like the modest synthesis suggested here, leaders of the field will have to think of themselves, above all else, as students of information, communication, and media per se, and of how and why these matter in all of their various manifestations, contexts, and applications. Considering the complex interrelationships inherent in such subject matter, it is easy to see the folly of balkanizing it in separate studies—like those of interpersonal, organizational, and mass communication—as the American field has increasingly done.

A more comprehensive view of the subject matter of communication, compatible with the one implied by the Four Rs, is that of the four-volume *International Encyclopedia of Communications* (Barnouw, Gerbner, Schramm, Worth, & Gross, 1989), which treats the field in 569 articles representing 30 major areas covering virtually every discipline. Unfortunately for all of us who share similarly expansive views of the field, however, the bulk of research, publication, and conference activities in communication today falls disproportionately in only a few of the *Encyclopedia*'s subject areas. Among these 30 areas, relatively little work in the American field has been devoted to—or even informed by—research on, for example, animal communication, computers, education, or folklore, largely because of the historical accident of entrenchment by specialties now a half-century old.

If communication scholars truly intend their field to represent the ambitious subject for which it is named, they must seek a more rational distribution of labor toward that end. Because virtually anything can be used to represent something else—and increasingly is in postmodern cultures—we must transcend the current concentration on perhaps a half dozen media in favor of descriptive, analytic, and comparative studies of the virtually unlimited numbers and types of media—wherever in societies we might find them. Why could it not be routine to find interpersonal, organizational, and mass communication treated in a single study, for example, or to conduct comparative analyses of widely diverse media as well as contexts? Few institutional changes could better improve the field of communication, or attract brighter students to the field, than the more dedicated pursuit of communication itself—the very idea—regardless of the subjects that might lead us to study.

This approach to building a discipline seems manifestly superior to its opposite extreme, what the "Call for Papers" describes as a focus "on socially relevant research" with the aim "to influence either the practice of journalism and communication or the formulation of communication policy." Nothing is more relevant to real-world concerns than theory and methods, as fields ranging from mathematics to economics can attest.

No one person could be expected to study all or even most aspects of communication, of course, no more than a single psychologist could be expected to master all of neuroscience and cognitive, developmental, social, and clinical psychology. Just as the ultimate status of academic psychology depends on its integration of this entire range of phenomena, so too does the future of American communication depend on its ability to synthesize its own subject matter in some way.

For this to happen in communication, however, its leaders must turn their collective gaze away from the past half century to the future—away from the field's now ossified subject divisions toward the other disciplines that increasingly usurp its claims to the academic study of communication. Most important of all, the field must fully embrace the subject of communication itself—the very idea—regardless of what that might entail for its current organization. For surely information and communication will continue to be the focus of intellectual excitement and integration into the foreseeable future, regardless of whether the academic field that today just happens to be named communication—largely through historical accident—manages to survive or not.

Reference

Barnouw, E., Gerbner, G., Schramm, W., Worth, T. L., & Gross, L. (1989). *International encyclopedia of communications* (4 vols.). New York: Oxford University Press.

Why Are There So *Many* Communication Theories?

by Robert T. Craig, University of Colorado at Boulder

Present trends in communication theory suggest a paradox. Even as the field has grown and matured, even as communication researchers have contributed more and better original theory in recent years, confusion, uncertainty, implicit dissension, and to a lesser degree, explicit controversy about the proper foci, forms, and functions of communication theory have markedly increased.[1] Yes, the field remains in "ferment" (*Ferment in the Field,* 1983) and more than ever requires "rethinking" (Dervin, Grossberg, O'Keefe, & Wartella, 1989). What communication theory should look like, what it is for, the boundaries and internal divisions of the field, how communication research should be conducted and how such research properly relates to the development of theory—all must be regarded as open questions at the present time.

So, even as we *do* more theory, we become (collectively if not individually) less certain of exactly *what* we are doing or should be doing. I do not intend to propose this correlation as a scientific "covering law." Doing more theory has not chiefly *caused* our present state of disorientation, nor has our uncertainty about the nature and purpose of theory *caused* us to do more theoretical work. To continue the metaphor in the idiom of quantitative analysis, the correlation, I believe, is largely spurious: Both our increased theoretical productivity as well as our increased confusion about theory are best explained by the influence of a third variable: a general transformation of the human sciences, which is transforming communication along with many other disciplines.

---

[1] Recent communication theory cannot be reviewed in detail in this short essay. For diverse examples, see Dervin, Grossberg, O'Keefe, and Wartella (1989, especially Vol. 2) and current issues of *Communication Theory,* which began publication in 1991. "More" and "better" are relative terms: One is not yet tempted to doubt whether additional improvements to the field are still possible.

Robert T. Craig is an associate professor in the Department of Communication, University of Colorado at Boulder.

Definitions of Theory and Other Curios

As recently as the 1970s, communication researchers were quite sure they knew what theory was, even though they had to admit that they themselves had not yet produced a lot of it. Optimism prevailed. Philosophers of science and sociologists specializing in "theory construction" had pointed the way to a social science of communication and we had only to follow their guidance. Theory, we knew, comprised a body of scientific generalizations describing functional relationships among empirically measured or inferred variables. The goals of scientific theory were explanation, prediction, and control of phenomena. Following Merton (1957), we knew that what we needed were "theories of the middle-range"—theories that would yield rigorously testable hypotheses concerning a delimited range of phenomena. Such theories (Festinger's [1957] theory of cognitive dissonance was a highly influential example) were considered superior both to speculative "grand theories" like Parsonian social theory, and to isolated empirical generalizations concerning, for example, the effects of fear appeals on attitude change. We had learned from Popper (1959) that the *sine qua non* of scientific theory was *falsifiability,* and that speculative grand theories inherently lacked this essential quality. Isolated empirical generalizations or "sets of laws," on the other hand, lacked the organizing and heuristic advantages of conceptually integrated middle-range theories. Following (however inconsistently) Kuhn's (1962) postpositivist history and philosophy of science, we knew, finally, that our communication science was in a "preparadigmatic" state as we searched for a paradigm.

Thus, in the 1970s we thought we knew what theory was. Indeed, if we believe publications such as the *Handbook of Communication Science* (Berger & Chaffee, 1987) and current communication theory textbooks, which continue to define theory largely in the traditional terms discussed in the preceding paragraphs, we still think we know what theory is. Those who continue to understand theory exclusively according to the received view of scientific theory may think preposterous my claim that communication theory is flourishing today as never before. Berger (1991), for example, writing in response to the question "Why are there so few communication theories?" bemoans the relative *lack* of original theory in the field and even cites Craig (1988) in support of the view that communication science has produced no major, seminal theories. This remains true, although as both Berger (1991) and Craig (1988) point out, the field has generated some notable efforts in that direction in recent years. The present essay, however, takes a broader view of the field, looking well beyond the boundary of communication science as surveyed, for example, by the *Handbook of Communication Science* (Berger & Chaffee, 1987; cf. Craig, 1988).

Berger (1991) clearly presupposes that we know what theory *is*: "Some researchers," he writes, "simply have never learned what a theory is" (p. 106); but, in a good graduate seminar in theory construction, "by the end of several weeks, most students have a good grasp of what a theory is" (p. 109). According to this view, we know what theories *are* but have not yet created nearly enough of them. More such theories are needed, no doubt, and Berger's analysis of our field's shortcomings points out some really serious problems. But in a broader perspective it becomes apparent that communication scholars are already doing a great deal of work that they regard as theory, but that would not qualify as such according to the received view of theory that Berger takes for granted.[2] Hence, only if we are willing to admit that we no longer know what theory is are we able to see that communication theory is now flourishing.

Conventional definitions of theory have lagged far behind practice; they no longer reflect the actual range of theoretical work in the field. This is our present state of confusion. We lack even a coherent vocabulary with which to discuss the great variety of ideas that currently announce themselves as communication theory. Different lines of theoretical work are premised on fundamentally different views of theory. The very notion of theory is radically contested.

Why are there now so *many* communication theories? This development is best understood in the context of trends that have transformed not only our own discipline but the human sciences generally.

Blurred Genres

The essential transformation of the human sciences that best explains both the recent proliferation of communication theory as well as our present confusion about theory was aptly named by Geertz in his famous essay "Blurred Genres" (1980). As Geertz pointed out, the formerly clear boundary between the social sciences and the humanities has become indistinct (hence the increasing usefulness of a term like "human sciences," which includes both). Scholars in many disciplines "have become free to shape their work in terms of its necessities rather than received ideas as to what they ought or ought not to be doing" (Geertz, 1980, p. 167).

---

[2] Redding (1992) and Purcell (1992) also comment on the narrowness of Berger's implicit definitions of theory and the field of communication. In response, Berger (1992) complains that he "cannot see how either single individuals or departments can possibly embody completely and competently the full spectrum of content included in the field" (p. 103). This is a reasonable view, but it does not address the point at issue, which concerns the range of intellectual products that will be admitted to constitute legitimate forms of communication theory, a field that no individual or department, of course, can hope to master in depth across its full breadth. How to cope with the blessing and curse of specialization in graduate programs is another problem.

The humanities have recently become, in a certain sense, more scientif-ic. By this I do not refer primarily to the use of computers and statistics for documentation, textual analysis, and the like, although such techno-logical developments have certainly occurred and are perhaps not with-out their own importance. What I wish to emphasize is that the humani-ties have become increasingly *theoretical*. If in the past we thought that the humanities were concerned with the understanding and appreciation of historically particular acts and artifacts while the proper business of philosophy and the sciences were questions of a general or theoretical nature, this definitely is no longer the case. Interdisciplinary theory in the humanities now engages philosophy, social theory, rhetoric, and cultural studies as well as literary and artistic critical theory in a common, though heterogeneous, discourse. Through this interdisciplinary discourse, strands of postmodernism, deconstruction, reader response theory, his-toricism, feminism, Marxism, psychoanalysis, and so forth, emerge, split, recombine, and spread across disciplines and continents.

Now it may be a little misleading to say that the humanities, in becom-ing more theoretical, have become more scientific, for the new schools of interdisciplinary theory show little resemblance to traditional forms of theory in the empirical sciences. Instead they have mounted a serious challenge to received notions of scientific theory. They have sensitized us to the inescapable literary and rhetorical dimensions of *all* language and *all* discourse, including all academic discourse, especially scientific theo-ry itself. Increasingly, schools of theory in the human sciences are thought of as *literary genres* in their own right. In what Simons (1990) has called "the rhetorical turn," theory is conceived as practical, historically situated discourse, not altogether different from those discourses of the arts and public life that were formerly the special province of the humani-ties. For those who understand theory to be an essentially *literary-rhetor-ical* enterprise, the relation of theory to such *epistemological* criteria as falsifiability or scientific explanation becomes problematic in the ex-treme. Theory-as-discourse has no clear relation to theory-as-knowledge. The traditional vocabulary of scientific theory construction may thus be largely irrelevant to the new forms of theory. However much one may like or dislike this situation, to ignore or deny it can only serve to worsen our present state of confusion about theory.

While the humanities have become more scientific in this admittedly rather special sense of the term, the humanistic elements and tendencies within the social sciences, influenced by the same international-interdisci-plinary trends, have flourished anew. It should be emphasized that the "harder" approaches to the social and behavioral sciences have also con-tinued to flourish. Their vitality apparently has not diminished, but it would be pointless to deny that their relative dominance in some of these fields has declined as other styles of social science have gained adherents and recognition. Postmodernism, feminism, and other such interdiscipli-nary movements have infected the social sciences no less than the humanities. Moreover, the practice of broad, speculative social theory—inspired by such prominent exemplars as Habermas, Foucault, Giddens, and others—has regained the prominent place from which it had apparently been dethroned by the vogue of middle-range theory in the decades after mid-century.[3] Increased attention to the *historical* dimension of social processes; increased interest in textual, discourse analytic, and ethnographic methodologies; increased appreciation and use of qualitative methods more generally—these trends are much in evidence across the social sciences. Each of these trends, along with the emergence of interdisciplinary theory in the humanities, contributes to the blurring of boundaries among the human sciences and calls into question the metatheoretical vocabulary of explanatory scientific theory in social science, according to which the social sciences can advance only by becoming *harder,* more quantitative, more like the physical and natural sciences.

Theory and Practice

Thus, as Geertz (1980) claimed, the boundary between the social sciences and humanities has become blurred. But the transformation of the human sciences has involved more than just an emergence of theory in the humanities accompanied by a resurgence of "softer" approaches to the social sciences. A blurring of the boundary between *theory and practice*—a development that raises basic questions about the *purposes and goals* of theory—also has been important to that transformation. The goals of social scientific theory have traditionally been formulated as explanation, prediction, and control, but the logic of explanation that frames this epistemological holy trinity has become deeply problematic as attention has been drawn to the potentially *constitutive* role of theory in social life. Theory, on this view, influences how people in society think and talk about their own activities and thereby shapes those activities and the emergent social structures produced and reproduced in them. But, according to the logic of scientific explanation, a theory that actively shapes the very phenomena it purportedly explains is essentially untestable and thus irreparably unscientific. Any observations that might be adduced to test predictions derived from such a theory are hopelessly contaminated, and the logic of explanation falls into a vicious circle.

Faced with this situation, social scientists might adopt either of two basic strategies. First, they could insulate society as much as possible from any influence of social scientific theories in order to preserve the

---

[3] Note, for example, the title of Quentin Skinner's (1985) anthology: *The Return of Grand Theory in the Human Sciences.*

logic of explanation. But the futility (at least in relatively open, liberal societies) as well as the ethical absurdity of such a stance is obvious. A second approach would be to embrace the constitutive potential of theory and to take due responsibility for its consequences. Among those consequences, however, is the fact that explanation, prediction, and control as goals of social scientific theory become insufficient. Given a constitutive rather than just an explanatory function, theory must now address other issues and pursue other goals in order to justify itself as an activity. As Hesse (1980) points out, social scientific generalizations, to the extent that they are undecidable empirically according to the logic of scientific explanation, are warrantably—indeed, inescapably—judged by other values, whether aesthetic, moral, or political.

Again we find that theory must be regarded, by those who embrace its potentially constitutive role, as something other, or more, than *knowledge* only. Theory becomes not the crowning achievement of an objective social science but rather an integral component of an engaged social practice. Not only epistemic but broadly practical criteria enter into the discourse of theory.

Whether values other than epistemic (and, supplementarily, aesthetic) ones should serve as criteria for theory remains highly controversial. Many self-identified communication theorists remain strongly committed to the ideal of an epistemically grounded, empirical social science. Moreover, even among those committed to the cultivation of a closer relationship between theory and practice, precisely *how* our moral or political concerns should bear upon theory is subject to a great many different interpretations—another index of our present confusion about theory.

In the spirit of the dictum attributed to Lewin, that "there is nothing so practical as a good theory" (cited in Thayer, 1982, p. 21), some might advocate the application of traditional social scientific methods to address socially significant problems, even while continuing to adhere to the received view of scientific theory. Social action in the service of moral and political ends, according to this view, should be informed by epistemically sound, scientific knowledge. Explanation, prediction, and control remain the essential goals of science as such, but science should be used (without compromising its methodological and ethical principles) to advance toward moral and political objectives. In contrast, to advocate social actions based on false or empirically untestable beliefs about the consequences of those actions, beliefs contaminated by ideological dogma or wishful thinking (i.e., theories based on moral and political predilections rather than scientific research), would be morally irresponsible.

Traditional social science might acquire, then, a moral rationale. Although this argument is not without merit, especially in regard to the moral implications of scientific method, it fails to take full account of the potentially constitutive role of theory in society as discussed above. My own view is that we should largely set aside the traditional models of scientific *theory* even while continuing to grant empirical research methods an important place. A "practical" communication theory would be essentially normative, and we would continue to conduct empirical and other types of studies, to investigate the empirical assumptions and consequences of practical theory (Craig, 1989).

Both of the views just sketched derive largely from a single school of philosophy: American pragmatism. They are by no means the only current views on how to merge theory and practice. Critical theorists, feminists, social constructionists, and postmodernists display a bewildering variety of approaches to this question. Hence, considering the state of theory on the whole, we cannot conclude that theory and social practice have now merged in any definite way, but that the boundary between them, like the boundary between the social sciences and the humanities, has become blurred, indistinct. This blurring of boundaries has opened our own field to many new, hybrid varieties of theoretical work, all now calling themselves communication theory, and all creating a relative abundance of communication theory today—and its equally impressive disarray.

The Future of Communication Theory

Does our present moment, one in which the boundaries of theory and communication seem to be in flux, offer us an opportunity to reconfigure the future of the field so as to breach old barriers and open up new possibilities for communication research?

Perhaps ways can be found by which the various, apparently incompatible or unrelated modes of communication theory that now exist can be brought into more productive dialogue with one another. Perhaps an updated scheme of paradigms for communication research will reduce our confusion and broaden our awareness of alternatives. Perhaps communication science can be understood as an integrated "practical discipline" in which critical, interpretive, and empirical research as well as philosophical reflection and applied work have deeply related, essential functions to perform (Craig, 1989).

However broadly the *discipline* is understood, however confused about theory we may collectively be, each of us who works at communication theory must do so in some specific way, according to some relatively definite, even narrow, notion of theory (cf. Berger, 1992). Even so, we may also hope that our work might speak to larger problems of the discipline. Basic questions about theory and research are now open and unsettled. Every contribution to communication theory, including one's own, exemplifies an approach to those questions—and the biases and limitations of that approach. Dialogue in the discipline will be advanced as we reflect on the various modes of theory and their characteristic biases and limitations. Situated within such a dialogue, work in our field cannot fail to engage with issues of broad concern throughout the human sciences.

References

Berger, C. R. (1991). Communication theories and other curios. *Communication Monographs, 58,* 101–113.

Berger, C. R. (1992). Curiouser and curiouser curios. *Communication Monographs, 59,* 101–107.

Berger, C. R., & Chaffee, S. H. (Eds.). (1987). *Handbook of communication science.* Newbury Park, CA: Sage.

Craig, R. T. (1988). The *Handbook of Communication Science*: A review. *Quarterly Journal of Speech, 74,* 487–497.

Craig, R. T. (1989). Communication as a practical discipline. In B. Dervin, L. Grossberg, B. J. O'Keefe, & E. Wartella (Eds.), *Rethinking communication; Volume 1; Paradigm issues* (pp. 97–122). Newbury Park, CA: Sage.

Dervin, B., Grossberg, L., O'Keefe, B. J., & Wartella, E. (Eds.). (1989). *Rethinking communication* (2 vols.). Newbury Park, CA: Sage.

Ferment in the Field [Special issue]. (1983). *Journal of Communication, 33*(3), 1–368.

Festinger, L. A. (1957). *A theory of cognitive dissonance.* New York: Row Peterson.

Geertz, C. (1980). Blurred genres. *The American Scholar, 49,* 165–179.

Hesse, M. (1980). *Revolutions and reconstructions in the philosophy of science.* Bloomington: Indiana University Press.

Kuhn, T. S. (1962). *The structure of scientific revolutions.* Chicago: University of Chicago Press.

Merton, R. K. (1957). *Social theory and social structure.* New York: The Free Press of Glencoe.

Popper, K. R. (1959). *The logic of scientific inquiry.* London: Hutchinson.

Purcell, W. M. (1992). Are there so few communication theories? *Communication Monographs, 59,* 94–97.

Redding, W. C. (1992). Response to Professor Berger's essay: Its meaning for organizational communication. *Communication Monographs, 59,* 87–93.

Simons, H. W. (Ed.). (1990). *The rhetorical turn: Invention and persuasion in the conduct of inquiry.* Chicago: University of Chicago Press.

Skinner, Q. (Ed.). (1985). *The return of grand theory in the human sciences.* London: Cambridge University Press.

Thayer, L. (1982). What would a theory of communication be *for? Journal of Applied Communication Research, 10,* 21–28.

The Past of Communication's Hoped-For Future

by Klaus Krippendorff, University of Pennsylvania

In this essay I am suggesting that much of communication scholarship to date has been message driven and that this dominant form of explaining what communication is or does is slowly being challenged by what one may call reflexive explanations. This discursive perturbation offers researchers the choice of either narrowing their domain of inquiry to where message-driven explanations can be enforced, or embarking on an exciting path of reconstructing our field.

Let me begin by stating what I see as three defining features of message-driven explanations and then consider some of their fruits.

First, messages are *objectively describable* compositions, texts, or events. They are created to be *moveable* from one physical location or context to another or *reproducible* at different places or times. They thus exist in an objective reality and independent of anyone receiving them. References to intertextuality, message systems, or situational structures do not substantially alter the subject-independence of this starting point of message-driven explanations.

Second, messages *affect,* persuade, inform, stimulate, or arouse those exposed to them. Whatever messages cause or bring to their receivers, their contents, the symbolic qualities they have, are believed carried in their composition or structure and must therefore be explained or be theorized *as a function of these objective properties*. Cognition is simplified to a linear process of responding to or interpreting given messages.

Third, *exposure to the same messages causes commonalities* among senders and receivers, and, in the case of the mass media, among audience members. This gives "communication" its social significance and a standard for evaluating its success. So, deviations from expected commonalities become individual failures, misinterpretations, ideological or cognitive biases, noise, systemic distortions, and so on.

Thus, message-driven explanations are both *objectivist* and implicitly *normative*.

Klaus Krippendorff is a professor at the Annenberg School for Communication, University of Pennsylvania.

The End of Theories We Grew Up With

Message-driven explanations have ancient roots—for example, in the rhetoricians' search for linguistic forms that make arguments compelling. Their modern incarnation comes from journalism. Journalists see themselves as writing newspapers and magazines that were created to be mass produced and *uniformly comprehensible* to their readers. When new media such as radio and TV came along, and when interpersonal encounters, political events, and organizational processes came to be seen as communication as well, the printed message quickly became the dominant metaphor for conceptualizing them. The fact that discrete messages were not so obviously, if at all, identifiable in these new media; that differences in interpersonal skills, accessability, and authority had no place in these explanations; and that definitions of community or of a public based on common exposure to messages became empirically untenable, did not prevent communication researchers from refining message-driven explanations.

In fact, Lasswell (1948) codified the field, its research questions and explanations, by *defining communication research* as providing answers to the five questions: "Who, says what, in which channel, to whom, and with what effects." To date, his formula defines many communication research designs. In the same year, Berelson and Lazarsfeld (1948) finished their conception of content analysis as an "objective, systematic, and quantitative description of the manifest content of communication," promising scientific accounts of what messages carry to everyone with access to them. Also in 1948, Shannon (Shannon & Weaver, 1949) first published *A Mathematical Theory of Communication*. Many communication researchers immediately embraced his terminology, probably for the scientific legitimacy a mathematically founded concept accorded to inquiries into communication, including the mass media (Schramm, 1954, 1955). However, only Shannon's diagram and Weaver's popular commentary on the theory entered the bulk of communications literature. Although the theory extended our vocabulary—adding such terms as *redundancy* and the *encoding* and *decoding* of messages—Shannon's statistical and relativist measure of information quickly became equated with news, or the stuff that messages "objectively" contain. After these basic notions were in place, message-driven explanations mushroomed. Without reviewing the many and more increasingly sophisticated versions of message-driven explanations that developed from these early beginnings, let me simply suggest that they now permeate the examplars in our field: studies correlating message variables and effects, inquiries into the effectiveness of different message designs, use of mathematical theories to predict attitude changes from media exposure, and so forth. None of these regard the human participants in the progress as capable of making up their own meanings, negotiating relationships among themselves, and reflecting on their own realities.

Where message-driven conceptions of communication entered serious empirical tests, they turned out to be of limited explanatory value. For example, Katz and Lazarsfeld (1955) found evidence that led them to replace the hypodermic needle conception of mass media effects with a *two-step flow model*. The first step involved exposure to the media and the second an informal opinion-creating process mediated by opinion leaders. Klapper's (1960) massive review of the effects literature concluded that the mass media had rather limited abilities to shape their audience members' lives. His conclusions were criticized because (a) his review was sponsored by the networks who, being under public scrutiny, had an interest in its outcome; and (b) industry would not continue to finance the mass media through its advertising without reasonable expectations of a return on its investment. A more likely explanation for Klapper's findings is that message-driven conceptions just don't work. Obviously, the effects researchers, their reviewer Klapper, and his critics' responses to the mass media and to each other could hardly be explained in causal terms.

Faced with these apparent failures, scholars came up with new conceptions. In the beginning these conceptions appeared to be mere stopgap measures, designed to keep linear causal explanations in place. But they also provided the seeds for alternatives to the dominant accounting practices. Let me mention some of these.

One is the *uses and gratifications approach*. This approach can be traced to propaganda effects studies during World War II and to Berelson's (1949) study of what missing the newspaper meant during a strike in New York City. Inquiries into the social and psychological needs, sources of expectations, and gratifications derived from mass media attendance led proponents Katz, Blumler, and Gurevitch (1974) to turn the message determinism the other way. How audience members *used* these messages was found to be far from uniformly distributed among audience members. There was no obvious message determinism of effects.

Another and far further going approach can be seen in the *information-seeking paradigm* (e.g., Donohew & Tipton, 1973). Here, "objective" contents of messages are largely irrelevant. Individuals are seen as actively engaged in diverse information seeking, avoiding, and processing strategies, which turn out to be explainable in terms of their "image of reality," their "goals, beliefs, and knowledge." Information no longer is explainable from the properties of message alone. Senders or producers no longer play the central role that message-driven explanations assigned to them.

In organizational communication research, a so-called *interpretive approach* (e.g., Putnam & Pacanowski, 1983) has become increasingly appealing. It centers on the way individuals make sense of their world through communicative behaviors, and it attempts to explain choices in terms of prevailing "organizational cultures" or working climates to which members of an organization come to be committed. It holds that *mean-

ings are created and negotiated, neither objectively given nor assignable by a scientific authority. Individual participation in a social network of interaction, not the messages, become the explanatory basis of outcomes and effects.

Probably the most productive demonstration of the not so minimal effects of mass communication is the research on the media's ability to *create issues* and *set agendas* for public opinion and action (e.g., McCombs & Shaw, 1972). Clearly, issues, agendas, and controversies exist neither outside language nor without political actors' understanding. They are part of the very public discourse within which mass media institutions and mass media audiences constitute themselves. Their *reality resides in the playing of a public game* of, albeit unequal, participation. One aspect of this unequal participation has been theorized as the so-called "spiral of silence" (Noelle-Neumann, 1984), which adds to the setting of agendas participants' perception of each others' opinions on issues; it attempts to predict the emergence of certain political realities by processes analogous to self-fulfilling prophecies.

Sources of Breakdown and Alternatives

Actually, communication research is comparatively late in experiencing such breakdowns of message-driven explanations which, while still rampant in public and everyday discourses, have been dismissed in other disciplines for different reasons.

The breakdown of the popular notion of language as descriptive or representative of an objective world external to us and separate from language has been slow in coming, but it now enters the study of communication at numerous not so obvious entry points. It started with the Wittgensteinian notion of *language as a game people play,* was paralleled in the Whorfian hypothesis of *linguistic relativity,* and has recently led to the search for more adequate accounts of meaning in terms of the *cognitive schemas* underlying the understanding of linguistic constructions (Johnson, 1987; Lakoff, 1987). Here, the objectivism in message-driven explanations is quite explicitly and thoroughly discredited and replaced by an experientialist alternative.

Social *constructivists* have shown that "facts"—from emotions to persons, gender, language, and cultural institutions—are *socially constructed,* in the sense of having been *invented,* perhaps at a time no longer accessible to individual memories (Berger & Luckmann, 1966) or from behind the facade of political institutions (Edelman, 1977), now being habitually reproduced by its participants. Constructivists can be divided into three camps. The first maintains the belief in an observer-independent reality relative to which constructions by the media and by ordinary people could be compared and verified. In the opinion research literature, Lippman was an early proponent of this view. Boorstin (1964) still believes he can distinguish pseudo-events from real ones. Tuchman (1974) considers news as constructed by the way the mass media are organized, and Gitlin (1979) demonstrates how hegemonic processes account for TV entertainment. Efforts to deconstruct social phenomena by showing how real social institutions, hegemonic forces, and power structures are responsible for them belong here as well. This approach—some call it *trivial constructivism*—is unable to take institutions, structures of domination, ideologies, and so forth, including the reality referred to in explaining these phenomena, as *the analysts' constructions.*

The second group, the *social constructionists,* tie themselves to the later Wittgenstein and subsequent natural language philosophers by arguing that all social phenomena can be explained by reference to language. Foucault (e.g., 1989) exemplifies a grand semiotic version of this view. Gergen (1985), his collaborators, and several discourse analysts—few of which build on Foucault's work—have shown how persons, emotions, gender, (self-)identity, taboos, and so forth are constructed and negotiated in language. They see no need to refer to facts outside of language. From their perspective, the mass media do not merely construct a public reality, they also construct themselves into it.

Finally, *radical constructivists* (Glasersfeld, 1991; Watzlawick, 1984), joining hands with second-order cyberneticians (Foerster, 1974; Mead, 1968) and with biological cognitivists (Maturana & Varela, 1987), go beyond language determinism by insisting that internal and external *reality* is omnipresent but *not knowable without constructive participation* by its observers. This seriously challenges the claim of privileged access to reality as a basis of scientific authority, questions the use of this metaphysics by scientists to justify their role as intellectually superior observers of less sophisticated others, and criticizes the failure of researchers to reflect on their own cognitive participation in the phenomena they claim to describe. It radically doubts anyone's ability to provide objective accounts of the meanings messages have for others and thereby removes the ground of message-driven explanations. Radical constructivists also embrace an important demand of feminist theory (e.g., Belenky, Clinchy, Goldberger, & Tarule, 1986) to treat *knowledge* not as abstract and freely transmittable, but as *embodied in a knower* who supplies his or her own terms for understanding, embracing both intellectual and emotional experiences. This means that no knowledge can exist outside knowers and that all facts have their factors, their makers. This constructivism is radical because its conceptual framework grants *no epistemological exceptions* to scientific observers, constructivists included.

Let me offer just one example of the kind of message-driven research whose unreflected claims I find increasingly offensive. Recently, I attended a workshop on the effects of television on children. A good part of it was devoted to *children's understanding.* Proceeding from commitments to message-driven explanations, the researcher exposed children to selected TV images and tested for what they could recall and correctly identify. The findings of these experiments were graphed and presented as showing how children's understanding improved with age. A constructivist critic might ask how and whose understanding is being articulated here: Clearly, the commitment to message-driven explanations was the researcher's, not the children's. What counted as messages (what the TV images depicted) was decided by the researcher, not by the children. And although children live, at least in my experience, in a very imaginative, fantastically rich, and certainly more varied world than adults do, the researcher allowed as data only what he could cast into the categories of his own operationalized understanding. The children's did not matter. The researcher observed no more than how well children's (unobserved) understanding conformed to his well articulated expectations of what children should see or do if they were more like him and less like the children they were. He acted as the self-appointed agent of an objective, shared, and adult world in which and to which children are expected to adjust, and explored no more than *his own preconceptions,* using children, much as they are used in society—as convenient props. Message-driven studies obviously *disrespect others' understanding.* The claim to have studied children's understanding is not sustainable in the face of the apparent intellectual imperialism.

One alternative to the above is the anthropologist Wagner's (1981) conception of culture. In the minds of objectivists, culture usually ends up being a causal agent of overwhelming power. For Wagner it becomes the anthropologists' way of explaining their encounters with people other than their own. Respecting, yet not grasping, the emergence of otherness in conversation, feeling the loss of certainties that everyday communication does afford, experiencing breakdown in the taken-for-grantedness of common sense, leads the analyst to invent and the interlocutors to coconstruct something both can live with. For reflexive anthropologists, this entails *reinventing their own culture.* Applied to the research example, Wagner might suggest listening to the children's stories with wonder and trying to make sense of why they tell us, if they do, what they see in terms of our understanding of their understanding of us. In such a reflexive loop, we might learn at least to appreciate children's ways of seeing. We might also come to understand something about our own understanding of, for instance, how constrained we have allowed ourselves to become.

In the above, I see a new convergence of natural language philosophy, ethnography and cognitivism in linguistics, social and radical constructivism, second-order cybernetics, reflexive sociology, and the above mentioned responses to the failures of message-driven explanations (not just in mass communication), to which one might add efforts to understand the new interactive media (computer interfaces, hyper-media, virtual reality) that have so far defied traditional theorizing. The epistemology of this new constructivism calls into question whether we could have communicated the way we said we did.

Consider the rather sketchy history of communication research, its failings and the emergence of alternative paths, that I constructed *as a story*. It began with familiar but simple-minded accounts of how messages drive humans into compliance. But in its unfolding it is obvious that this simple beginning *contradicts the very experience of constructing, communicating, and listening* to (or reading) it. Here, our story confronts its own reality, which resides in its present telling. It suggests that the reality we sought to approximate by our scientific accounts always was *of our own making,* and it now calls on us to bring into focus the very communication practices we use in inquiring and writing about communication.

To me, this realization marks a bifurcation point for communication research. I don't anticipate that message-driven communication research will disappear. People in positions of authority are all too eager to embrace deterministic reality constructions that can offer them the prospect of forcing predictability and controllability onto others. Witness the use of message-driven vocabulary in the mass media, politics, education, advertising, public relations, and management. Communication researchers can withdraw into this comfortable niche where message-driven explanations are enforced and the handmaidens of manipulatory interests are rewarded. This would surely be the end of our story.

Becoming aware of the reality in telling our story of communication is a way of getting out of the trap of message-driven explanations. But it also means accepting the notion that reality is a social invention. Surely, we could not otherwise explain the experience that *Reality Isn't What It Used to Be* (Anderson, 1990) and how our constructions of communication could be evolving, as they do, in the very process of inquiring and communicating about them. The revolution that this new understanding of reality can set in motion could be of a Copernican magnitude. However, while Copernicus's theory challenged only the location of the center of the then known astronomical universe and left the hierarchical organization of social and religious life and the objectivist construction of the universe pretty much intact, the epistemology of this new constructivism challenges the privileged role of disembodied knowledge and reveals its complicity in the emergence of hierarchical forms of social and political authority and its attendant requirement of submission.

Constructability of and in a Virtuous Future

I am suggesting that the strands of scholarship mentioned above could be woven into a radically *new and virtuous synthesis,* seeing humans first as cognitively autonomous beings; second, as reflexive practitioners of communication with others (and this includes social scientists in the process of their inquiries); and third, as morally responsible interveners in, if not creators of, the very social realities in which they end up living. To embrace this new epistemology, let me end this essay by suggesting that communication scholars recognize the social constructibility of reality, with all of its consequences, and make commitments on each of these three points.

First, the commitment to respect the *cognitive autonomy* of those observed and theorized. This presupposes the recognition that language, communication, indeed all *social phenomena* exist only in the knowledge their participants have of them. Specifically, there can be no scientific or everyday understanding of human communication without an understanding of the understanding of those involved in communication. Storytellers can attest to this. Scientists know it when writing for their peers. I am merely suggesting that we grant those we seek to theorize like abilities of understanding. In contrast, message-driven explanations equate scientists' understanding with objective truth and therefore cannot respect others' understanding of communication, unless they all agree. Nor can they acknowledge that anyone's understanding of communication is reflexively embedded in communicating about it. *Cognitive autonomy* resides in the (my) fact that (a) individuals cannot be forced or caused to understand something as intended, as it exists, or as it should be; (b) that nobody can directly observe someone else's understanding; (c) that all individual actions are dedicated to preserve individual understanding, and (d) that understanding is never final, even in the absence of external stimulation.

Respecting this autonomy *prevents abstract and disembodied communication theory constructions* and encourages explanations of communication phenomena (and of other social constructions) from the bottom up, from the knowledge and practices *embodied in its participants*. This contrasts with top-down explanations that attribute determining forces to someone else's (usually the observing scientists') super-individual constructions—for example, ideologies, hegemonic forces, cultural determinisms, rules, or objective meanings. Respecting this autonomy also means *abandoning the idea of creating general theories without obtaining, as far as possible, the consent of those theorized.* If people do hold different theories of communication and practice them with each other, a general theory of communication may not do justice to either. Indeed, there are plenty of eminently practical folk theories people live by—for instance, communication as imparting knowledge, as maintaining or creating relationships, as domination or control, as healing wounds, as dance, and so forth. For inquiries in communication, I prefer a conversation metaphor because it respects the cognitive autonomy of others (Krippendorff, in press, a).

Second, communication scholars should commit themselves to *reflexive theory constructions* by means of which they can enter others' understanding into their own understanding. As understanding is never finished, this means that a reflexive reality cannot remain fixed either. It is continually created, tried out and tested each time it is being talked about. This is so for social scientists, whose analytical categories, originally invented for mere analytical purposes, can become real (Giddens, 1991, pp. 40–41); for politicians, whose campaign promises can change political practices; for engineers, whose inventions keep technology on the move; and so it is in the everyday life of communication. All social theories must also be communicable, at least among scientific peers, and may reach and affect those theorized therein. Neither can they escape the self-reference this entails, nor can their stability be assured in being communicated. Denying the reflexive nature of human communication (theory) sets communication researchers apart from their subjects and creates reality constructions that aid technologies and can support oppressive social structures. Reflexivity is perhaps the most outstanding feature of human communication. I have proposed (Krippendorff, in press, b) that human communication scholarship redefine itself in terms of the discourse that embraces itself.

Third, we need a commitment to what one might call *a distributive ethics* for social inquiry. In the preceding, I claimed that knowledge, especially social scientific knowledge—communication theory, for example—can hardly be prevented from entering the phenomena it addresses. Whether it is intended to be critical or merely descriptive, it can delegitimize what exists or contribute new social constructions. The changes thus brought forth encourage the emergence of radically distributed realities, a multiverse of reflexive constructions, that no general theory can capture. I believe that the increased awareness of our role in the socially (and hence communicationally) constructed, distributed, and emerging nature of contemporary realities has brought us, as social scientists, to a point where truth is secondary to the responsibilities we bear for our constructions. To be consistent with this new multiverse means to distribute this responsibility. I made the methodological proposal to invite those affected to participate in the construction of communication theories concerning them (Krippendorff, in press, a). Living such an ethics may not be easy. However, practical difficulties should not deter us from developing methodologies that assure respect for those theorized.

No story of our message-driven past can tell us what the future has in store. But its present telling demonstrates the constructed nature of our field, and the awareness of this demonstration affirms our role in inventing and reconstructing the social realities we work in. If this is so, we might as well take the poetic licence to construct, and put into a story, the most desirable realities we can imagine. Understanding this understanding could be a moment of liberation.

References

Anderson, W. T. (1990). *Reality isn't what it used to be*. New York: Harper Collins.

Belenky, M. F., Clinchy, B. M., Goldberger, N. R., & Tarule, J. M. (1986). *Women's ways of knowing: The development of self, voice, and mind*. New York: Basic Books.

Berelson, B. (1949). What missing the newspaper means. In P. F. Lazarsfeld & F. N. Stanton (Eds.), *Radio research 1948–49* (pp. 111–129). New York: Basic Books.

Berelson, B., & Lazarsfeld, P. F. (1948). *The analysis of communication content*. Chicago & New York: University of Chicago & Columbia University.

Berger, P., & Luckmann, T. (1966). *The social construction of reality*. New York: Doubleday.

Boorstin, D. J. (1964). *The image*. New York: Harper & Row.

Donohew, L., & Tipton, L. (1973). A conceptual model of information seeking, avoiding and processing. In P. Clarke (Ed.), *New models for mass communication research* (pp. 243–268). Beverly Hills, CA: Sage.

Edelman, M. (1977). *Political language*. New York: Academic Press.

Foerster, H. von. (1974). *Cybernetics of cybernetics or the control of control and the communication of communication*. Urbana: University of Illinois, Biological Computer Laboratory.

Foucault, M. (1989). *Power/knowledge: Selected interviews and other writings, 1972–1977*. New York: Pantheon.

Gergen, K. J. (1985). The social constructionist movement in modern psychology. *American Psychologist, 49,* 266–275.

Giddens, A. (1991). *Modernity and self-identity*. Stanford, CA: Stanford University Press.

Gitlin, T. (1979). Prime time ideology: The hegemonic process in television entertainment. *Social Problems. 26,* 251–266.

Glasersfeld, E. von. (1991). Knowing without metaphysics: Aspects of the radical constructivist position. In F. Steier (Ed.), *Research and reflexivity* (pp. 12–29). London: Sage.

Johnson, M. (1987). *The body in the mind*. Chicago: University of Chicago Press.

Katz, E., Blumler, J. G., & Gurevitch, M. (1974). Utilization of mass communication by the individual. In J. G. Blumler & E. Katz (Eds.), *The uses of mass communications* (pp. 19–32). Beverly Hills, CA: Sage.

Katz, E., & Lazarsfeld, P. F. (1955). *Personal influence*. New York: Free Press.

Klapper, J. T. (1960). *The effects of mass communication*. New York: Free Press.

Krippendorff, K. (in press, a). Conversation or intellectual imperialism in comparing communication (theories). *Communication Theory*.

Krippendorff, K. (in press, b). Trying to talk the Russellian Ghost back into the bottle it escaped from. In D. Crowley & D. Mitchell (Eds.), *Communication theory today*. Cambridge, MA: Polity Press.

Lakoff, G. (1987). *Women, fire, and dangerous things*. Chicago: University of Chicago Press.

Lasswell, H. D. (1948). The structure and function of communication in society. In L. Bryson (Ed.), *The communication of ideas* (pp. 37–51). New York: Harper.

Maturana, H. R., & Varela, F. J. (1987). *The tree of knowledge*. Boston: Shambhala.

McCombs, M. E., & Shaw, D. L. (1972). The agenda-setting function of the mass media. *Public Opinion Quarterly, 36,* 176–187.

Mead, M. (1968). Cybernetics of cybernetics. In H. von Foerster et al. (Eds.), *Purposive systems* (pp. 1–11). New York: Spartan.

Noelle-Neumann, E. (1984). *The spiral of silence*. Chicago: University of Chicago Press.

Putnam, L. L., & Pacanowski, M. E. (Eds.). (1983). *Communication and organization: An interpretive approach*. Beverly Hills, CA: Sage.

Schramm, W. (1954). How communication works. In W. Schramm (Ed.), *The process and effects of mass communication* (pp. 3–26). Urbana: University of Illinois Press.

Schramm, W. (1955). Information theory and mass communication. *Journalism Quarterly, 32,* 131–146.

Shannon, C. E., & Weaver, W. (1949). *The mathematical theory of communication*. Urbana: University of Illinois Press.

Tuchman, G. (1974). *The TV establishment: Programming for power and profit*. Englewood Cliffs, NJ: Prentice Hall.

Wagner, R. (1981). *The invention of culture*. Chicago: University of Chicago Press.

Watzlawick, P. (Ed.). (1984). *The invented reality*. New York: Norton.

Verbing Communication: Mandate for Disciplinary Invention

by Brenda Dervin, Ohio State University

Most of the polarities that divide our field—universalist vs. contextual theories, administrative vs. critical research, qualitative vs. quantitative approaches, the micro vs. the macro, the theoretic vs. the applied, feminist vs. nonfeminist—are symptoms, not the disease. They are shallow indicators of something more fundamental. Because that which is fundamental eludes us, we see both tolerance (a comfortable acceptance of theoretical pluralism) and dissent (ideological and methodological contests) everywhere. It is as if we are all studying a very large elephant. Without addressing the question directly, we seem to assume that we are studying the same elephant, while comfortably relegating ourselves to our own parts. But every once in a while we bump into each other.

Our contradictions are used both as a measure of our tolerance (after all, she does x while I do y) and a measure of our dissent (but she is doing x the wrong way, or her work has these negative consequences). While caught in these ricochets between tolerance and dissent, we can pontificate on why media effects remain a black box or why our research seems irrelevant to practice or why disciplinary status eludes us. It's because "they" use the wrong methodology, wrong theoretic perspective, wrong ideology, wrong . . . They should become more like "us." What we have is dissent mythologized as tolerance.

At root here is the issue of difference—both the differences between different sectors of our field and the differences that are at the heart of what we study—the differences that characterize human beings, their symbolic lives, and their symbolic products. I would propose that it is how we treat the latter differences that confounds our own differences.

Our field and the social sciences in general have for the most part handled difference in ways that are not fundamental. Because of this, our theories are weak and we end up attending with much energy to artificial, symptomatic differences, squabbling over turf and status. We end up try-

Brenda Dervin is a professor in the Department of Communication, Ohio State University, Columbus. The author thanks the following for their useful comments on the draft manuscript: Sam Fassbinder, Robert Huesca, Priya Jaikumar-Mahey, Tony Osborne, Peter Shields, and Vickie Shields.

ing to use the summation of the products of our current work as if they showed the way out. In our periods of tolerance, we call for meta-analyses, hoping these will point to ties that bind. When they don't, we move into one of our periods of dissent. Being unable to deal with difference in a way that fundamentally makes a difference, we make no difference.

Ironically, in grappling with their own substantial and/or illusive polarities, most of the other social sciences point to the phenomenon of our field—communication—as the way out (e.g., Giddens, 1984; Habermas, 1987). Bruner clearly does this when he suggests that it is the making of meaning that is the "proper study of man [sic]" (Bruner, 1990, chap. 1). One sees calls for the study of communication everywhere. In a recent speech at Ohio State University, an anthropologist publicly challenged our field. Anthropology, he pronounced, has found communication and will do it better.

The anthropologist is optimistic. And, some in our field are too pessimistic about the potential of our field for disciplinary coherency (e.g., Beninger, 1990; Schramm, 1983). There is no reason to expect that the other social sciences will change easily, cast as they are in unresponsive disciplinary frames. Nor is there reason to expect that we, upon confronting our own disarray, cannot do something about it if we can recognize that in fact we have yet to capitalize on our strengths. In one sense, it might be said that we can never be a discipline because when the disciplinary frames fall, what must rise are process-driven alternatives based on fundamentals. Clearly, communicating is a fundamental. Everyone may rightly claim it. But, in a second sense, it can be said that if we were not so busy modeling the very disciplinary structures that blind us, we might find our strength. Our field does no better than other academic fields concerned with human beings in bringing the practical together with the theoretic. Yet no one contests the bounty of practical wisdom embodied in our ranks. Even some of our most theoretic and critical scholars are called upon to make practical judgments in arenas ranging all the way from media design and practice to policy and legal considerations to the conduct of everyday personal, relational, and organizational lives. More often than not there is at least a disparity and sometimes an enormous discontinuity between the guiding academic project and the practical wisdom offered. The gap is filled with the consciousness of the individual communication academic. We call this the theoretic vs. applied contradiction and accept it as a given of our field. What we don't seem to understand is that this contradiction may mask our greatest strength. It is not that our work ignores theorizing for practice. Rather, we subordinate it to the more pressing academic mission. To theorize the practice of communication would require that we focus on communication theorizing of communication. We focus, instead, on other kinds of theorizing—sociological, psychological, anthropological, and so on. This essay asks: What if we were able to develop communication theory for communication practice—if we could bring our practical and our theoretical activities to-gether?

The immediate response—from within the many caves in our field where our contests are waged—is that theory for practice is not possible. The reasons would themselves form an array of contradictions. On the one hand, for example, theory for practice would be challenged as too oppressive, prescriptive, modernist, and/or totalizing, or erroneously uni-versalist, leaving important cultural, contextual, and personal factors be-hind. On the other, it would be challenged as too ambitious and/or too removed from experimental control.

This listing does not exhaust the challenges that can be mounted. Re-acting to each would require an essay in itself. The important point I wish to make here is this: These challenges rise out of the same kinds of theo-rizing about difference that beget the unproductive dualisms that encum-ber our field. Our field's hidden strength is that our phenomenon of inter-est—communication—is positioned at the very cutting edge of the study of the human condition. We already know much of what it would mean to develop a communication theory of communicating. To capitalize on that strength, we must let go of the theoretical strategies that prevent other fields from looking at communication communicatively and look to our hidden strengths, our foundational interest in how communicating is done.

We already have within our grasp a variety of coherent theories of com-municative practice, but the clarity of our vision is clouded with debris we have imported from other fields. If we can clear this debris away, we may be able to reach for a core that in no way will eliminate our contests, but will give them productive meaning. Our differences would become informative.

Metaphors for Difference: Nouns and Verbs

The debates in our field and in the social sciences generally rest on a ro-tating axis of polarities. The polarizations have between them something common and something uncommon. Thus, for example, the universalist vs. contextualist debate focuses on positions that adhere to and challenge the idea that universal theories of human situations can be developed. In contrast, the quantitative vs. qualitative debate rages between those who accept and reject quantification. Those who accept quantification are also more likely to accept the quest for universalist theory while those who use qualitative approaches are more likely to accept the quest for con-text-bound theorizing.

Because academics use normative narrative practices to make advances by first defeating enemies, there is some utility in looking at how innova-tive theoretical and methodological work is often built on fortresses of critique. This is as true of advances within literature genres—where, for example, one postmodernist tears down another, or one quantitative study proceeds by challenging another—as it is between genres—as, for example, in the critiques of so-called positivist approaches by advances in qualitative research, ethnomethodology, and cultural and postmodern studies (e.g., Hall, 1989; Lather, 1991).

Having identified an enemy is not, however, the same thing as having diagnosed a disease. Unfortunately, the metaphors get mixed and intertwined. If the enemy is called positivism, anything that has any related symptoms (i.e., quantification, analytic methods, statistical tools) is automatically also called enemy. As a result, the polarizations on the rotating axis of contest proliferate. At one end of the polarity we most often find fundamentalism, totalization, modernism, authoritarianism, structuralism, and master narratives. At the other end of the polarity we most often find relativism, postmodernism, contextualism, culturalism, and poststructuralism. "Isms" proliferate, and in the context of the debate (and the publish-or-perish mandate that fuels it) words get used so facilely and glibly that it becomes difficult to understand what all the fury is about, particularly when the results of the fury do not seem to advance significantly our individual or collective projects.

It is a simplification, but one useful for purposes here, to suggest that at the center of all these contests is the issue of difference—where to locate it, how to define it, what to call it, and how to look at it. In our field—and, it appears, in most of the social sciences—difference is most often defined simply as that which is not the same. The approach is not to identify what difference will make a difference but rather to identify a difference that is not yet claimed as another's turf, thereby claiming it as one's own. It would be unfair, of course, to relegate all of this solely to turf war, for it often represents a genuine concern for untapped difference. It is in this context that scholarships of the disenfranchised have had such important force, for each has impelled a new voice to the fore as a voice relevant for scholarly attention, a voice heretofore ignored or marginalized.

It is not my purpose here to trace the treatment of difference in our literature. The intent is to be suggestive of our history and pertinent to our present. To do this, I shall focus on the methodological moves involved in locating something we define as difference. This is a methodological concern that embodies within it acts of defining, labeling, and looking. It constitutes a fundamental methodological construction, resulting from a synergy of moves.

Our projects are ultimately about difference. We search for pattern and for deviation from pattern regardless of whether we define ourselves as in the business of description, explanation, or prediction or whether we reject, as I do, this too facile division of labors. Some of us search for pattern in a straightforward prescribed manner—via statistical tools, for example, or authoritative readings, or master narrative theorizings. Others of us suffer qualms of uncertainty for fear our search for pattern disrespects difference. As we get more sophisticated in understanding discourse, we begin to understand that even the methodological act of locating difference—the act of differencing—is itself an imposition of pattern.

In short, when we difference, we must put difference somewhere. In our field now there are two primary sites—one is in culture; the other is in agency. On the surface these look like quite different methodological moves. But from the standpoint of this essay they are construed as being fundamentally identical. They both deal with difference without dealing with difference.

Culture is a wonderfully rich term, "warmly persuasive" as Williams (1976) characterized it in his useful archeological dig into the term *community*. When efforts to describe, explain, and/or predict human communication proved alarmingly limited using structuralist frameworks (e.g., class, organizational, personality, language, and text structures), scholars reached for nonstructuralist ways of embodying the differences that were implied by but eluded earlier efforts. Culture has in effect become one of the latest catchalls. Difference resides elusively there. This is manifested, for example, in quantitative work when cultural factors are added to predictive formulations. It is manifested in qualitative work when, for example, discourse is analyzed as at least culturally anchored if not culturally prescribed.

The culture metaphor for difference is quite simple. Culture becomes a box into which groups of entities (i.e., people, texts) are slotted. The relationship is part–whole. Culture is the whole. Humans and texts are the parts. Culture is frozen at least for that moment, conceptualized as a noun. The humans and the texts are also conceptualized as nouns. Pattern is framed in these terms, as is deviation from pattern. Difference becomes defined as discrepancies between entities conceptualized statically. Culture is at one moment homogenizing structure, at the next resistant difference. Individuals are positioned as homogenized or discrepant.

The structure vs. agency distinction is most often identified with Giddens (1984). My use of the term here addresses the amalgamation that has become prevalent in our field—an amalgamation that has captured some of the substantive emphasis in Giddens's project but has failed to bring forward as well the important aspects that attempted to avoid a structure–agency dualism. Most of the references in our field have not sidestepped this difficulty.

As it is commonly used in our field, the structure vs. agency distinction is a step forward in that it does not on the surface define agency as deviation from structure. It is important, however, to note that structure still is for the most part conceptualized as static. Structure is noun; agency is verb. Constraint and homogenization rest in structure; freedom and variation rest in agency. Although it is often acknowledged that there are contradictions and spaces within and between structures, at any given level of analysis the structure–agency dualism serves as a rigid methodological blinder constraining formal theoretical work.

Critiquing False Dichotomies

The difficulty with both of these sites for locating difference—culture and agency—is that they still invite the ricochet between fundamentalism and anarchy, authoritarianism and relativism, modernism and postmodernism. They still posit structure as noun and thus discursively static. Anything that is fluid must thus be in opposition: culture vs. individual, structure vs. agency, power vs. freedom. These ricochets allow the methodology of the moment to advertise itself cloaked in false dichotomies. These ricochets confuse method with methodology and theory with ideology. They assume, for example, that those who prefer qualitative approaches don't observe or analyze while those who use quantitative approaches don't think (Bruner, 1990). They resort to a technological determinism in assuming that methods are entirely constrained by ideology.

Even when we attempt to move away from these false dichotomies or to move away from the very strategy of polarizing as a way of defining our regard for each other, our contests bind us to a brute portrait. Difference is free; homogeneity is bound. We deny it, often vigorously, but at one end of our axis of polarities is structure, constraint, power, homogenization, order, subject as object; at the other, freedom, variety, chaos, subjectivity. We are forced to choose an entry point. Even in the most recent efforts to bring together different viewpoints—as, for example, the critical with the postmodern—the choice usually remains (e.g., Best & Kellner, 1991). We are unable to stop taking sides and start moving toward multiple perspectives that might inform each other in a dialogue of differences.

There are many avenues for critiquing the false dichotomies that bind our field. Most damning for us is that these dichotomies often lead us away from the study of communication. Pattern gets located in society (sociology), culture (anthropology), individual (psychology); political and economic processes are defined as homogenizing, serving pattern; and interpretive freedom is defined as the only place where difference can safely reside. This narrative structure is as dominant in administrative work as it is in critical work.

Because we continue to embody difference as that which is in opposition to structure, we fail to fully capitalize on (even though we give lip service to it) our understanding of the role of communication in the implementation of order as well as disorder, structure as well as agency, constraint as well as freedom, homogeneity as well as difference. In forcing ourselves to choose one end of the polar axis or the other, we allow our own phenomenon of interest to elude us. Because we define both constraint and freedom essentially as nouns, we fail to see that both are made, maintained, reified, and changed in communicating. We fail to fully conceptualize difference as differencing, as a communicating move, as a fundamental condition of human experiencing.

The irony of this situation manifests itself most clearly in the current debates on theories of the subject. The question at hand is: How shall we conceptualize our human being? Shall this human entity be cognitive, emotional, spiritual, physical, desiring, unconscious, conscious, discursively created, empowered, disempowered, or some combination? This listing does not exhaust the possibilities nor does it represent any single corner of the debate. Rather, it is designed to represent the diversity in the array and suggest that here again we are focusing on choosing a particular static way (or set of ways) with which to characterize humans. The characteristics are defined as adjectives, attributes of the human nouns.

We have begun to discuss the ever-changing subject, but for the most part, we posit the subject as moving from one state to another. The emphasis is on the states, not the moves. Given the nature of the polarities from which we come, this avoidance is understandable. How does one explain a subject that is ever-changing if one has only the conceptual tools of structure (homogeneity) or freedom (difference) with which to work? How can one focus on the moves when all one has is nouns with which to work?

What the idea of the ever-changing subject has accomplished is that difference is now being conceptualized as both across time (e.g., one entity differing across time) as well as across space (two entities being different at the same time). What is important about difference across time is that it begins to force us to attend to difference as fundamental, not as noun but as verb, as differencing. In doing that we can begin to genuinely capitalize on the study of communication. With such a change, our field might come into its own—because difference makes a difference *in* communication; differences come into existence *in* communication; differences rigidify *in* communication; differences are bridged *in* communication; and differences are destroyed *in* communication. Likewise, structures that attempt to homogenize difference as well as those that attempt to display it come into existence *in* communication; maintain, rigidify, and disappear *in* communication.

Homogenizing and differencing are reconceptualized as communicatings. Among other possible communicatings are: idea makings and idea repeatings, thinkings and emotings, listenings and arguings, positionings and vacillatings, cooperatings and contestings, polarizings and nuancings, categorizings and hierarchializings, nounings and verbings, and a host of other ways in which we humans individually and collectively make and break order. The clumsy verbings of nouns are intentionally used here to make this point: Our strong suit is our understanding of communication as process in myriad contexts. But our understanding of process has been relegated to second-class status—we do that when we testify, when we consult, when we teach. When we do our scholarship, we focus primarily on entities, not process; on nouns, not verbs.

My major point is this: If our field can refocus on communicatings instead of communication, we can begin to conceptualize in such a way that we can find more relevance (which does not imply more agreement) in each other's work. Further, the move will allow us to begin to transcend the false dichotomies that prevent us from theorizing communication as practice, as the verbings that humans, collectively and individually, use to construct bridges across gaps—self and other, self and community, structure and individual, self at time 1 and self at time 2, one aspect of self at time 1 to another aspect of self at time 1, chaos to order, order to chaos, homogeneity to difference, difference to homogeneity.

Even though we tend to be ashamed of our pervasive and foundational practical side, our field is the field that has always dealt with difference *in communicating* and that has always accepted both structure and difference and conceptualized communicatings as that which energizes the in between. We already have the theoretic potential for which this essay makes a call. Sometimes this potential shows clearly, sometimes through a fog. Sometimes this potential has been a point of major emphasis (e.g., Carter, 1991; Craig, 1989; Dervin, 1991).

In short, we know a lot that we don't know we know. Communicating is where the micro becomes the macro, the macro the micro. It is the in between, the doing, the making, the experiencing. No matter what stripe the scholars in our field wear, at some point all of them can be heard talking *communicatings* rather than *communication,* verbs rather than nouns. We are where structure and agency meet, both implemented in communicatings. It might be more useful to say we are where structure and agent meet, both implemented in the agency of communicatings. We are where individual as object and individual as subject meet, both implemented in communicatings. We are where the conscious and the unconscious meet, both implemented in communicatings. We are where hegemony and resistance meet, both implemented in communicatings.

Denouement: Whence the Species

Our field has already done more to move the social sciences from noun theories and methodologies to verb theories and methodologies than any other field. One difficulty here is that our contributions to other social sciences have been more methodological than substantive. In essence, we have propelled a communicative way of looking at things. This is a double problem. First, the social sciences—particularly U.S. social sciences—don't value methodology per se. Second, the social sciences journals are organized substantively. Our contributions get hidden in the cracks. We are also so busy taking potshots at each other and valuing those in other fields that we have not garnered our strengths or resources. Our progress is impeded by extant disciplinary structures, so our movement toward the verbs of communicating is not refined, gracious, or easy. It is being propelled in contradiction and failure. It is impeded by structures heavily in

place (e.g., publish or perish) and made worse by the economic and ideological encroachments on the academy that characterize our time.

But the bottom line is this: From the beginning we have stood more in
between—the humanities and the social sciences, the social sciences and
the physical sciences, the fields within the social sciences—than any
other field. While other fields worry about long-time fractionalizations,
our disarray is characterized by often disarming fluidity. Cast in the mirror
of current disciplinary structures, all this makes us appear weak. But
more than any other field we have been unable to escape the mandate of
difference. Psychology can find stable patterns in individuals across time;
sociology can find stabilities in societies; anthropology can characterize
culture as entity. But we are left with the sternest test of all—what happens in the elusive moments of human communicatings.

While others may be rushing in to claim the ground we have tread,
from the beginning we have had to deal with theory and practice, micro
and macro, structure and agency. And from the beginning we have had to
deal with process. We have praised process, we have even offered it to
the world as practical wisdom. We have only recently begun to acknowledge it and develop it intellectually. It is process, however—the verbs of
communicating—where we have something to offer that is, if not ultimately unique, at least for now ahead of the others. Because of this, we
can lead the way, if only we will.

References

Beninger, J. (1990). Who are the most important theorists of communication? *Communication Research, 14*(5), 698–715.

Best, S., & Kellner, D. (1991). *Postmodern theory: Critical interrogations*. New York: The
Guilford Press.

Bruner, J. (1990). *Acts of meaning*. Cambridge, MA: Harvard University Press.

Carter, R. F. (1991). Comparative analysis, theory, and cross-cultural communication. *Communication Theory, 1*(2), 151–158.

Craig, R. T. (1989). Communication as a practical discipline. In B. Dervin, L. Grossberg, B.
J. O'Keefe, & E. Wartella (Eds.), *Rethinking communication: Volume 1—Paradigm issues*
(pp. 97–124). Newbury Park, CA: Sage.

Dervin, B. (1991). Comparative theory reconceptualized: From entities and states to
processes and dynamics. *Communication Theory, 1*(1), 59–69.

Giddens, A. (1984). *The constitution of society*. Berkeley: University of California Press.

Habermas, J. (1987). An alternative way out of the philosophy of the subject: Communicative versus subject-centered reason. In J. Habermas (Ed.), *The philosophical discourse of
modernity: Twelve lectures* (F. Lawrence, Trans.) (pp. 194–326). Cambridge, MA: MIT
Press.

Hall, S. (1989). Ideology and communication theory. In B. Dervin, L. Grossberg, B. J.
O'Keefe, & E. Wartella (Eds.), *Rethinking communication: Volume 1—Paradigm issues*
(pp. 40–52). Newbury Park, CA: Sage.

Lather, P. (1991). *Getting smart: Feminist research and pedagogy with/in the postmodern*. New York: Routledge.

Schramm, W. (1983). The unique perspective of communication: A retrospective view. *Journal of Communication, 33*(3), 6–17.

Williams, R. (1976). Community. In R. Williams (Ed.), *Keywords: A vocabulary of culture and society* (pp. 75–76). New York: Oxford University Press.

Images of Media: Hidden Ferment— and Harmony—in the Field

by Joshua Meyrowitz, University of New Hampshire

As of now, media scholars have a rather limited shared vocabulary to describe exactly what it is they are studying about media or about a particular medium. This situation is not necessarily a serious problem for the
scores of fields and research traditions whose concepts and vocabularies
are brought to bear on media research questions, but it is a glaring problem for media studies because, even apart from other differences, we
have no common understanding of what the subject matter of the field is.

In comparing and contrasting one work with another, scholars sometimes rely on rather ambiguous dichotomies such as "content vs. structure," "content vs. form," or "manifest vs. latent." Yet as I will describe
below, terms such as *structure, form,* and *latent* are used so differently in
different media studies that many researchers misunderstand or talk past
each other, when they bother to speak to and listen to each other at all.
More typically, overviews of the field draw on a long laundry list of terms
and approaches specific to particular research camps. It is often unclear
how the findings of these different camps relate to each other or build
into some larger corpus of knowledge about media.

This essay argues that a fair amount of confusion in media studies has
resulted from the lack of explicit treatment of the most basic of questions:
"What are media?" Such a question has generally appeared too elementary to merit a serious response. Perhaps the widespread use of modern
media, such as the telephone, movies, radio, television, computers, and
tape and disk technologies—which has been a major stimulant to the
rapid growth of media studies in the first place—has fostered the belief
that everyone knows what media are and that one can therefore move immediately to other research questions. Yet even when researchers have
not confronted the issue of the nature of media explicitly, they have had

Joshua Meyrowitz is a professor in the Department of Communication at the University of
New Hampshire. Portions of this research were supported by summer fellowships from
the Graduate School and the Center for the Humanities at the University of New Hampshire. An early version of this article was presented as a paper at the Seventh International
Conference on Culture and Communication in Philadelphia in 1989. The author wishes to
thank Ed Wachtel, Beverly James, Donna Flayhan, and Michael Pfau for their helpful suggestions on earlier drafts.

to address it implicitly in order to conduct studies. And an examination of media scholarship with the question "What are media?" in mind reveals that different researchers have answered the question quite differently.

As with all attempts to comprehend complex phenomena and processes, we rely, often subconsciously, on metaphorical thinking to simplify and clarify our conceptions of media. I believe the field of media studies can be strengthened over the coming years by more attention to what is common and different, limiting and liberating, about the various metaphors for media. In this article, I attempt a preliminary meta-metaphorical analysis, by suggesting that the scores of surface metaphors that are used to describe media are manifestations of a handful of even simpler metaphorical constructs.

Media Metaphors

One does not need to dig too deeply to see that both popular and scholarly media analysts draw on an abundance of metaphors. Television alone, for example, has been described in terms of dozens of metaphors, including: companion, new state religion, plug-in drug, Big Brother, window on the world, baby-sitter, teacher, instrument of terror, network of social relations, thief of time, pulpit, shared arena, cultivator, agenda setter, white noise, new language, glass teat, electronic wallpaper, anthology of texts, and nineteen-inch neighborhood. Although media metaphors abound, they are sometimes treated as unproblematic descriptions of aspects of media or, more commonly, they are seen merely as figures of speech that have aesthetic rather than epistemological implications. Yet different metaphors flow from and foster different perceptions of media and lead to different research questions and findings. Metaphors are potent tools for seeing clearly, but they also blind us to other ways of seeing (Lakoff & Johnson, 1980).

I suggest that virtually all the specific questions and arguments about a particular medium, or media in general, can be linked to one of three underlying metaphors for what a medium is. Although various terms could be used to convey the general sense of these three metaphorical constructs, I summarize them here as *media as conduits, media as languages, media as environments.*

Media as Conduits
By far the most common image of a medium is that it is a sort of *conduit* that is important insofar as it delivers *content*. The conduit metaphor leads to such questions as: What is the content? What social, political, economic, organizational, ideological, and other factors influence the development and perception of content? How accurately does media content reflect reality? How do various audiences interpret the content? What effects does the content have? What alternative types of media content are possible?

This metaphor is so common because content is the first thing we react to when we use a medium. A message appeals to us or repels us. We respond with belief or disbelief. We are moved by a news story on starvation in Africa, uplifted by a heroic rescue, or troubled by the reported state of the economy. We wonder whether to buy an advertised product. We want our children to learn some intellectual skills from one television program, or we are worried about what social behaviors they may learn from another. And when we communicate through a medium, we usually are concerned about "getting our message across." We all have a sense that there is a difference between one truth claim ("Iraqi invaders pulled Kuwaiti babies from incubators") and a very different truth claim ("The incubator story was promulgated by a public relations firm hired by Kuwaitis to incite Americans to war"). While there are differences among the various channels through which content can be conveyed—such as newspaper, telephone, radio, television—the differences in messages stand out as the first thing to see, respond to, and study.

Although some researchers draw on more complex definitions of content, which include aspects of media made visible through other metaphors, the study of content that is stimulated by the conduit image of media is largely *medium-free*. That is, the focus on media content tends to minimize the attention given to the nature of the particular medium that holds or sends the message. Television content is an extremely popular topic of study, for example, simply because so many people attend to TV's messages. Yet most of the questions asked about television content deal with behaviors and communications that do not necessarily require the existence of television. Violence, sexism, sexuality, and government disinformation, for example, all exist without television; indeed, they exist without the use of any particular medium (at least in the most typical use of the term to refer to an impersonal mechanical communication device, in contrast to considerations of the vocal chords, tongue, ear, brain, ground, air, or culture as media).

Although it sounds strange to say that one can study media content without paying much attention to media, it is something that most people do daily. (Content researchers simply do it more rigorously.) When we miss a favorite television program, we may ask a friend or spouse to tell us orally "what happened." We accept that a written phone message tells us at least something about an oral telephone call; or to push this further, we accept at the start that the electronically reproduced sounds over a telephone yield a "conversation." We talk about movies being made from books ("faithfully" or "unfaithfully"). We read a transcript of a recorded interview or discussion and assume that it retains something from the original discussion and from the recording. These and other daily experiences and practices suggest that it is common in our culture to believe that there is some *content essence* that can be transported relatively unchanged from medium to medium—or from face-to-face interaction to medium, or from medium to face-to-face interaction.

Of course, those who draw on the conduit metaphor generally provide much more sophisticated analyses than a spouse's description of a missed TV program or telephone call. In addition to quantified and statistically analyzed studies of manifest content, scholars look at underlying cultural value systems and gender assumptions; examine the ways in which media narratives are shaped by political, economic, psychological, and organizational factors; probe the ways in which long-term exposure may cultivate certain attitudes among audience members; analyze topically or thematically defined genres; look at the ways in which different audiences engage in different "readings" of media "texts" (in effect, co-creating their own content); and so on. Many of these more sophisticated explorations point to the most common uses for terms such as *structure, codes, form,* and *latent* in media studies: structure of the content, content codes, form of the content, and latent content. For even in its most complex forms, research that grows from the conduit metaphor tends to look at some aspect of content and to ignore other latent aspects of the structure of mediated communications.

The conduit metaphor is widely shared in both the popular and scholarly arenas. It underlies broadly held concerns over children's imitation of antisocial behaviors seen on TV. It helps frame debates over news bias, gender portrayal, cultural elites, and family values. It is a stimulus for concerns over the public's susceptibility to propaganda. Ironically, this metaphor is one shared among competing social activists and among research camps that barely speak to each other, such as conservative Reed Irvine's Accuracy in Media (AIM) and progressive Jeff Cohen's Fairness and Accuracy in Reporting (FAIR), Feminists for Free Expression (FFE) and the Moral Majority, traditional content analysts and most critical theorists. Thus, many people who claim to share little with each other in terms of media study, actually share a fundamental image of what media are and what one should examine about media. They also often share a neglect of at least two other important conceptions of media.

Media as Languages

Another core metaphor that has generated much media scholarship (especially in film studies) is that media are *languages*. Unlike the conduit conception, the media-as-languages metaphor, as I am using the image here,[1] has tended to focus attention on the unique *grammar* of each medium. Those who draw on this metaphor have explored the particular expressive variables, or production techniques, within each medium or

---

[1] The images I analyze here are actually my metaphors for what I claim are the implicit conceptions underlying various forms of media inquiry. My three metaphors, therefore, do not necessarily match the explicit usage of similar terms in the literature, which is often very inconsistent. Sometimes, for example, the notion that each medium is a unique language is used to refer to the third conception of media analyzed below (e.g., Carpenter, 1960, (cont.)

each general type of media (film and video, for example, share many similar variables). Rather than viewing the medium as a relatively passive conduit, grammar analysts look at the plasticity of the medium in altering the presentation and meaning of content elements.

The language metaphor leads to questions such as: What are the variables that can be manipulated within each medium? What are the effects of such manipulations in terms of perception, comprehension, emotional reaction, and behavioral response? To what extent are the *grammatical codes* for each medium shaped by the physical nature of the medium, by the culturally variable codes of face-to-face communication, and/or by early production conventions? What political and ideological factors affect typical grammar variable choices? How do different audiences react differently to similar manipulations of production variables?

While the conduit metaphor leads one to analyze content that crosses easily from medium to medium and from live interaction to medium and back, the language metaphor tends to focus attention on those variables that function only within a specific medium or within a particular type of media. When a singer multitracks a vocal, for example, she is making a decision that cannot be made in real life or in still photography. And no matter how upset we are with a rainy afternoon, in real life we cannot "dissolve to a sunny morning."

Of course, one cannot discuss grammar choices without also considering content. In print, for example, one needs words before one can vary the sizes and styles of type; in visual media, one cannot have a long shot or a closeup of nothing; in aural media, one must have some sound content to employ equalization filters or to create sound perspective (the sense of relative distance fostered by different microphone placements).

Nevertheless, although grammar studies must include consideration of content, grammar questions are generally quite different from typical content questions. A content analyst exploring women's images in media, for example, may be concerned with elements such as the roles held by women (housewives or executives, for example), women's treatment (are they respected as equals by men, worshipped as madonnas, or viewed as sex objects, for example), whether women characters are punished in some manner for exhibiting personal or professional independence, and so on. A grammar analyst, in contrast, might examine the particular structuring of these roles, relationships, and behaviors within the particular medium. In television, for example, grammar concerns over women's images would include whether the women are framed in intimate, personal,

---

(cont.) p. 162). Similarly, Altheide (1976, p. 155) suggests that TV news practices are themselves media for filtering reality, yet his excellent analysis of "emphasis, omission, interpretation, and presentation" does not deal with what I discuss below as relatively fixed features of a medium, but with what in my model here are choices in the structuring of media content.

or social space; whether women are made to look weak through high-angle shots or strong through low-angle shots; whether filters are used in closeups of some female characters to create a soft, ethereal glow; whether shot structure focuses attention on a woman's body parts; and whether the overall action is viewed from a male or female perspective (such as in the all too common sequence of a woman passing a man, followed by a cut to a shot of her rear end). Thus, even the seemingly clear terms *image, portrayal,* and *genre,* tend to have very different meanings within different media metaphors.

The contribution of grammar to the overall message is made most apparent when one actually or hypothetically holds content elements constant as grammar variables are changed. Of course, in naturally occurring media artifacts the specific content generally shifts along with the grammar, but often one can still sort out the different strains of impact. A simple but striking example is offered by Henry Hampton (1989), producer of the award-winning documentary on the American black civil rights movement, "Eyes on the Prize." Hampton and his staff studied hundreds of hours of TV news footage. They found that one basic grammar element in the coverage changed dramatically over time, and that this element seemed to reflect the degree of identification with the protestors that journalists felt and promoted. Hampton describes how in early demonstrations the cameras take an outside, white perspective, observing the black demonstrators confronted by white racists. But as time passes, the cameras move "behind the march leaders and look outward at the hostile sheriffs and their deputies" (p. 39). With the calls for "black power," the view again shifts outside. And, finally, with the 1967 riots, the camera's point of view is from behind the police lines.

These shifts in camera position parallel the manipulations that are used in some fiction films, most blatantly in war movies, to encourage audiences to identify with one "side" as opposed to another. Manipulations of grammar variables also partially explain why in some movies audiences tend to identify with the criminals (a content concept), as in "Bonnie and Clyde" and the Godfather movies, and in others with the police (or more typically, with *one* or *some* of the police). In addition to general camera perspective, the vicarious distance established between audience and performer encourages various degrees of emotional involvement. It is easiest to react personally (both positively and negatively) to characters who are seen often in closeup. Indeed, we may feel that a movie has a happy ending, even though hundreds of people are killed, as long as those we have been vicariously "close to" escape largely unharmed.

As the above examples suggest, the examination of media grammar variables involves a second, quite different set of meanings for the terms *structure, form, latent,* and *code.* These terms have yet another group of meanings that grows from a third image of media.

Media as Environments

A third answer to the question "What are media?" is that each medium is a type of *environment* or *setting* or *context* that has characteristics and effects that transcend variations in content and manipulations of production variables. This leads to what I call *medium analysis*. I use the singular *medium* because those who draw on this metaphor examine the relatively fixed features of *each* medium.

Medium features are an implicit subject of study in both content and grammar research. After all, when one studies the content of TV *images* or the content of a *paragraph*, one is implicitly studying what is unique to TV and to print, respectively. Similarly, an analysis of the effects of microphone pickup patterns on the resulting "landscape of sound" in an audio recording clearly deals with variables that do not exist in many other media or in live interaction. In addition, the special features of a medium are sometimes explicitly used to justify the significance of studying the medium's content or grammar (as when the basic accessibility of TV images to young, preliterate children in contrast to the relative opaqueness of print is used to justify analyses of the content of TV programs). But medium research goes further: It focuses specifically on advancing our understanding of the ways in which the differences among media make a difference.

Broadly speaking, the environment metaphor leads one to ask: What are the characteristics of each medium (or each type of media) that make it physically, psychologically, and socially different from other media and from live interaction, regardless of content and grammar choices? How do the features of a medium influence content and grammar choices for that medium? What social, political, and economic variables encourage the development and use of media with some features over media with other features? How does the addition of a new medium to the existing matrix of media alter the function and use of older media? How does the rise of new forms of media alter social roles and institutions whose structure and functions were dependent in some way on the characteristics of previously dominant media? How do the characteristics of each medium interact with cultural codes and customs?

Of course, it is impossible to analyze the features of the medium without in some way recognizing the existence of content and grammar choices. To whatever extent there is an "environment of television," for example, it still needs programs to become visible. Indeed, medium analysts may use systematic tendencies in content and grammar choices as partial evidence of the different contexts for communication fostered by different media. Transcripts of telephone conversations contrasted with letters, for example, may be used to point to the relatively informal, bidirectional, and less linear nature of the telephone; and TV producers' tendency to rely more than film producers on the closeup may be related to the greater intimacy fostered by various aspects of the current form of the TV medium.

At the same time, typical medium questions are quite distinct from typical content and grammar questions. Analysts of both content and grammar focus on variables that can be manipulated after the medium of communication is chosen. With medium analysis, the focus is on those *environmental features* of the medium that are largely out of the control of users once the medium is in use. One can give in to the tendencies of the medium (such as the relative informality of the telephone) or one can resist them (by not having a phone, for example) or try to work around them (by buying an answering machine). But they are there, and one must contend with them in some way. With medium analysis, then, the key decision is whether or not to invent, adopt, or employ the medium in the first place.[2]

Looking at media *as* contexts is often confused with looking at media *in* social context. But the larger social context is relevant to all three images of media. Social, economic, political, and organizational variables influence, and are in turn influenced by, (a) the production and perception of media message content, (b) manipulation of media grammar variables and the reaction to such manipulations, and (c) the development and use of different media settings.

One can study media settings on both the micro, single-situation level and the macro, societal level. On the micro level, medium questions explore the implications of choosing one medium over another in a given situation. What, for example, are the medium-related implications of a job applicant choosing to write a *letter* of introduction as opposed to making an introductory *telephone* call, or of a child choosing to relax after school by reading a *book* rather than watching *television,* or of a business using the *radio* rather than the *newspaper* to advertise a new product?

On the macro level, medium analysis deals with the larger social implications of the widespread use of a medium. Thus, macro-level issues would include the impact that the telephone has had on business and social interactions in general, such as its impact on the art and function of letter writing. With regard to TV, school, and children, a macro-level analysis might examine the ways in which TV may undermine print conceptions of education and childhood. As for businesses and advertising, a sample macro-level medium concern might be how a political and eco-

---

[2] Changes in technology can alter the setting of a medium, even when it goes by the same name. The social context of the telephone, for example, has been altered by advances in switching equipment, tone dialing, voice mail, and most recently by "caller ID," which allows subscribers to see the phone number of the caller before answering the phone. Similarly what we call "television" has been an evolving environment of broadcast, cable, satellite, and, soon, high-definition TV—each with different implications. In effect, then, the names we call various media often refer to a cluster of similar, but not identical, *subcontexts* of communication. Technological evolution also changes the range of grammar variables available within a medium, but grammar and medium variables remain analytically distinct: During particular communications, medium characteristics are fixed while grammar variables can be manipulated.

nomic system that focuses on selling products and promoting a single vision of democracy may encourage the development of broadcast radio and television (which are unidirectional, centrally- and mass-distributed, and allow for relatively little local input, feedback, and discussion), while other more interactive and community-based technologies receive relatively little support or encouragement.

Macro-level medium questions address potential large-scale changes, such as the ways in which different medium environments may foster different thinking patterns; alter the dividing line between public and private life; stimulate changes in child-adult, male-female, and leader-follower role relationships by altering who-knows-what-about-whom and who-knows-what-compared-to-whom; increase or decrease opportunities for participatory democracy; change the social significance of physical location and physical barriers; affect the criteria that are used to evaluate political leaders; alter the relative status of various social institutions; and so on.

The impact of the medium's setting is most visible when content variables are actually or hypothetically held relatively constant and when one medium is contrasted with another medium or with live interaction. Consider, for example, the content element of "advice to parents about what to tell or not tell young children about sex." When placed in a book, the context of print supports the content of such advice and the authority of parents, because most young children cannot read an adult book and do not even learn about the existence of this parental concern. When placed in the setting of broadcast TV, however, a paradox arises, because thousands of children may be listening in, hearing about the things parents are advised not to tell children, as well as sensing the anxiety and confusion of parents. In effect, then, even when we try to hold the content constant, the functional *message* often changes along with the medium.

As the above examples suggest, the image of media as environments leads to a whole other set of meanings for the concepts of media *structure*, *codes*, and *form*. To say, for example, that the basic access code of TV, as a medium, is less complex than the access code of print, is not to say that a young, preliterate child who is able to watch TV necessarily understands the codes of particular thematic genres (content) or of shot structure (grammar).

Like content and grammar studies, medium analysis tends to ignore as much as it explores. A full consideration of any media-related issue, therefore, requires exploring questions that grow out of all three images of media.

Re-Imaging Media Studies

When taken together, the three images of media outlined above offer one way of defining the current subject matter of media studies and of comparing, contrasting, and synthesizing research findings. If my analysis here is correct, at least some of the confusions and disagreements in the field have stemmed from the fact that functionally there have been *three different* "media studies," plus various hybrids, based on three different conceptions of media.

These three competing images of media foster hidden ferment and hidden agreement in the field for several reasons. Since the subjects of all three forms of inquiry are referred to by the same general terms—such as *media effects, media control,* or *perception of media*—the very different assumptions underlying each are obscured. Further, because content, grammar, and medium elements of the same mediated communication offer their own *thrusts* of influence—which may or may not be in the same direction—potentially complementary and additive studies are often misconstrued as competing and contradictory.

A scholarly or popular analysis that suggests that a particular TV series contains positive images of blacks or women, for example, may not necessarily contradict another study that argues that blacks or women are negatively portrayed in the series. One needs to look at what aspects of the portrayal are being examined—content and/or grammar.

Just as content thrusts may be in tandem with or opposed to grammar thrusts, medium thrusts may support or undermine content and grammar decisions. A look at traditional television content, for example, may suggest that TV has been oppressive to women, but a medium perspective could argue that TV, regardless of its portrayal of women *characters,* has exposed women *viewers* to a wide array of previously all-male arenas and has therefore encouraged greater blending of male and female roles in everyday life. It is not necessary to accept either of these particular content and medium claims to see that they are each addressing a different aspect of mediated communications. Yet they may be incorrectly viewed as simply two contradictory answers to the same question: Does TV support or undermine a feminist world view?

A lack of examination of the metaphorical base of media inquiry may also mask significant disagreements. For example, an analysis that suggests that the medium features of TV weaken adult control over what information children have access to may be confused with popular and scholarly concerns over the content of children's programming, but they are actually very different types of analyses with very different implications for social policy, media regulation, and childrearing practices.

When researchers ignore the range of metaphors, there may simply be confusion over what has been found in a study. Cultural and subcultural variations in perception of a TV show, for example, are usually explained in terms of content elements (roles, narrative, action, etc.). But such variations may also be linked to culture-specific perceptions of grammar variables (such as the spatial zones symbolized by camera shots) or even cultural variations in interaction with the medium of television.

Conclusion

Of course, the separate consideration of content, grammar, and medium variables grows partly from an analytical fiction. Analytically, one can separate media processes into those elements that transcend any particular medium (content), those elements that involve manipulations of a particular medium's production variables (grammar), and those aspects of each communication environment that are relatively constant, regardless of content and grammar choices (medium). But the fact is that any use of media involves all three dimensions simultaneously.

Nevertheless, in research practice and popular thought, the metaphors have usually operated in relative isolation. While some media analysts draw on more than one image—some even on all three—the vast majority of popular and scholarly discussions of media, including most of my own, draw primarily or exclusively on only one of these conceptions. Even when two of the metaphors are bridged within a single study, rarely is the third introduced.

This situation has been fostered by the fact that the underlying media metaphor—which helps to form a researcher's question and shapes a good measure of the answer—is most often left unstated and unexamined. Without exposure, the latent conception acts as a source of seemingly boundless vision. Once analyzed, however, the edges of each image begin to show, and the desirability of drawing on other images of media becomes clearer. This is not a particularly pleasant experience for us as researchers. For one thing, it initially draws our attention away from the issues that have been our main focus and asks us to consider the underlying images of media that feed a variety of research questions. This is about as pleasant a task as trying to savor a meal in our favorite restaurant while listening to a lecture about the strange foods eaten by members of other cultures. It also forces us to consider the possibility that questions we have not yet considered and are not sure how to approach answering may be as central to our claimed topical concern (children and television, gender and media, audience analysis, political persuasion, analysis of news, hegemony, and so forth) as the specific questions we have spent so much time investigating. For some of us, this is akin to an unsettling discovery that a loved one whom we thought we were taking care of nicely has a whole set of problems that we did not know about and have developed no strategies for addressing. Nevertheless, I believe the future of the field will be enhanced by confronting the metaphors outlined here because a full exploration of any media-related topic requires a bridging or a new synthesis of all three images.

References

Altheide, D. L. (1976). *Creating reality: How TV news distorts events*. Beverly Hills, CA: Sage.

Carpenter, E. (1960). The new languages. In E. Carpenter & M. McLuhan (Eds.), *Explorations in communication* (pp. 162–179). Boston: Beacon Press.

Hampton, H. (1989, January 15). The camera lens as two-edged sword. *The New York Times,* Section 2, pp. 29, 39.

Lakoff, G., & Johnson, M. (1980). *Metaphors we live by.* Chicago: University of Chicago Press.

The Consequences of Vocabularies

by Joli Jensen, University of Tulsa

Epistemological upheavals knock us for a loop. What was taken for granted under one system of thought becomes problematic in another. What was once invisible appears at center stage; what was once of crucial importance now seems beside the point. We live in exciting epistemological times, and the kinds of things that seem given about communication inquiry are obviously being renegotiated.

Attempts to map this process of negotiation are attempts to regain our balance, and get our bearings. What are our central questions? Our chosen methods? Our crucial theories? As we construct answers, we can teach students and colleagues about what we call "the field," and we can locate our work in historic and more legitimate academic traditions. Such is the purpose, I believe, of this particular collection of essays, as was its predecessor, *Ferment in the Field* (1983). They represent attempts to map, explore, or define what our field is up to, so that we can do research, teach students, and justify our work with more ease and certainty.

In doing such mappings, explorations, and definitions, we create what we pretend to merely describe. That is, of course, the bracing conclusion of symbolic interactionism, cultural studies, and now post-structuralism—the world that we live in is a world we ourselves create, via something that can be called language or discourse or culture. From these perspectives, communication inquiry is simply one more interpretive world that we have created, and now live in—the field is something we make in common, and it is constantly being remade by the social, cultural, and interpretive practices in which we engage.

If inquiry is inescapably definitional (not merely descriptive or explanatory), then we must pay attention to what it is we are making when we engage in particular modes of inquiry. What is the nature of our interpretive worlds? What are their consequences? Who do we become when we engage in the study of communication? If we accept that inquiry constitutes a reality, rather than merely describing it, then we can and should ask questions about the nature and worth of that which we are constituting.

Joli Jensen is an associate professor of communication at the University of Tulsa. The author gratefully acknowledges conversations about these matters with students (especially those in her theories of communication class) and colleagues, particularly John Pauly and James Carey.

The Field Today

Our understanding of what we are up to in communication study has been based in a belief in a neutral "world out there" waiting for us to figure it out. This is our epistemological heritage; it has shaped our understanding of what constitutes good inquiry, appropriate evidence, sound conclusions. It has, less obviously but no less deeply, shaped our beliefs about ourselves, our teaching, our research, our common enterprise.

Academic disciplines developed in relation to assumptions about categories of knowledge and thus of inquiry: botany distinct from zoology, physiology distinct from philosophy. Universities developed in relation to this same epistemology, a belief in the knowability of a world that was already diversified and categorizable. Knowing these different chunks of reality, through experiment or measurement or close study, would allow particular groups of people to have a deep and true picture of the world. With systematic study in institutional settings, knowledge could be deepened and widened, the chunks reconstituted, the world better—even fully—understood.

Let us bracket, for now, whether categories of reality are out there. It is clear that the lines between academic disciplines are humanly constructed and maintained, visibly so when our maintenance practices seem to break down or be in flux. Turf battles over particular courses, grant applications to particular agencies, submissions to particular journals are all moments when the need to sustain distinctions between fields becomes crucial. Similarly, there are moments when what is within the boundaries of the field is clearly being constructed: Whenever introductory courses or comprehensive exams or curricular revision or faculty searches or tenure reviews are undertaken, the field matters greatly.

Academic journals and professional organizations offer the most obvious markers for what constitutes the field boundaries and content. In professional meetings and in article reviews, concerns over a field's unity or fragmentation, its dominant paradigms, its disciplinary status, and its methodological battles are most obvious.

Why does all this matter? Who cares if what we're up to is a field or a discipline or an area of inquiry, or if we can agree on certain theories or procedures? Well, we care because the answers to these questions shape our livelihood—not only what we do, but how we are perceived by others. Having a coherent field with coherent paradigms, disciplinary legitimacy, and methodological certainty makes our academic life infinitely easier—we can teach by transmitting doctrine; do research by extending or filling in extant lines of inquiry; revamp curricula, hire colleagues, give tenure, and justify our existence to colleagues and administrators in unambiguous terms.

In short, we can flourish in the university environment when we become what the institution needs us to be: a field of circumscribed and transmissible knowledge, arrived at with agreed-upon methods, whose worth can be ascertained by agreed-upon procedures. To be fully legitimated in our academic setting is to be able to say we are members of a coherent discipline who study a particular set of phenomena with particular methods.

Communication has never had that full legitimacy, although only a brave few have suggested we abandon the quest. We cannot let go of questions about the field's unity, paradigms, and methods because they are status-related questions, based in institutional necessities. They are therefore questions that will bedevil us, but not because they are, in themselves, important questions.

Let me first suggest that institutional logic is not, necessarily, worth our loyalty. Universities developed their practices based on previously held, now radically questioned assumptions about the knowability of the world, the transmission of that knowability to others, the usefulness of knowledge accumulation by particular groups of people in particular relationship to one another. Practices based on those assumptions may not be particularly valuable, or humane, or useful. If epistemological upheaval has any point, it is to help us question the meaningfulness of previous practices, in the light of shifts in what is given about the world.

Universities may be using a language that is harmful or inadequate, and for us to replicate their logic may waste our time and energy. Our efforts to give communication inquiry the status of a field commit us to a logic that may, itself, need an overhaul. If we let go of questions of the unity or fragmentation of our field, seen as containing particular warring paradigms, we might find ways to take up more important questions, like "What are the consequences of what we are doing?"

The Great Divide

Questions about the consequences of actions are seen as ethical questions, and when our field is divided up into qualitative or quantitative, interpretive or social science, expressivist or objectivist approaches, the humanist types get to ask them, while the scientist types can leave them alone. The recognition of deep epistemological differences gets simplified almost beyond recognition or use: cold and calculating scientists vs. warm and fuzzy humanists; the facts vs. the values; the messages vs. the meanings; the details vs. the big picture.

I want us to get past such simplistic oppositions. We cannot escape the endlessly interesting epistemological divide between objectivism and expressivism, between belief in a neutral world out there that waits for us to know it and belief in a world that is constituted in our knowing of it. But we can recognize it is unresolvable, and pay attention instead to the consequences of our ways of knowing.

That division too readily gets flattened into a way to map the field—World View I vs. World View II, scientific vs. humanistic, qualitative vs.

quantitative, administrative vs. critical; and thus we attempt to teach students both methods, to hire people from both traditions, and to write essays that talk about how to live with, or overcome, or make the most of, both ways of doing communication inquiry.

Orientations to Inquiry

I want us to engage questions that spring from somewhere other than the field or the divide. *The Field* offers us questions about legitimacy and status; *The Divide* offers us questions that become endless comparisons, or endless standoffs, or endless calls for rapprochement. The questions I think we should ask instead are about who we are, and who we become, when we engage in certain kinds of communication inquiry—ontological questions about epistemological consequences.

A preliminary suggestion: Let us assume that anyone who claims to study communication is up to something we recognize and honor. Thus membership in the field is by self-definition, not institutional allegiances, or sites of publication, or professional organizational membership. The Field, thus defined, would insistently remain contradictory, in flux, with irresolvable differences, and permeable boundaries. It is constituted by practice, a shared interest in phenomena called communication, demonstrated through public conversation.

Let me also suggest that we leave The Divide alone, at least temporarily, and look instead to the modes of inquiry we deploy when we engage in study of phenomena-called-communication. These can be seen to fall into four modes: experimental, descriptive/analytical; ethnographic; critical/theoretical. When we examine these modes, we find that they mobilize different assumptions about who we are and what we are up to, and thus they have consequences for how we act in the world.

The experimental mode presumes that processes and structures that underlie reality can be revealed, under appropriate circumstances. Situations can be constructed (the experiment) that allow these processes and structures to be revealed. Scientific procedures are tools that can reveal the laws and structures underlying and determining communication.

The descriptive/analytic method can use quantitative and/or qualitative research techniques; the distinctive feature here is that description—via survey, statistics, close readings, reconstituted narratives—is not yet explanation. This mode does not claim to get at what is real underneath—it tells instead what is there to be understood.

The ethnographic mode foregrounds the experience of the investigator, and the inextricable connections, in understanding, between the knower and the known. It is about what happens when things are perceived, experienced, addressed. It too does not claim explanation, but rather a particularly intense, revelatory form of description.

The critical theory mode offers previously constructed perspectives as illuminative procedures: Applying theory reveals the structures or discursive processes that determine and constitute the world. Thus theory becomes, like the experiment, a procedure that shows, and thus can even explain, what is really going on in the world.

These four modes differ, then, in their claims about level of analysis (underlying processes or ongoing processes) and outcome (description or explanation). Notice that the critical theory mode (explicitly antipositivist in heritage) resembles the experimental mode in these particular claims.

The quest for certainty is the heritage of the objectivist epistemology. If there is an anchoring, bedrock reality, something that underlies or determines surface characteristics, then it can be revealed via experiment. Critical theory claims also to reveal what is really going on, but it does not need a constructed experimental situation. Instead, the theoretical apparatus reveals when it is applied to circumstances—certainty develops as particular Marxist or post-structuralist or semiotic or reception theories demonstrate, repeatedly, their explanatory power.

Notice that, in the experimental method, certainty can be shaken by findings—evidence can demolish experimentally based theory, leading to doubt, new hypotheses, new tests, new theories. The experimental mode seeks certainty, but through a seeking of (contained) refutation. This is not the case in critical theory: There is no direct way that evidence can demolish theory-based theory—it inevitably proves itself by recurrent, fruitful application.

Descriptive/analytical and ethnographic modes do not claim to explain what is really going on, and thus do not claim certainty. Attempts are made to remain open to situations and circumstances, to remain capable of surprise. These modes are doubt-filled, chronically hesitant to make even the most specific claims.

Tellingly, researchers in these two traditions are often exhorted to ground their claims in experimental evidence or in theory—these are offered as guarantors of the findings. Failure to ground research in previous experimental work, or in previously developed theories, makes the work appear illegitimate—unsystematic or lacking in rigor. These terms reveal their origins in positivist notions of knowledge, faith that good scholarship evidences a firm, systematic, solid edifice of reality. If we question positivist notions, so too must we question the terms on which we evaluate research, and our expectation that research findings be secured by the familiar moorings.

These differences in certainty can be connected to differences in loyalties. The experimental mode is loyal first to evidence—no matter what the hypotheses, theory, or paradigm, if evidence challenges it, then the evidence wins. The evidence is trusted as being stronger than the techniques and languages used to uncover it.

Loyalty in the descriptive/analytical mode is to what is being looked at—the setting or text is seen as offering or determining meaning, with the researcher attempting to best recognize, record, or retell those meanings. In the ethnographic mode, this recording or retelling is problematized—meaning is believed to be made, not found, so loyalty is to the encounter with evidence that creates meaning, with constant attempts to be fair to those elements of the encounter that trouble, surprise, mystify, or confuse.

In the critical theory mode, loyalty is to theories themselves. Researchers define themselves as particular kinds (or combinations) of theorists, and their work is designed to demonstrate the power of theory, and their facility in using it. Evidence is merely the substrate for theoretical action; meaning is revealed through theory, not direct observation; and the role of critics is unproblematic—they are theorizing.

Consequences of Expertise

These variations in levels of analysis, claims about goals, degrees of certainty, and location of loyalties have consequences for who we think we are, and how we act toward one another, in communication inquiry.

Experts are people who have faith in their own special understanding or mastery of the world. Use of research techniques that claim to offer special understanding, even explanations of what is hidden from others, supports belief in one's own expertise. Thus those who do experiments and those who do theory are more likely to see themselves as experts than those who do descriptions or ethnographies.

Belief in one's own privileged knowledge can be tempered, in the experimental mode, by the ever-present possibility of being contradicted by research findings—someone else can and will come along to refute you and your beliefs, based on new kinds of experiments, new findings, or new kinds of data. Maintenance of expertise requires constant updating, and an acknowledgment that the findings will surpass the efforts of any one individual.

In the theoretical mode, contradiction comes through the passage of time, and findings depend on the expertise of the theorist. The worst that can happen is that one becomes outdated. New theories come into fashion, and favorite theories look out of date. But, in this mode, expertise can be maintained simply by keeping up with new theories and their attendant languages.

Such ease of expertise is not possible in the descriptive/analytical or ethnographic mode. One can be accomplished methodologically, but what one knows is local and particular, contingent on one's efforts in specific circumstances. One does not have or generate explanations that apply broadly or deeply, one does not speak in languages that others ("subjects") do not readily understand. One is not, in these modes, an expert. One is still, like the experimental researcher, subject to refutation by experiential evidence.

What are the consequences of expertise? Experts see themselves as possessing special knowledge not shared with others. Thus teaching becomes a process of transmitting that expertise, and learning the process of becoming more and more like the teacher in knowledge possession. As experts we risk becoming arrogant, since we are always surrounded by people who seem to know less than us, and we risk becoming smug, because we think we know what's really going on. Our relationship with our students is likely to be one that is autocratic and controlling, since our expertise is what we think they need.

This danger is greatest, it seems to me, in the theoretical mode, because loyalties are to theories, not empirical evidence or lived experience. Reality, be it experimentally proven or individually apprehended, is humbling. It is complex and contradictory when apprehended as experimental data, and poignant and mysterious when apprehended by observation or ethnographic experience.

Teaching, in modes other than the theoretical, is usually about "the stuff"—data, observation, and experience—not about "theories about the stuff." Experimental findings, observation, and experience are things that the student and teacher can confront together. Especially in the descriptive and ethnographic mode, teachers and students share the expertise of close and careful observation. Teachers are not in possession of an exclusionary language or specialized methods. This means teachers will listen to students even if they don't have a mastery of theories or research methods. Teaching becomes less of an autocratic encounter and more of an exploratory enterprise.

This is a model of teaching, and of scholarship, as conversation. It suggests that what we do, when we teach and when we research, is engage in a series of discussions with each other about what we think is happening with others and with us. We are less interested in presumed processes that underlie or explain, and more interested in events and experiences that happen and mean. Such discussions are conducted with a sense of humility in the face of the complexities of the world, and with loyalty to the attempt to illuminate, comment on, rethink what we find when we explore it.

In some of these modes, then, we make a world that values and promotes the participation of others. We do so because we are listeners as much as speakers, students as well as teachers. This is more likely to occur in modes of inquiry that foster uncertainty over certainty, and understanding over explanation. It is more likely to occur in modes that respect the possibility of surprise. It is less likely to occur in those that seek certainty and explanation, and in those that have no direct loyalty to the likelihood of being wrong.

There are other ways of talking about the ontological consequences for epistemological vocabularies–some are more likely to be coopted by institutional forces, to insist on massive footnoting, to result in turgid prose; some support more direct considerations of ethical dilemmas; some deflect passionate engagement; others foster more curiosity, others more respect.

My purpose here is simply to suggest that a conversation about such consequences is necessary and fruitful, and to suggest some terms in which it can begin. We can start, as we do in this forum, by asking ourselves what we are up to, and why. But we get further if we ask these questions not of The Field but of ourselves, as people who live and act in worlds for which we are responsible. I believe we can act in ways that make those worlds better when we recognize and make explicit the consequences of our chosen vocabularies.

Against Theory

by Barbara O'Keefe, University of Illinois

Recent studies and analytic essays testify that the field of communication is and has been composed of a set of fairly discrete subdisciplines, a complex structure that promotes recurring disciplinary reflection. Previous discussions have focused on the tension between rhetorical studies and communication research, between behaviorist and antibehaviorist conceptions of interpersonal processes, and between competing agendas for the study of mass communication. Currently, attention is drawn to relationships between mass and interpersonal communication research.

Studies of citation patterns by So (1988) and Rice, Borgman, and Reeves (1988) provide persuasive evidence that scholars in interpersonal and mass communication draw on different intellectual sources and seldom cite each other—they function as discrete intellectual traditions. Commentary by such scholars as Berger and Chaffee (1988), Reardon and Rogers (1988), Wiemann, Hawkins, and Pingree (1988), and Dervin, Grossberg, O'Keefe and Wartella (1989a, 1989b) has sought to provide a framework within which scholars from interpersonal and mass communication studies can understand their differences and address them.

Interest in understanding relations between interpersonal and mass communication research is undoubtedly a product of many interacting forces, but one of the most important is the increasing integration (institutional, not theoretical) of the two fields. These two traditions in communication research historically have been segregated in separate academic departments, with interpersonal communication research primarily found in speech or speech communication departments and mass communication research located within journalism (Delia, 1987; Wiemann et al., 1988). But development of new programs, consolidation of old programs, and shifts in theory and research agendas are creating pressure toward integration. These developments not only create a new propinquity between formerly disparate groups but also create pressure to rationalize integration.

While many commentators have represented the current situation as one of increasing fragmentation in the field, I am struck by increasing interconnections between previously separate disciplines and by projects that deliberately attempt to forge connections (e.g., Berger & Chaffee,

Barbara O'Keefe is an associate professor in the Department of Speech Communication, University of Illinois, Urbana–Champaign.

1988; Dervin et al., 1989a, 1989b; Hawkins, Wiemann, & Pingree, 1988). Hence, rather than seeing the discipline of communication research as subject to increasing fragmentation and specialization, I see several distinct intellectual traditions attempting to accommodate each other in a new, integrated, institutional context. Some have argued that where once there was unity, there is now discord, which should be resolved by finding a unifying theory or philosophy of communication (see, for example, Berger, 1991; Reardon & Rogers, 1988; Wiemann et al., 1988). Others see in the current instability opportunities for change in the research agendas of both disciplines (Baxter, 1992; Bochner & Ellis, 1992; Lannamann, 1992; Leeds-Hurwitz, 1992). But I see different fields, brought together by institutional forces rather than considered choice, trying to see how they might, and might not, support each other.

In this essay I consider two issues relevant to our collective predicament. First, I comment on the attempt to construct a coherent intellectual viewpoint to span the disparate projects of interpersonal and mass communication research. Second, I consider how the discipline of communication can construct functional representations of itself despite diversity.

The Case Against Coherence

Bochner and Eisenberg (1985) have documented the powerful appeal that visions of theoretical coherence seem to hold for communication scholars despite the fact that theoretical unification is neither desirable nor attainable. The subdisciplines of communication have different projects; different issues dominate their agendas and different literatures provide intellectual resources. Since we have different viewpoints for good reason, imposition of one common theoretical viewpoint would simply mean displacing some important work from the field.

The idea that communication research is a single discipline is belied by the history of communication research and by the facts of disciplinary organization. As Delia (1987) has shown in his analysis of the history of communication research, studies of communication have no single intellectual origin, nor has communication research ever been a unified enterprise. Certainly, groups of researchers within subdisciplines have, for short periods of time, maintained common focus and theoretical consensus.

The occurrence of local moments of coherence has not contributed materially to the development of broad theoretical consensus. For a host of reasons, research areas prosper and then die, leaving behind a legacy of concepts and findings as well as nostalgia for a period of time in which the field purportedly moved toward true disciplinary status.

This sort of nostalgia exacerbates the perception that communication research is becoming increasingly fragmented. At present, no area of the field exhibits the kind of consensus and common focus of attention that characterize such periods of coherence. In comparison to a distantly re-called Golden Age, the current state of things appears quite chaotic. And there is little prospect for the emergence, any time soon, of relatively siz-able and stable research communities.

Various analysts have suggested that one or another recent theoretical development is likely to produce an organized consensus. One candidate that is often mentioned in this context is cultural studies, because the in-tellectual sources on which it draws are having such a wide impact on the field at the moment. This suggestion, however, overlooks the powerful centrifugal forces within the cultural studies movement itself. Key figures in cultural studies are explicitly antidisciplinary (Hall, 1992; Nelson, Tre-ichler, & Grossberg, 1992). As an intellectual movement, cultural studies is distributed across the academic landscape and into general intellectual culture. The very success of cultural studies as an intellectual movement, combined with the resistance within cultural studies to identification with a subject matter, makes it unlikely that cultural studies will stabilize as a coherent research tradition.

Moreover, while cultural studies has transformed the study of mass communication, it is as yet unclear what this movement has to contribute to the study of interpersonal processes. A second group of theorists, who style themselves as "social constructionists" or the "social approach" to communication, represent themselves as carrying forward the cultural studies project in studies of interpersonal communication. They have sug-gested that the social approach one finds in mass communication and, to some extent, in the study of organizational cultures, offers a distinct alter-native to the standard approach to the study of interpersonal processes (see, for example, Lannamann, 1992; Leeds-Hurwitz, 1992).

However, in the process of characterizing social approaches broadly enough to encompass work in interpersonal, organizational, and mass communication, social constructivism has lost its distinctiveness as a the-oretical position. Although these theorists draw on much of the same lit-erature that informs other approaches to cultural studies, the social ap-proach is not clearly differentiated from fairly standard approaches to the study of interpersonal processes. Consequently, it is unclear whether so-cial constructionism ought to be treated as the application of a cultural studies approach to interpersonal communication (which would repre-sent cultural studies in a way that most in the movement would find ob-jectionable) or as a separate movement involving the assimilation of liter-ary and cultural theory (particularly Bakhtin, Derrida, and Geertz) to the received view, which is heavily influenced by the work of Goffman.

Those who advocate social approaches to interpersonal communication fail to recognize that social constructivism (at least as it is characterized in recent discussions) has been a dominant viewpoint in interpersonal com-munication for more than two decades, as has a view of the individual as articulated only within relationships (see Dance, 1982). The view that meaning is socially produced rather than privately encrypted was introduced into the area with the absorption of ethnomethodology and conversation analysis (see, e.g., Coulter, 1979; Jacobs, 1985). Notably, recent borrowings from social psychology, often taken as evidence of individualism, are drawn from approaches that treat personality and emotion as semiotic categories and investigate their invocation in everyday explanations of conduct (O'Keefe & Delia, 1989).

The one thing that is distinctive in the *new* social constructivism is an implacable hostility toward quantitative and laboratory-based research. Ironically, this methodological cleansing is being advocated for the study of interpersonal processes at precisely the moment that other areas are beginning to develop a more sophisticated view of method. As Dervin et al. (1989b) note, the assumption of a tight connection between social approaches and qualitative research and between classic individualism and experimental research cannot be sustained in light of the complexity of contemporary research practice. Increasingly researchers have come to realize that a range of questions arises within every research topic, that different kinds of questions require different methods of gathering relevant evidence, and that the best approach is to match the question with the method rather than to insist on using a single method regardless of the question being addressed.

As Sigman (1992) has pointed out, the field of interpersonal processes is not well served by the construction of social constructionism. Within the study of interpersonal communication, the theoretical landscape is much more finely organized; differences among theorists such as Goffman, Garfinkel, and Geertz, who tend to be conflated within a general social constructionism, are understood and appreciated. The attempt to create an overarching social constructionist position in interpersonal communication thus involves a retrograde movement toward imprecision in theory and rigidity in method.

A final candidate approach that is often mentioned in discussions of disciplinary integration is cognitive science. For example, Berger (1992) suggests that a view of communication research as the study of message production and comprehension (understood within a traditional mentalistic analysis of goals, plans, and communication) could frame all research on communication processes. Certainly our field is making and will continue to make contributions to the study of cognition and communication, but such a cognitive framework is unlikely to be appealing to any but researchers working on very narrowly defined problems. Within this viewpoint, message production and comprehension are treated as private mental processes, functions of the individual in isolation. Yet researchers in both interpersonal and mass communication have increasingly come to recognize the ways in which virtually all messages reflect the collaboration of a community in both their production and in the meaning they come to have. A view of communication as private mental activity is unlikely to have the broad appeal necessary to attract the allegiance of a large community of scholars.

Moreover, the study of communication and cognition is itself quite unstable at the moment. Much of the standard research on this topic (and almost all of the research on cognition in the field of communication) is conducted within a framework that employs the intentional vocabulary of folk psychology, reifies a view of the individual as constituted prior to culture, presupposes that rational explanations are literal descriptions of mental processes, and equates meaning with subjective representation. This received view has come under increasingly trenchant criticism within a number of contemporary intellectual traditions, including the philosophy of mind (e.g., Dennett, 1987; Stich, 1983), ethnomethodology (e.g., Coulter, 1979), and cognitive science itself (e.g., Clark, 1989). These criticisms of mentalism have not yet become prominent in the published literature on communication, but they are increasingly cited in our literature (e.g., Greene, 1987; O'Keefe, 1992). Within cognitive science, the standard approach now competes with views of cognition as situated (e.g., Lave, 1988), views of performance as locally managed (e.g., Suchman, 1987), and views of representation as distributed in patterns of association between features (Clark, 1989; Lambert, 1992). In the study of discourse processes, the standard view of communication as intersubjective agreement now confronts views that problematize standard models of intersubjectivity (Taylor, 1992).

In light of the lack of theoretical consensus regarding the nature of cognition and communication, it seems unlikely that studies of cognition and communication will achieve any kind of stability in the near future. Neither of the two great cross-disciplinary movements of the last 15 years, cognitive science and cultural studies, appears to hold much promise from the standpoint of creating broad or stable consensus in either mass or interpersonal communication. Obviously, then, they are even less likely to provide useful integrative frameworks for the field as a whole.

Given the very real limitations of the theoretical apparatus offered by current approaches, it follows that efforts to enforce theoretical coherence on the field can only work to our collective disadvantage. Even so, we have seen a flood of recent essays arguing for a return to grand theorizing as well as colloquia and edited collections devoted to metatheoretical issues. The impulse behind these developments (the exigence of institutional integration) is fully understandable, but as a field we need to adopt a more reflective stance toward our situation. We do not have a problem that can be solved by arguing for our own viewpoints; in fact, heightened dissensus is precisely what we should avoid.

The Case for Cohesion

Bochner and Eisenberg (1985) have argued that the need for integration should be solved by cultivating cohesion rather than coherence. They make a persuasive case against the common view that disciplines, especially successful disciplines, are characterized by theoretical and method-ological coherence. In fact, they argue, successful disciplines—those that are able to garner power and resources in the academy—are character-ized by cohesion rather than coherence. In cohesive disciplines, the sib-ling subdisciplines adopt a posture of respect and protection toward each other, presenting a common front to the rest of the academy.

This kind of common front is of crucial importance during periods of self-criticism and shrinking resources within higher education. Recogniz-ing this, the integration of interpersonal and mass communication studies can be seen both as an opportunity and a danger. The great opportunity offered by integration is the possibility of making a common cause. Rather than competing in separate units, the communication disciplines can provide support for each other. Since larger units are less vulnerable than smaller ones, integration offers increased stability and security to programs that are chronically understaffed.

The great danger, of course, is that faculty within these units or in the discipline at large will treat theoretical coherence as a prerequisite for community and become locked in self-destructive squabbles. If scholars in the field cannot agree to value the work being done, how can scholars outside be expected to appreciate our achievements? In the current con-text, especially, we should place a low value on dissension, especially about fundamental theoretical commitments. We should place a high value on enterprises that contribute to cohesion and to the field's ability to point to clear achievements.

The first such enterprise is problem-centered research. Whereas recent commentary has often held up problem-centered research (research on health communication, family communication, new technologies, chil-dren and television, and the like) as evidence of harmful fragmentation, this kind of research offers the greatest promise of generating results that are demonstrably useful to the communities we serve. By forcing theory to confront the details of a specific application area, this kind of research also promotes refinement in communication analysis. And finally, by shifting the researcher's commitment to the problem rather than the ana-lytic framework, such research fosters theoretical integration in a context where it is meaningful and appropriate.

The second enterprise that should be encouraged is systematization of our field's contributions to theory and method in the analysis of commu-nication. My work in interdisciplinary teams has convinced me of the unique and valuable contributions we have made in developing methods for message analysis; yet little of the knowledge involved in this research practice has been presented in systematic form, either in articles or books. Similarly, anyone who teaches either beginning or advanced sur-vey classes discovers that there are very few good research reviews avail-able in our literature. It is difficult to represent the field well, either to stu-dents or external reviewers, because we have failed as a community to organize our contributions in a systematic fashion.

Finally, we should reclaim the curriculum for the community and organize it explicitly to foster cohesion. Survey courses at all levels should avoid a comparative theory orientation and focus instead on substantive issues and problems. Students should be encouraged to develop a broad acquaintance with basic facts about language, interaction, culture, and media and a positive view of the field as a whole.

To sum up: There is little justification for the continuing search for a unifying theoretical consensus. Even if there were frameworks available with the requisite broad-based appeal (and I have argued that there are not), the field is better served by promoting theoretical and methodological tolerance and disciplinary cohesion.

References

Baxter, L. (1992). Interpersonal communication as dialogue: A response to the "social approaches" forum. *Communication Theory, 2,* 330–336.

Berger, C. R. (1991). Communication theories and other curios. *Communication Monographs, 58,* 101–113.

Berger, C. R. (1992). Curiouser and curiouser curios. *Communication Monographs, 59,* 101–106.

Berger, C. R., & Chaffee, S. H. (1988). On bridging the communication gap. *Human Communication Research, 15,* 311–318.

Bochner, A. P., & Eisenberg, E. M. (1985). Legitimizing speech communication: An examination of coherence and cohesion in the development of the discipline. In T. Benson (Ed.), *Speech communication in the 20th century* (pp. 299–321). Carbondale: Southern Illinois University Press.

Bochner, A. P., & Ellis, C. (1992). Personal narrative as a social approach to interpersonal communication. *Communication Theory, 2,* 165–172.

Clark, A. (1989). *Microcognition: Philosophy, cognitive science, parallel distributed processing.* Cambridge, MA: MIT Press.

Coulter, J. (1979). *The social construction of mind: Studies in ethnomethodology and linguistic philosophy.* London: Macmillan.

Dance, F. E. X. (1982). *Human communication theory.* New York: Harper & Row.

Delia, J. G. (1987). Communication research: A history. In C. R. Berger & S. H. Chaffee (Eds.), *Handbook of communication science* (pp. 20–98). Newbury Park, CA: Sage.

Dennett, D. (1987). *The intentional stance.* Cambridge, MA: MIT Press.

Dervin, B., Grossberg, L., O'Keefe, B. J., & Wartella, E. (Eds.) (1989a). *Rethinking communication, Vol. 1: Paradigm issues.* Newbury Park, CA: Sage.

Dervin, B., Grossberg, L., O'Keefe, B. J., & Wartella, E. (Eds.) (1989b). *Rethinking communication, Vol. 2: Paradigm exemplars.* Newbury Park, CA: Sage.

Greene, J. O. (1987, November). *The language of action: Symbolic representation in the output system.* Paper presented at the annual meeting of the Speech Communication Association, Boston, MA.

Hall, S. (1992). Cultural studies and its theoretical legacies. In L. Grossberg, C. Nelson, & P. Treichler (Eds.), *Cultural studies* (pp. 277–294). New York: Routledge.

Hawkins, R. P., Wiemann, J. M., & Pingree, S. (Eds.) (1988). *Advancing communication science: Merging mass and interpersonal processes.* Newbury Park, CA: Sage.

Jacobs, S. (1985). Language. In M. L. Knapp & G. R. Miller (Eds.), *Handbook of interpersonal communication* (pp. 313–343). Newbury Park, CA: Sage.

Lambert, B. L. (1992). *A connectionist model of message design.* Unpublished doctoral dissertation, Department of Speech Communication, University of Illinois.

Lannamann, J. W. (1992). Deconstructing the person and changing the subject of interpersonal studies. *Communication Theory, 2,* 139–147.

Lave, J. (1988). *Cognition in practice: Mind, mathematics and culture in everyday life.* Cambridge, England: Cambridge University Press.

Leeds-Hurwitz, W. (1992). Forum introduction: Social approaches to interpersonal communication. *Communication Theory, 2,* 131–138.

Nelson, C., Treichler, P. A., & Grossberg, L. (1992). Cultural studies: An introduction. In L. Grossberg, C. Nelson, & P. Treichler (Eds.), *Cultural studies* (pp. 1–22). New York: Routledge.

O'Keefe, B. J. (1992). Developing and testing rational models of message design. *Human Communication Research, 18,* 637–649.

O'Keefe, B. J., & Delia, J. G. (1989). Communicative tasks and communicative practices: The development of audience-centered message production. In B. A. Rafoth & D. R. Rubin (Eds.), *The social construction of written communication* (pp. 70–98). Norwood, NJ: Ablex.

Reardon, K. K., & Rogers, E. M. (1988). Interpersonal versus mass media communication: A false dichotomy. *Human Communication Research, 15,* 284–303.

Rice, R., Borgman, C. L., & Reeves, B. (1988). Citation networks of communication journals, 1977–1985. *Human Communication Research, 15,* 256–283.

Sigman, S. (1992). Do social approaches to interpersonal communication constitute a contribution to communication theory? *Communication Theory, 2,* 347–356.

So, C. Y. K. (1988). Citation patterns of core communication journals: An assessment of the developmental status of communication. *Human Communication Research, 15,* 236–255.

Stich, S. (1983). *From folk psychology to cognitive science: The case against belief.* Cambridge, MA: MIT Press.

Suchman, L. A. (1987). *Plans and situated actions: The problem of human–machine communication.* Cambridge, England: Cambridge University Press.

Taylor, T. J. (1992). *Mutual misunderstanding: Scepticism and the theorizing of language and interpretation.* Durham, NC: Duke University Press.

Wiemann, J., Hawkins, R. P., & Pingree, S. (1988). Fragmentation in the field—and the movement toward integration in communication science. *Human Communication Research, 15,* 304–310.

Building a Discipline of Communication

by Gregory J. Shepherd, University of Kansas

In the "Call for Papers" for this issue of the *Journal of Communication* the editors wrote, "Communication scholarship lacks disciplinary status because it has no core of knowledge. Thus institutional and scholarly legitimacy remains a chimera for the field."

For a variety of reasons, I find the premise and consequent of this proposition—that the field "lacks disciplinary status" and thus lacks a certain "legitimacy"—to be self-evident. What seems less obvious is the cause of our undisciplined and illegitimate status. The posed proposition suggests that our disciplinary troubles are the result of having "no core of knowledge." This essay will argue, however, that disciplines are defined not by *cores of knowledge* (i.e., epistemologies) but by *views of Being* (i.e., ontologies). Disciplinary status for a field rests on the ontological status of that field's "idea"—disciplines represent various foundational ideas. Understanding the lack of disciplinary status accorded communication requires an understanding of the lack of ontological status granted the idea of communication in modernity—understanding how the foundationalist sense of a discipline is antithetical to the nonfoundational, modern sense of communication.

The Nature of Disciplines

An etymological consideration of the word *discipline* suggests it to be a character of ontology and characteristic of modernity's faith in foundations. *Discipline* is derived from the Latin *disciplina*: instruction of disciples. Disciples, in turn, are instructed in a doctrine (and by "doctors")— they are "indoctrinated." Thus, Berkeley could write that to be "undisciplined" is to be "nurtured to no doctrine" (Oxford English Dictionary, 1971, pp. 741, 3496).

Doctrines provide disciples with foundations for beliefs and action, but those foundations are views of Being more than cores of knowledge. The religious connotation that accompanies the word *disciple* is illuminating:

Gregory J. Shepherd is an associate professor in the Department of Communication Studies at the University of Kansas, Lawrence.

To be a disciple is to adhere to a particular faith in the nature of existence and, furthermore, to promulgate that ontology. Disciples are defined more by faith than knowledge; their beliefs and practices depend on views of Being which they witness, not cores of knowledge that they claim.

Taking this ontological and foundational sense to the academy allows for understanding academic disciplines as something other than fields of study built around knowledge of particular phenomena or practices. Academic disciplines, in this view, are distinguished not by the parcels of existence that they study, but by the views of existence they afford. Anthropology, art, biology, chemistry, economics, history, philosophy, physics, political science, psychology, religion, sociology, and others, each offer a particular view of Being. Is existence best understood as cultural, creative, or chromosomal? Is the foundation for all best thought of as the molecule, a commodity, or time? Is ontology best viewed as rational, material, or governmental? Is Being best seen as self, soul, or society?

It may seem curious to characterize this ontological question in terms of what it is *best* to believe. After all, many might argue that existence is multidimensional. But it is precisely the nature and purpose of disciplines and their disciples to forward a unique view of Being among all the alternatives and say, "There is something primary, or essential, about this particular view." Disciplines depend on disciples acting as advocates for the ontology they forward, making implicit and explicit arguments that their view "matters" (i.e., is of the "mother-stuff," from the Latin *mater*). Perhaps this ontological sense of disciplines can be illustrated by the views of an object different disciplines might suggest. In the room where I am working is a chair. How is this object most essentially viewed? The discipline of physics teaches a doctrine that suggests the chair to be a swirling mass of subatomic particles. The essential nature of the chair, says a disciple of physics, is an affair of matter and energy. The discipline of biology, however, offers an alternative view of the chair founded in an understanding of the organic nature of existence—the cotton backing, the wooden legs, etc., understood as cellular structures. And, of course, the discipline of art urges an understanding of the chair as essentially aesthetic, while the discipline of economics argues for viewing the chair primarily as an asset, and so on. Thus I can well imagine the physicist, the biologist, the artist, and the economist speaking in turn to me of the chair in my room: "How sturdy!" "How natural!" "How handsome!" "How much?"[1]

It is the forwarding of a unique foundational ontology that grants fields of study their disciplinary status. Correspondingly, it is communication's failure to articulate a unique, foundational ontology that has kept it from achieving the status of a discipline and not its failure at articulating an

---

[1] That there is no agreed-upon noun for naming a disciple of communication—nothing like physicist, biologist, artist, or economist—may be an interesting bit of circumstantial evidence for the undisciplinary nature of the field of communication.

agreed-upon core of knowledge. After all, fields of study we consider to be "established disciplines" also lack epistemological coherence (Becher, 1981). What the fields of study we call "disciplines" have that communication does not are not more narrow and knowable subjects, longer histories for the establishment of knowledge, shared methodological commitments for testing knowledge, and the like, but unique ontologies they forward as materially essential to Being, and a corps of disciples committed to the foundationalist nature of their beliefs.

This discussion of the "nature" of disciplines has been grounded by an ocular metaphor: Disciplines offer *views* of Being. As Rorty (1979) has shown, this metaphor is pervasive in modernity, representing faith in the existence of foundations. There is a material world to be viewed, and there are essences to be seen. Modernity has defined disciplines as eyes on existence. Understanding our field's lack of disciplinary status requires an understanding of how modernity defined communication as incapable of seeing foundations.

A Nonontological Foundationless Idea

Bruce Kimball (1986) has written a compelling account of the history of liberal education that shows the proponents of a communication-centered view of education engaged in a losing struggle for supremacy with the proponents of philosophy and their subsequent allies of science. Although the story dates to the time of Isocrates and Plato, the crucial battle in this war took place at the dawn of the Enlightenment, when communication was stripped of its force. Understanding this story's decisive developments in the 17th century is key to understanding communication's antithetical relationship to the foundationalist sense of what it means to be a discipline.

The 17th century witnessed the publication of works by Shakespeare, Descartes, Newton, and Locke that would mark, respectively, the establishment of modern English, philosophy, science, and social theory. All of these works displayed the making of a great division between the material and immaterial, what matters and what does not. It was, in fact, this essential bifurcation that grounded the many dualisms modernity came to offer, and the character of the bifurcation was such that communication and its associated terms were taken to define the inessential side of the great division. Thus, in the 17th century, the world was firmly divided into linguistic and nonlinguistic spheres, and the strategy of claiming materiality for something by contrasting it with "mere" words, talk, or rhetoric was fixed in literature, philosophy-science, and social theory until, by the close of the century, communication was rendered immaterial, an idea void of ontological force.

In many of his plays, but particularly those of his latter period, Shakespeare penned an ironic sense of the powerlessness of words. As a man of words, Shakespeare forwarded a tradition in English literature for trivializing the importance of words. It was a perfectly modern thing to do. In *Troilus and Cressida* (circa 1601–1603), for example, Shakespeare gave the association of "mereness" to words: "words, words, mere words, no matter from the heart" (1936, p. 857).[2] And in *King Henry the Eighth* (circa 1612–1613) he gave voice to what came to be the most popular of dualisms meant to trivialize words: "words are no deeds" (1936, p. 1350).

Shakespeare was not, of course, the only author among the 17th century's literati to forward a bifurcated sense of the world, in which things linguistic were contrasted with things that mattered. In what was to become a renewed avalanche of the ancient Socratic critique and characterization of rhetoric as sophistry, for example, Richard Braithwaite wrote, in 1615, "Heere is no substance, but a simple peece Of gaudy Rhetoricke" (Oxford English Dictionary, 1971, p. 2535). And in 1645 a new term was coined (first used by Joseph Hall) to capture the immaterial nature of communication: *verbality,* "the quality of being (merely) verbal; that which consists of mere words or verbiage" (Oxford English Dictionary, 1971, p. 3609).

The rise of modern science in the 17th century had much to do with the teaching of the lesson on display in the century's literature—the foundational insignificance of communication. In the first part of the century, Galileo recognized that his troubles with the Church were tied to the apparent insignificance he attached to communication from God—that is, the words of the Bible. Sensing the charges of heresy that were forthcoming, Galileo tried to reconcile the facts of Copernican theory with the truth of the scriptures by suggesting a certain linguistic dualism. He argued that there are two languages, the language of words, as found in the Bible, and the language of nature, as seen through his telescope. Both languages "proceed from the Divine Word," but when truth is expressed vernacularly, as it is in the Bible, it is subject to interpretation. In contrast, nature's expression of the truth "is inexorable and immutable" (see Boorstin, 1983, pp. 322–323). As science took hold in the 17th century, Galileo's argument was generalized into a commitment to "the language of nature" as the only language that mattered.[3]

---

[2] Attributing the first collocation of "mere words" to Shakespeare is not entirely fanciful. Prior to the 17th century, and well into the 18th century, the adjectival sense of *mere* meant the opposite of what it means to us today and what it meant to Shakespeare in his use of "mere words." Where once *mere* meant "nothing less than," it now means, of course, "nothing more than." And the earliest use of this modern sense of *mere* appears to have occurred no more than two decades prior to Shakespeare's, and was then apparently reserved or typically used for denoting the insignificance of humankind, as in the common current use of "mere mortals." (See the Oxford English Dictionary, 1971, p. 1773.)

[3] Galileo's creative but contrived and doomed defense was popularly echoed by many writers later in the 17th century (including Locke), who distinguished two types of discourse: a pure and direct internal sort derived from sensation, and a tainted and ancillary external type for public consumption (see Peters, 1989).

Descartes was, of course, the 17th century's most important bifurcator, and his work more than any other provided the philosophy that was to ground modern science. And modern science, as realized in the 17th century, defined itself, in large measure, in opposition to the world of words. It is telling, for example, that when the Royal Society was chartered in 1662, its founders took as their motto the idea that was becoming proverbial as history accelerated toward the Enlightenment: *Nullius in Verba.* Words are nothing. There is nothing in words. Thomas Sprat, in writing his history of the society that Newton was to preside over for nearly a quarter of a century, noted that the group was to be organized around an interest in "not the Artifice of Words, but a bare knowledge of things" (cited in Boorstin, 1983, p. 394).

As Peters (1989) has pointed out, the modern conception of communication was introduced by John Locke in 1690. As a man of the 17th century and a fellow of the Royal Society, John Locke shared his contemporaries' view of language as insignificant. Words, to Locke, were insubstantial and untrustworthy nothings. Words, in fact, stood in contrast to the things that mattered—individuals, their selves, and especially their "ideas." But in his attempt to build the social theory of liberalism, Locke was confronted with a dilemma: How was a *social* theory based on the primacy of *individuals* to succeed? A mechanism for connecting individuals was required lest liberalism fall prey to solipsism. Locke answered this conundrum by borrowing words as empty nothings, putting them in the service of "ideas," and inventing the modern vehicular view of communication.

In Locke's liberalism, communication was invented as a conveyance, a conduit, a mere medium for the transfer of thoughts, an ancillary mechanism for mediating individuals. This invention was premised on a by then well-established belief in the inessential character of words. It was because Locke shared his century's disdain for the immaterial nature of words that he took them as free to be filled with matter—that words could be employed as the messengers of material ideas—and communication could be born in modernity as a foundationless and Being-less idea.

Communication's struggles with obtaining disciplinary status should now be clear. Disciplines forward unique ontological views; they tell us what matters about Being and they represent essentialist ideas. From modernity's point of view, then, how can there be a discipline of communication? *Nullius in Verba.* How can one be a disciple of nothing? As a mere vehicle, communication has no existential status in modernity. In a sense, communication may carry Being, but in and of itself, communication is Being-less. What unique view of Being can a Being-less idea forward? Modernity said of communication what Gertrude Stein said of Oakland: There is no there there. As scholars of communication, how are we to respond?

Coping with A Bifurcated World

Our challenge is to respond to modernity's vision of a bifurcated world and view of communication as inessential in a way that will legitimate our interests. Our choices are basically three: (a) We may accept modernity's bifurcation and view of communication, but attempt to obtain legitimacy through association with and service to other disciplines (the undisciplinary response); (b) we may reject modernity's bifurcation by accepting communication as nonfoundational as we argue against the legitimacy of any essentialist ideas (the antidisciplinary response); (c) we may deny modernity's bifurcation by asserting that communication is foundational and attempting to forward a unique communication ontology (the disciplinary response). Each of these responses is associated with a unique set of challenges that will prove consequential to the future of the field.

As a 20th-century field of study, communication has had to cope with modernity's 300-year-old conception of communication as nonfoundational. The field's traditional coping response has been *undisciplinary,* a response of surrender: Accept modernity's conception of communication and tacitly endorse a view of the world bifurcated into the linguistic and the nonlinguistic. With this response, legitimacy for the field is sought primarily through association with and service to "real" disciplines and the material stuff of the other side of the bifurcation that those disciplines represent.

The nearly necessary consequence of accepting modernity's conception of communication as a nonontological vehicle is a field of study devoted primarily to the investigation of skill, practice, and use, (dis)organized by context. The study of communication, in this view, is primarily the study of how a vessel can be manipulated and shaped, in particular contexts and under certain circumstances, in order to best transmit the material it contains, the essences that established disciplines reveal. This view encourages a defense of communication as an academic enterprise which is not in itself disciplinary, but is cross-disciplinary. Because communication may be thought, for example, to carry individual selves, transmit social rules, and convey culture, disciples of the ontologies of psychology, sociology, and anthropology have an interest in studies of this vehicle. As a cross-disciplinary field of study, communication thus becomes a place where disciples of various disciplines congregate. The research products of this field may tell us much about the existence of the self, the essence of society, the foundation of culture, and the like, but they cannot, by virtue of the definition the field has accepted, tell us much about communication—for little can be told of nothing.

The challenges that come with accepting the modern bifurcation are well known, and primarily institutional. As the traditional response, surrender to modernity's view has led to the perceived set of problems that undoubtedly motivated the proposition framing this essay. Accepting communication as an inessential vehicle makes the straightforward development of a communication discipline impossible, and leaves the field in a precarious position of illegitimacy. We will be, from this position, forever caught trying to obtain legitimacy through association with "real" disciplines, forced into borrowing existence from their ontologies in an attempt to obtain life for our vehicle by virtue of the passenger cargo it carries. We admit that communication does not matter, but will claim significance for communication as a carrier of things that do matter. We can argue for the importance of maintaining such a nondisciplinary field, but will constantly struggle against charges of incoherence and triviality.

The most popular postmodern response to modernity has been *antidisciplinary*: reconcile the bifurcated world by rejecting foundationalism. There are no foundations, nothing is essential, and ontological views are without meaning. Rather ironically, this postmodern response seizes upon and celebrates modernity's conception of communication as the foundationless idea. The nature of existence is the nature of the immaterial idea: Words do not matter (in the Latin sense of lacking any "mother-stuff" or essence), but then again, nothing matters.[4] Therefore, all must be words. Everything is symbolic. There are only appearances (or, in the postmodern vernacular of the Canon camera ad campaign, "image is everything"). This postmodern response reconciles the antithetical terms of discipline and communication by denying legitimacy to the idea of disciplines. There are no foundational doctrines, or views of Being, so there can be no disciplines. Communication is not disciplinary, but neither is philosophy, physics, or any other of those "merely" rhetorical constructions modernity had us believing were "established."

This antidisciplinary response returns the world to sophistry, as Richard Rorty, a spokesperson for this choice, makes clear: We ought to be "where the Sophists were before Plato . . . we shall be looking for an airtight case rather than an unshakable foundation" (1979, p. 157). In a world with no essence to be revealed, technique and effectiveness are all that remain. Modernity's grand bifurcation is obliterated by elevating the modern conception of communication as sophistry to a position as the all-encompassing idea.

A field of communication organized by this response will know no boundaries. Scholars of communication will act as postmodern Sophists, taking their intellectual wares to all corners of the academic world. As Rorty (1979) can be read to imply, the goal of communication scholarship would be "edification" and the field would be reactionary. Communication's task would be to deconstruct the constructed and engage in "abnormal discourse."

---

[4] The irony here is quite layered: Nothing matters; communication, in modernity, is nothing (*Nullius in Verba*); therefore communication, in this ironic postmodern sense, matters (as the nothing, it is the only thing that can matter).

The problems associated with adopting the antidisciplinary response are largely moral. The ancient arguments against the Sophists will also be brought to bear on communication, should this be our choice. Accepting a view of communication, shorn of any ontological foundation, and making it paradigmatic of existence entails celebrating the insignificance of Being itself. As MacIntyre (1981, p. 23) has argued, such a view is immoral in the Kantian sense; it leads to treating others as means to an end rather than as ends in themselves, with persuasive technique as our only guide. Such "base" pragmatism displaces notions of "good" and "right" with unashamed effectiveness and opens itself to charges of relativism.

A third response also means to obliterate the grand bifurcation, but involves denying, rather than accepting, modernity's view of communication as inessential. This response entails arguing for a definition of communication as foundational and attempts to make communication *disciplinary* through the development of a unique ontology. Communication is not simple skill; it is more than a mere vehicle for the conveyance of substance; it is, in itself, material.

This response, unlike the previous two, refuses to accept modernity's view of communication. As such, it is the only one from which a discipline of communication can be built. Scholars of this response will act as disciples of, advocates for, a communication-based view of Being. These disciples will argue that existence is essentially symbolic, but that there is nothing "mere" about that. In such a view the "sticks and stones" that break bones in a child's nursery rhyme would no longer be contrasted with "words," which by their immateriality, "can never hurt." Rather, words would viewed as *the* ontological force, where language constitutes existence, and communication makes Being be; where the essential character of sticks, stones, bones, and the chair in my room is "communicationally" constructed; where communication rather than cellular structure, energy or mass, aesthetic quality or commodiousness, is the foundation for Being. This is the view of Being that communication disciples would offer as alternative to the views of other disciplines.

The disciplinary field of study built from this response would look quite unlike the fields suggested by the traditional modern and popular postmodern responses. It would not be focused on effectiveness, nor organized by context. Rather, the disciplinary field would research the general grounding of Being in communication, and query the ways in which particular manifestations of existence (e.g., individuals, societies) are "communicationally" constructed.

The troubles that will visit should we choose the disciplinary response are those that come with uncertainty and complexity. It is difficult to know how or whether the details of a communication ontology might be worked out. It is one thing to state the doctrine of communication; it is something else again to provide compelling theory based on that ontology. A communication ontology will be unlike those of other disciplines. Communication as the essence of Being suggests a sort of self-generating foundation: Communication is both the builder of Being's foundation and the foundation itself. The amorphous nature of this ontology poses a severe challenge to would-be disciples of communication. Gadamer (1989, p. 378) has captured the character of this challenge: "What language is belongs among the most mysterious questions that man ponders. . . . it seems to conceal its own being from us. . . . we are endeavoring to approach the mystery of language from the conversation that we ourselves are." However daunting the challenge of rejecting modernity's bifurcation by "ontologizing" communication, that is the challenge that must be met in building a discipline of communication.

References

Becher, J. (1981). Towards a definition of disciplinary cultures. *Studies in Higher Education, 6,* 109–122.

Boorstin, D. J. (1983). *The discoverers*. New York: Random House.

Gadamer, H. G. (1989). *Truth and method* (2nd rev. ed., J. Weinsheimer & D. G. Marshall, Trans.). New York: Crossroad.

Kimball, B. A. (1986). *Orators and philosophers: A history of the idea of liberal education*. New York: Teachers College Press.

MacIntyre, A. (1981). *After virtue*. Notre Dame, IN: University of Notre Dame Press.

Oxford English Dictionary (compact ed.). (1971). Oxford: Oxford University Press.

Peters, J. D. (1989). John Locke, the individual, and the origin of communication. *Quarterly Journal of Speech, 75,* 387–399.

Rorty, R. (1979). *Philosophy and the mirror of nature*. Princeton, NJ: Princeton University Press.

Shakespeare, W. (1936). *The complete works of William Shakespeare: The Cambridge edition text* (W. Aldis Wright, Ed.). Garden City, NY: Garden City.

Perspectives on Communication

by Kurt Lang and Gladys Engel Lang, University of Washington

Ten years ago, when invited to write about the "ferment in the field," more than one skeptical communication scholar must have wondered if there really was any ferment; yet, anyone really looking managed to find signs of it. Now, in responding to the request to assess the future, one cannot resist a prior question: Just what are the boundaries of the field whose fate we are to prognosticate? Does it qualify as a recognizable academic discipline or is it just the product of fortuitous contingencies that now place certain scholars and practitioners in schools, departments, or institutes under the rubric of communication? We take this issue to be implicit in the larger question.

The easy way to define the field is by whatever those "in communication" do. Though obviously unsatisfactory, such an operational definition may express the reality better than anything else. Even the most cursory glance at course titles, papers read at meetings, or publication lists reveals an astounding lack of unity. Perhaps this is inevitable, especially when one takes stock of some older and more firmly established social sciences. Political science, anthropology, and sociology are still plagued by a similar, even if less problematic, diversity. As a late arrival (except for rhetoric) in the groves of academe, communication is still in the process of carving out a niche for itself. In the process, it has, by necessity, drawn on the other social sciences as well as on the humanities and several relevant professional fields (Beniger, 1990). However, we see no reason to cut the umbilical cord to these parents by searching for a separate "disciplinary" identity or, as some would have it, a specific communication paradigm to organize and integrate the existing body of what passes for substantiated knowledge.

One can hardly anticipate the future without first taking a look backward to see how we got to where we are today. We will do this from the perspective of sociology, an approach to the study of mass communication with which we have been identified, but without limiting ourselves to the work of sociologists.

Research and scholarship (we use these terms interchangeably) have grappled with three sets of issues: (a) the nature of human communica-

Kurt Lang is a professor of sociology and communication and Gladys Engel Lang is a professor emerita of communication, political science, and sociology at the University of Washington in Seattle.

tion, (b) the effects of mass communication, and (c) the connections between the media system and the society in which it is embedded. Each has its own approach to the subject.

The first, by asking how communication is possible, treats the process itself as problematic. Investigators can focus on transmission (information theory), on interpretation (hermeneutics), on the acquisition of language (social psychology), or on the processes and structures through which meanings are assimilated (cognitive science). However basic this set of issues, we choose not to address them in this brief essay.

Regarding the effects of mass communication, scholars concluded by 1948—if concluded is the right word—that "some kind of *communication* on some kinds of *issues,* brought to the attention of some kinds of *people* under some kinds of *conditions,* have some kinds of *effects.*" This tongue-in-cheek statement by Bernard Berelson (1950, p. 451) did indeed reflect the conventional wisdom in the "golden age" when media research in the dominant center was preoccupied with pinpointing effects, typically through surveys or experiments, on individuals, usually within some specified time span and with the full understanding that any measurement of change due to media exposure over extended time periods would be contaminated by other influences. Progress came to be measured in terms of the number of barriers or facilitating conditions, also called mediating factors (Klapper, 1960), that research had been able to identify.

And progress there was. Claims about the significance of findings, however, often rested on false dichotomies that made the break with the past appear sharper than it actually was. One view, supposedly laid to rest by the studies of the 1940s and '50s, was of the media as "all powerful." It was being replaced, so people said, by another (though equally one-sided) view of media power as severely limited, a view often characterized as the "minimal effects model." Both views are parodies that functioned as straw men in a nonexisting controversy that distracted, and may unfortunately continue to distract, scholars from investigating issues that deserve our full attention.

We have never found even a suggestion in what passes as social science literature, not even in the writings of a crowd psychologist like Gabriel Tarde, that the media are all-powerful—quite the contrary. Tarde anticipated by more than a half century the discovery of personal influence by American scholars, a precedence that Berelson, Lazarsfeld, and McPhee did indeed acknowledge (1954, p. 300). However, the interest of Tarde lay in how newspapers helped transform individual opinion into social opinion, with journalists defining "what is or appears to be interesting" and, in the long run, "imposing the majority of their daily topics upon conversation" (Tarde, 1969, pp. 301f., 304). This is how people all over the country were made conscious of a similarity of ideas until those on one side managed to eclipse the other by sheer force of number.

If this sounds very much like agenda setting, so does a statement from

The People's Choice, the much cited landmark study in support of minimal effects: "Issues about which people had previously thought very little or had been little concerned took on a new importance as they were accented by campaign propaganda." This redefining of issues accounts for the few conversions in voting behavior observed (Lazarsfeld, Berelson, & Gaudet, 1944, p. 98). The same media effect showed up again in 1948 when "the image of what was important in the campaign . . . *did* change to a dominance of socioeconomic issues"—issues whose salience was brought home through the mass media (Berelson et al., 1954, p. 264). By associating himself with these issues, President Truman succeeded in winning potential defectors back into the Democratic fold. The last proposition is a striking illustration of the priming effects more recently produced by Iyengar and Kinder (1987) on laboratory subjects as a demonstration, along with agenda setting, of the power of the media.

Findings like the above have been hailed as a rediscovery of media power but are not quite the refutation of earlier work they may seem. In their formulation of agenda setting, McCombs and Shaw (1972) made a distinction—in our opinion too absolute—between the power of the media to tell us what to think versus what to think about. The power of mass communication in general, conclude Iyengar and Kinder, "appears to rest not on persuasion but on commanding the public's attention (agenda-setting) and defining criteria underlying the public's judgments (priming)" and even this is limited insofar as television does not create priorities or standards out of thin air (1987, p. 117). The difference between the formulation of these and earlier scholars is one of emphasis. Whereas the circle around Lazarsfeld never tired of pointing to the small number of conversions mass communication was able to achieve, others since then have pursued this same search for effects, often with better tools that offer a sharper focus on every kind of shift, however subtle or small, and techniques that allow them to tease out effects from more readily accessible and larger data bases. The difference between then and now boils down to whether the glass is half empty or half full, but the paradigm remains essentially the same.

As an alternative, one can move beyond the individual as the target of converging influences from other people, from the community, and from the media, toward a more sociological conception of effects, in which the media system is a subsystem within the society with many links to other sectors. In focusing on the third set of issues, long of concern among scholars, one starts with the time-honored axiom that society exists only in communication. Accordingly, the question that commands attention is how the presence of the various means of communication allows individuals, groups, and the various organizations and institutions in society to go about their business in the way they do. We live in a world of media events (Dayan & Katz, 1992). Without modern means of communication to sustain them, there could be no great society, with its spectacles and rituals, and no public opinion as we have come to know it.

Scholars from Gabriel Tarde to Ferdinand Toennies, from Walter Lippmann to Harold Lasswell, were intrigued by the media as harbingers and accomplices of social change. None has written more suggestively than Harold Innis (1950) about how the introduction of a new medium—be it by invention, immigration, or trade—can upset the precarious balance between sacred and temporal power, between culture and military force, between monopolies of knowledge rooted in an oral tradition and those built on writing with alphabets of varying difficulty. The issues Innis raises crop up time and again in different terms and in modern contexts.

This is where the real challenge lies. Although the volume of research on communicators, on media organizations, on news-making and culture production has increased significantly in recent years, we need to move toward a better integration of observations on the microlevel with systemic generalizations about macrolevel phenomena. Concepts are useful. They help us define problems in need of an answer but only when defined so as to make answers possible. Thus, the neo-marxist concept of hegemony, still fashionable, is saddled with too much ideological ballast to take us very far. Questions about how the media operate in relation to the centers of political and economic power can be formulated with equal sharpness as a test of First Amendment theory: Has the press been a consistent and effective adversary, or is it, because of its many connections, primarily an extension of the establishment? Research would examine situations of actual and potential conflict, their causes, and the process by which accommodation is reached (see, for example, Blumler & Gurevitch, 1981). The dynamics behind journalistic feeding frenzies would appear as special cases.

Looking to the future, we see the best prospects for advancing scholarship in an even more definitive reorientation of research away from the media behavior and responses of individuals and toward the cumulative consequences of media behavior over time. For instance, instead of locating vote changers in order to determine how their communication behavior differs from those with a consistent preference, why not concentrate on the larger picture: how trends in media usage common to voters on both sides of the political aisle may be changing the process of decision-making, how candidates adapt to new media developments, how this affects the kind of candidate considered electable. Similar questions can be asked about the media interface with the cultural sector: Is culture disseminated via the media inevitably trivialized? How well do the media reflect the diverse needs and mix of interests in the population? Have they stunted artistic creativity, elevated taste, provided an adequate outlet for new talent?

We are not suggesting that such questions have never been seriously probed. However, much of the discussion has been speculative and all too often totally uninformed by the more mundane findings of audience and effects research. Or, if based on content analysis, the results are often allowed to speak for themselves instead of being coupled with equally systematic observations of the producers and the recipients of such content. Nevertheless, certain communication events and developments are propelling us toward a better integration of the atomistic and the systemic level. They also give us a vision of the future.

These events and developments include, first of all, the technological advances that underlie what is often characterized as the communication revolution. A series of inventions *in toto* have extended the reach, rate, and speed at which content of ever greater fidelity, combining sight and sound as in *actualité,* can be distributed over a large area to a heterogeneous population. They ushered in the era of mass communication that dates back to the rise of the popular press roughly 150 years ago. Since that time, change has accelerated. The time lag between the first successful telecast and the near universal diffusion of sets is far less than it was in the case of the newspaper, the movies, or even radio—media that provided the initial stimulus for systematic empirical research on mass media. As a natural response, most research in recent decades has been aimed at television, and there is no doubt that this medium will continue to occupy an important place as a focus of study.

Still other inventions have changed, and will continue to modify, the nature of the media–audience relationship. The increased capability to target audiences, to store and make available for convenient retrieval vast masses of material, and to communicate in the interactive mode, by giving people more options, has tended both to undermine and to supplement the mass audience (Neuman, 1991). Miniaturization has further helped to restore to them some of the autonomy they had lacked as members of the mass audience. On the other hand, fears that the ascendency of such minimedia will fragment the great society into communicatively separate enclaves strike us as exaggerated. People remain interdependent, while the same flexibility from which audiences benefit also increases the power of communicators. Communicators can use the technology for personalized messages or, at least, for messages tailored to take advantage of the unique vulnerabilities of whatever types of people they are trying to influence. Authority may change its character but does not itself disappear. The trend is away from domination to manipulation, from direct control by a central authority to more indirect forms of control within some framework of common understanding. Communication research can no longer concentrate on a particular medium or set of messages in isolation from their overall distribution.

Third, the explosion of knowledge and an accelerated information flow across all boundaries has led to a greater appreciation of the importance of the communication sector, both as a condition for economic development and as the neural activity that holds a society together. Without communication, it would not be possible to coordinate the traffic in material goods essential for efficient production; advertising, especially for new products, affords consumers opportunities for critical selection they would not have if completely dependent on what was on the shelves.

Quite apart from such functions, many communications are salable commodities, valued for their content and supported by industries that run the whole gamut from the manufacture of newsprint to highly sophisticated electronics.

Finally, as the combined consequence of growing interest and the quantum jump in information processing capability, we have been accumulating vast masses of data for a more encompassing analysis of communication than was once possible. There are surveys, masses of them, on just about every imaginable aspect of audience behavior and communication response. Once novel operations have become routine. Someone is always in the field to catch the instantaneous reaction to every major event. With data banks housing material from surveys covering decades, we are now in a position to document trends in relation to technological innovation and changes in the general pattern of life. We could, for example, compare the impact of festivals, public disorder, the outbreak of war, and disasters on populations in different media environments. As for the flow of messages, machine-readable texts and software have greatly reduced the labor involved in content analysis, making it easier to monitor the flow of news in space (through the media and across the world) and its thematic emphasis over time. These same methods will be useful in other areas as digital storage becomes more common.

We are not so naive as to believe that data banks and statistical processing technology will solve all research problems. There is, for one thing, the Janus-faced problem of standardization. For comparisons to be valid, the data on which they are based have to be collected in a uniform way, which is not always the case. Definitions change as methods improve and there are always new circumstances that must be taken into account. Conclusions can be especially troublesome when the categories and method of collection of data used for crossnational comparison are not the same. Moreover, some countries lack the resources and infrastructure to collect data at a level of validity acceptable in the wealthier industrialized part of the world. In addition, standardization creates problems of its own. If powerful public and corporate agencies with the means of collecting information insist on their proprietary rights, the data will not contribute to our understanding as much as they could. Often these agencies gather, and analyze, only the information directly useful to them while scholars, in their disinterested pursuit of knowledge, rarely bother themselves over such cost effectiveness. There can, however, be problems even when such information is made freely available, or at a reasonable cost, as has been the practice of some. Data collected for alien purposes can limit what questions are asked. Nothing is more natural—for us as for the drunkard—than to start searching where the light is. Meanwhile, standardization in the interest of comparability tends to freeze things. Creative and suggestive work by scholars working independently and forced to rely, because of the nature of their research problem, on what is so often disparaged as "anecdotal" evidence is likely to be overshadowed, even more than it is today, by elegant analyses of impressive masses of data. We are not sure that the two will be able to prosper together or whether one will be driven out of circulation by some kind of Gresham's law.

To revert once more to the problem of integrating of the micro and macro levels of analysis: We discern an increase in the style of theorizing that, if not actually rooted in, certainly mirrors the reasoning involved in statistical data processing. The typical result is a model that links variables. In its more concrete version, it takes the form of a path diagram that arranges the determinants of some aggregate pattern of communication behavior. The model can also be presented as a schematic description of an entire communication system, in which variable labels cover empty boxes. Both look like flow charts. How well this kind of model simulates reality depends on the content that goes into it. But simulation is not yet interpretation, which can only come out of the mind of the analyst.

The alternative is to link communication research more closely to cultural studies, to style and content as affected by the structural bias of the media system. What image of society do the media cultivate? And how does this in turn affect the relations among people and groups within society? Such a perspective takes in a lot of what other social disciplines cover.

Looking into the future, we see the same basic options that many departments of communication are facing today. To simplify, there is the option of communication as a managerial science oriented to the explication of a specific set of "communication" problems versus the option of communication as a perspective that can shed light on issues shared with one or more of the social science disciplines. Both have a certain legitimacy, but the bigger question has to do with how much they have in common and how well they can coexist under the same label. The one probably belongs in a professional school. The other, given the range of data put into our hands by technology, affords new opportunities for creative synthesis—though, in our opinion, not for the development of a single paradigm. Fields are more likely to develop by opening up new areas and providing fresh perspectives than by drawing tight boundaries around themselves, and it is in this way that communication may yet define itself as a field.

References

Beniger, J. R. (1990). Who are the most important theorists of communication? *Communication Research, 17*, 698–715.

Berelson, B. (1950). Communication and public opinion. In B. Berelson & M. Janowitz (Eds.), *Reader in public opinion and communication,* (pp. 448–462). Glencoe: Free Press.

Berelson, B. R., Lazarsfeld, P. F., & McPhee, W. N. (1954). *Voting: A study of opinion formation in a presidential campaign.* Chicago: University of Chicago Press.

Blumler, J. G., & Gurevitch, M. (1981). Politicians and the press: An essay in role relationships. In D. D. Nimmo & K. R. Sanders (Eds.), *Handbook of political communication,* (pp. 467–493). Newbury Park, CA: Sage.

Dayan, D., & Katz, E. (1992). *Media events: the broadcasting of history.* Cambridge, MA: Harvard University Press.

Innis, H. A. (1950). *Empire and communication.* Oxford: Clarendon Press.

Iyengar, S., & Kinder, D. R. (1987). *News that matters: Television and American opinion.* Chicago: University of Chicago Press.

Klapper, J. T. (1960). *The effects of mass communication.* Glencoe: Free Press.

Lazarsfeld, P. F., Berelson, B., & Gaudet, H. (1944). *The people's choice: How the voter makes up his mind in a presidential campaign.* New York: Duell, Sloan and Pearce.

McCombs, M., & Shaw, D. (1972). The agenda-setting function of mass media. *Public Opinion Quarterly, 36,* 176–187.

Neuman, W. R. (1991). *The future of the mass audience.* New York: Cambridge University Press.

Tarde, G. (1969). Opinion and conversation. In T. N. Clark (Ed.), *Gabriel Tarde: On opinion and social influence,* (pp. 297–318). Chicago: University of Chicago Press.

The Legitimacy Gap: A Problem of Mass Media Research in Europe and the United States

by Paolo Mancini, Università di Perugia

Even if there are substantial differences between mass media research in Europe and the United States (Blumler, 1978), there are also certain features and problems that are commonly shared. One of these is the low level of legitimacy held by mass media studies in the academic world. This is not a new problem, but it is significant that 10 years after the publication of the *Journal of Communication* issue "Ferment in the Field" not much has changed in this regard and the editors of the present volume of *Journal of Comunication* have proposed this problem as one of the possible themes for new essays.

In the summer 1983 issue of the *Journal of Communication,* Wilbur Schramm wondered rhetorically whether "we have produced only ingredients of a communication theory" (Schramm, 1983, p. 14). The accent was evidently on the adverb *only* and on the substantive *ingredients*: Until that moment scholars in this field had been unable to produce an exhaustive, complete theory. All they had succeeded in accomplishing was to discuss and establish some of the elements for making up a broader scientific enterprise, still in large part to be constructed. In effect, Schramm, one of the "founding fathers" of the discipline, wanted to raise a very precise point. In part it entailed attempting an interpretation of the state of the discipline that would not be entangled in discussions of whether mass media studies enjoyed legitimacy or why these studies had failed to rise to the rank of an independent discipline on the level of sociology, psychology, etc. In Schramm's view that was not the main question. Rather, he believed we should ask whether mass communication researchers would continue to produce knowledge and to interact with researchers in other fields in the future to create new departments or centers of research.

Schramm's question implicitly signaled how much his fellow researchers felt a lack of legitimacy. Other articles in the same issue of the

Paolo Mancini is an associate professor at the Istituto di Studi Sociali, Università di Perugia, Italy.

Journal of Communication offered further material on the field's lack of standing (Rogers & Chaffee, 1983).

However, it is necessary to add that the lack of academic legitimacy is a more general problem which affects all the social sciences, albeit with different levels of intensity, based on the development, the history, the characterization, and the system of relations that each discipline shares with other scientific domains and with the larger social world. Still, it is true that media studies suffer this gap more than other social sciences because of the "publicity" of the debate over the mass media system itself.

The fact is that in the '70s and '80s the field of mass communication research was characterized by an alarming continuity: During this period the disciplinary field developed enormously, the number of specialized publications grew, and in Europe new faculties were created; but the lack of academic legitimacy largely remained. In this respect there is little difference in the situation on the two shores of the Atlantic. If anything, the difference is in the timing, in that the development that took place in the United States in the '70s occurred in Europe in the decade 1980–1990. During that same period in the United States, the discipline tended to consolidate and then compress. The problem has been noted in different moments and with different "dramatic force." In the United States the perception of communication's "minor status" has tended to remain within the field, limited to discussions among researchers. In Europe, by contrast, it has become a public question involving and regulating the relationships between researchers, mass media operators in general, and civil society as a whole. The dramatic nature of the European problem also has to do with the speed with which in the '80s the mass media system developed in the old continent and underwent a quick succession of changes. In this regard, Curran and Gurevitch speak of passing "from the age of channel austerity" to "an era of channel abundance" (Curran & Gurevitch, 1991, p. 8) while Blumler coins the captivating metaphor of "commercial deluge" to indicate the fast and often sweeping disappearance of public television and the establishment of commercial networks at the expense of the until then dominant public-service radio and television (Blumler, 1992). The changes in research have been just as fast and traumatizing as those in the mass media system: The number of departments, schools and faculties, and students interested in these subjects has grown rapidly. Thus the perception of the "minor status" of the discipline has become more evident and indeed "dramatic."

There is however something missing in the writings that we have cited to this point and in the proposal of the editors of this issue of the *Journal of Comunication*: If, as we have seen, "the lack of academic legitimacy" is rather common, what is missing is a clear indication of what is meant by this statement. Authors fail to specify what they mean when referring to it. They present it as a given, while perhaps it is not. It thus becomes necessary to fill in the gaps in the meaning of *the lack of academic legitimacy*. I can only do so taking as reference the European context and in particular the Italian. As a matter of fact, lack of academic legitimacy, while a rather generalized phenomenon, assumes different traits depending on the contexts of reference.

To start with, mass communication is weak compared to the strength of other disciplines that enjoy greater influence within the university, the research centers, and—in Europe—the ministries in charge of university development. This phenomenon is rather widespread and comes to light whenever a new faculty, a new course, or a new department of communication is being created, or when economic resources are being assigned. But there is more to it: Within the university world, mass media are looked down upon. It is considered a trivial subject, with no theoretical stature, no strong methodology, poor in tradition and history. Even within the field of mass media this low opinion of research and training prevails. The schools of journalism are an example: They are often judged negatively not only within the university but also within the world of journalism. Everette Dennis summed it up this way: "Journalism schools are 'the Rodney Dangerfields of higher education—they don't get no respect. . . . No one challenges the position of law and medicine in society, or the corresponding status of law and medical schools in universities. But a yawning chasm separates schools from the industries they serve" (Dennis, 1988, p. 3). The problems are many. As already mentioned, even in this case it is not easy to find a clear, explicit accusation against studies on mass communication. If anything, criticism is veiled, presented more as gossip than systematic reflection, and never in public. And yet this criticism hits the mark, and in a certain sense convinces media researchers of the theoretical gap. What is more serious and arouses personal concern among media researchers is that these criticisms are never thoroughly elaborated and, unlike their academic colleagues, media researchers find advancement of their university careers hindered and, when competitive examinations are involved, penalized.

Another problem is the undeniable "youth" of the field. The study of mass communication came into being in regard to a specific subject; around the second half of the 20th century the advent and consolidation of the system of mass media offered researchers of different scientific background a new subject to study. It is mainly related to the disciplines of the humanistic field, and among those the study of mass communication was the last to arrive. It was only a couple of decades ago that communication scholarship managed to create a space for itself, independent of the disciplines it derived from, earning its own university role, its own chairs, and its own scientific organization.

The Risk of Specialism

Why does communication lack disciplinary status? A question of time? Undoubtedly, but many other factors influence this negative perception.

Let us try to list them, once again considering essentially the European and in particular the Italian scene. In this framework the first factor that has undoubtedly delayed creation of a strong, complete scientific statute recognized abroad, is that of the specialism of the discipline. The study of mass communication, which was created in respect of a specific subject, has, even with the passing of time, only partially managed to find theoretical and empirical liaisons with other subjects of equal focus or with other disciplines in general. Attention has centered increasingly on the specific subject and has led to the creation of sophisticated, but because of this, extremely narrow instruments of methodology and interpretation, which are only partially exportable to other thematic fields and which only partially have permitted finding links, that are not just conjectural, with a more general theory of society. Mass communication scholars have undoubtedly improved their instruments and specific reflections, but by doing so have progressively confined themselves within their own field.

The study of media effects and, above all, content analysis may be presented as examples of the narrow specialism of the discipline, which has created a sort of *media centrism* in which attention and the results of the analysis have quite often had an essentially self-referential range, concentrating on one hand on the critical discussion of the instruments used and on the other on the subject of mass media seen in its different aspects. What has proved lacking or incomplete is the connection with other problems of society that are influenced by the mass media and with the sciences or approaches that analyze them. In short, only rarely have the instruments of mass communication and their different problems been analyzed as one of the many and different components of social change and therefore within the system of relationships between its components. "Quantitative content analyses are accused of having fragmented meanings that should never have been sundered" (Blumler & Gurevitch, 1987, p. 18). Indeed, in their introduction to the section of the 6th volume of *Mass Communication Review Yearbook* dedicated to the theories of agenda setting, Blumler and Gurevitch enumerate the components of a possible theory of *critical pluralism* that overcomes the specialism and monocasualism prevalent in U.S. research. The two authors suggest the need for multiple approaches capable of dealing with the problem "of characteristics that may significantly shape and constrain media performance, including its implications for other institutions and individuals" (Blumler & Gurevitch, 1987, p. 20). This affirmation, on the other hand, was but a continuation of Todd Gitlin's criticism of research on behaviorist effects in his widely discussed ariticle in *Theory and Society* (Gitlin, 1978).

The birth of further internal specialisms in the same field—the conducting of research on political communication, the study of electoral campaigns, psychological studies on communication, etc.—have led to accenting the separateness of the discipline. In some cases, as in the study of political communication and in the relationship with political science researchers, ways have been found for connecting with and expanding the thematic field. But in many other cases the birth of these further specialisms has exhausted not only theoretical reflection but also the capacity for interpretation of the more general social world. Elihu Katz noted in this regard that the process of institutionalizing the discipline has led to a strong "disconnection" between the new specialisms and sociology, from which it largely derives (Katz, 1987).

While in the United States communication's devotion to its subject matter is more compatible and organic with the organization of the scientific world, it is much less so in Europe. In the old continent the other humanistic disciplines to which the study of mass communication is related have a strong theoretical/abstract and generalist makeup, which sharpen their detachment from the methodologies and approaches utilized by students of the mass media.

The dark picture we have just drawn is applicable to only one part of the discipline, that is, to the part that can be definitely defined as "minor" and not to the great teachers, prestigious names, or important studies. Likewise, it can be affirmed that what Blumler defines as "professional experts orientation" (Blumler, 1978, p. 228) and which indicates both a sectorialization of the field and an exaltation of the more immediately operative and applicative aspects, represents above all—to return to the differentiations found between the two shores of the Atlantic—a peculiar feature of North American research. This is certainly true, but at the same time it can no longer be denied that in Europe too in recent years there has been a strong impetus in the direction of empiricism and applied research, which has strengthened and in some cases exasperated communication scholarship specialism.

There is no doubt that the trend has been toward research that succeeds in producing applicable results directly and immediately—in other words, Adorno's famous criticism of Lazarsfeld's Bureau of Applied Social Research—and that this limits the possibility of relating studies on mass communication to other disciplines and of overcoming its specialism. As we said above, in the last few years, but less so recently, in Europe too the demand for research from organizations, companies, and institutions interested in possessing knowledge that can be used to select correct strategic choices has increased enormously. This has meant a growth in applied research, marketing research, and research conducted by professional companies, but it has also affected academic research, it too having received financing for applied empirical research or for a more precise sectoring of pure research.

The case of demand by the Radiotelevisione Italiana (RAI) is emblematic of this atmosphere: Italian public television invests a lot in research that is prevalently applied research often entrusted to university researchers for satisfying certain obligations provided by the RAI reform law of 1975 and for operationalizing the notion of public service and involvement in promoting research, culture, and debate on themes of mass communication. The resulting applied research is confined to time limits imposed by the principal funder and only partially allows thorough examination. At the same time, on the other side, operators in the networks do not always accept the results and only rarely apply the suggestions.

Also the creation of professional schools—for example, schools of journalism—and therefore the inclusion in the university world of energies coming from this professional field has increased the trend toward "professional experts orientation." This has occurred in the United States and in many countries of Europe (i.e., France, Sweden, Great Britain). The schools of journalism have been inserted in the university but, as we have seen, very often in positions of open conflict with the rest of the academic world, partly because many of their professors come from the outside and partly because of the close relationships the schools have with the professional world. But what is most alarming, as stated before, is that this relationship is not always positively approved by the professional world. Skirmishes, petty resentments, and jealousies have arisen that, taken together, do nothing to facilitate integration of this type of training into the university system.

But there is a difference between the two shores of the Atlantic. In Europe, studies on mass communication have a weaker academic standing. While in the United States the discipline has become a viable part of the university, autonomous with its own departments, scientific organization, and specific PhD programs, the same has not occurred in Europe. In countries of the old continent, with the possible exception of Spain, departments of mass communication are usually staffed by members of other faculties, departments of sociology or linguistics or political science. This is represented to the extreme in Italy: Only in 1992 were courses for a degree in communication sciences established and these under strong influence of the departments of linguistics or letters. Except for the Department of Communication Sciences at the University of Bologna, no departments of communication exist nor are there scientific organizations for those specializing in mass communication who are mainly part of the academic field of sociology.

The Normative Dimension of European Research

The widespread *normative dimension* of the discipline is another element of difficulty in relationships with the rest of the university system. A strong trend in mass media research has brought it to light (McQuail, 1983) and is much more evident in Europe than in the United States for various reasons of a historical nature and because of established cultural and scientific traditions. The political culture of the "welfare state," which incorporates the idea of public broadcasting service and was supreme in Europe until the beginning of the '80s, works in favor of a close connection between the mass media and the needs and demands of society in its various organized expressions, including entrusting to mass media tasks of educating and promoting the cultural growth of their users. Consequently, scientific research is charged with tasks of control and orientation according to prefixed objectives. The norms of research on mass media are therefore part of this framework. Often the public radio-television service explicitly favors involvement by researchers and more generally by intellectuals in the network's strategic and cultural choices, including defining principles for directives and mass media regulations. This is the case of the different English committees and their consulting organs, which have contributed so much to changing the English television system. We are finding more and more frequently that the radio-television corporations' activity of supplying knowledge and more generally that of the organs assigned to control the overall system of mass media are based on contributions from university research centers, once again called upon to supply knowledge but also opinions on the work of the mass media production centers. On one hand this increases the specialization of the discipline and its professional orientation, but on the other it risks involving the researcher in judging or giving advice on operative choices and increasing the separateness of the researcher from the rest of the scientific community, which in a certain sense remains on the contrary closed in its splendid isolation also because of the absence of a strong demand for knowledge from the outside world. As already stated, this does not mean that the relationship with the professional world and media operators is idyllic, indeed criticism often comes from this area attacking studies on mass communication, and the completed studies remain a "dead letter."

The tradition of the researcher's social and political involvement is not associated solely with the idea of public service and its obligations but represents a constant that characterizes, with some differences between the different disciplines and different countries, the entire European scientific world. The researcher, the European intellectual, has always lived civil and political commitments personally. The researcher's intellectual activity and studies have always been accompanied by deep involvement in social questions, an involvement starting with the researcher's awareness of the political and ideological weight of didactic activity and proceeding to more direct participation in the life of the political parties and social movements. And within Europe, in this regard, there are significant differences: We go from the extreme of the southern countries (e.g., Spain, Italy), where the close relationship with the political system defines the researcher as a "public figure" whose actions and reflections strongly influence the world outside the university, to Great Britain where the researcher's social and political role is close to the separateness so widespread in the United States.

It is not just the more or less direct commitment in political life but the public dimension of theoretical reflection and empirical research that is so well established in the European academic world. Consider, for example, how important the signatures of European intellectuals in the world of printed paper and television are. When they write for newspapers or appear on television, they do not write or appear as simply experts in their field but as opinion leaders "authorized" to express their own views on the most diverse subjects. Moreover, the researcher who concentrates on subjects of mass media has more opportunities and interest in and for the "public" dimension of his field; he has more and closer relationships with operators in the system of information and easier access to the media, which divides him further from his colleagues.

For some European countries, where the above tradition had already been well established, the '70s and '80s marked a period of public involvement by mass media scholars that will be difficult to repeat. Once again, my scene of observation is Italian (Mancini, 1986; Mancini & Wolf, 1990), but a similar situation exists in France, Germany, and other countries. The debate on the concept of hegemony, on journalistic objectivity, and on the processes of concentration of the mass communication system have afforded occasions in which the already dominant critical theory has found similar subjects of discussion. Very often the abstractness of the debate and its normative dimension have prevailed not only over empirical research, relegated to a position of secondary importance as an expression of Americanization and neopositivism, but also over the methodicalness of theoretical reflection, very often subordinated to the imperatives of political commitment and therefore victim of a lack of precision—none of which has served to improve the external reputation or internal cohesion of the discipline.

The development of the mass media in the '70s and especially the '80s has greatly stimulated curiosity and demands for knowledge concerning it. The mass media have become a phenomenon of fashion. They have become a constant subject for public discussion and debate, they have filled the pages of newspapers, created news styles, and popularized figures of special editors. A market for knowledge and opinion has been produced by a self-referential process with a continuing demand for opinion leaders and intellectuals capable of expressing themselves successfully and articulately in this field.

Media researchers have been part of this new and growing attention: Their work has often assumed popular tones in which discovery and scientific innovation and the depth of theoretical exmination have become secondary. They have written and continue to write easy and appetizing texts; public success releases them from university recognition and is much more rewarding.

While media researchers have adapted to the leveling demands of the mass media circuits, the wealth of the field and the attention it arouses have also convinced many other intellectuals and researchers from other disciplines to try it. This has led to analyses of certain interest that cannot, however, become part of a school or tradition of theory and research in a field largely unknown to their authors. In many cases the debate on the subject has become more abstract and independent of the results of applied research, while the subject of mass media has continued to be ground for intervention by researchers coming from the most varied fields. The possibility of public success, and also sales, has certainly attracted them, but the entire disciplinary field has resented it, as the researcher-specialist popularizer has emptied studies of their content.

Because of the concern for values that has always been a feature of our discipline, the frequency of contacts with operators in the field, the wealth of applied research, and the interest its subjects arouse in general public opinion, the discipline suffers strange contradictions. On one hand, especially during the last decade, there has been a rich field of demand, which favors specialisms, sometimes to exasperation, that are inserted in a strong normativism which directs and places them into context. On the other there is critical/theoretical reflection, which privileges the abstractness of debate and ideological macroconstructions, often overlapping and confused with direct political involvement. Moreover, the broad interest the mass media arouse in public opinion determines a process of trivialization of the discipline, constantly exposed to incursions by intellectuals and opinion leaders of the most different backgrounds who are not good for the discipline itself in its relationships with the rest of the academic world. Specialism, abstract normativism, and frequent yielding to intellectual fashions and the demands of the mass media system are among the reasons for the lack of academic legitimacy in media research. All these reasons can explain, without risk of contradiction, the situation of mass media studies in Europe and their relationship with the academic world. As for the United States, it is mainly the specialism, together with the other two elements, that has caused the delay in effectively constituting a sufficiently recognized disciplinary field within the university world.

It is difficult to draw conclusions because the field of studies on mass communication is still being developed in both Europe and the United States. While the picture I have painted may appear even more pessimistic than what I am about to say, it seems legitimate to state that the relationship with the rest of the academic world is not idyllic and that there still exists a gap of legitimacy affecting this discipline in different ways and with different intensity depending on the contexts.

Probably every new field experiences similar problems and suffers not very different gaps. What seems certain to me is that we are in the presence of a development that has been perhaps too tied to the very system mass communication research intends to study. From this we have on one hand the narrow focus and on the other the normativism that weaken the discipline within a framework of different and often contradictory causes. Finally, the mass communication system itself contributes to trivializing the disciplinary field and creating around it a certain fascination but also a subtly harmful interest.

References

Blumler, J. (1978). Purposes of mass communications research: A transatlantic persective. *Journalism Quarterly, 2,* 219–230.

Blumler, J. (Ed.). (1992). *Television and the public interest.* Newbury Park, CA: Sage.

Blumler, J., & Gurevitch, M. (1987). The personal and the public. Observations on agendas in mass communication research. In M. Gurevitch & M. Levy (Eds.), *Mass communication review yearbook* 6 (pp. 16–21). Newbury Park, CA: Sage.

Curran, J., & Gurevitch, M. (Eds.). (1991). *Mass media and society.* London: Edward Arnold.

Dennis, E. (1988). Whatever happened to Marse Roberts's dream? *Gannett Center Journal, 2*(2), 1–23.

Gitlin, T. (1978). Media sociology: The dominant paradigm. *Theory and Society, 6,* 205–253.

Katz, E. (1987). Communication research since Lazarsfeld. *Public Opinion Quarterly, 51*(4), S25–S45.

Mancini, P. (1986). Between normative research and theory of forms and content: Italian studies on mass communication. *European Journal of Communication, 1*(1), 65–97.

Mancini, P., & Wolf, M. (1990). Mass media research in Italy: Culture and politics. *European Journal of Communication, 5*(2), 165–187.

McQuail, D. (1983). *Mass communication theory.* London: Sage.

Rogers, E. M., & Chaffee, S. H. (1983). Communication as an academic discipline: A dialogue. *Journal of Communication, 33*(3), 18–30.

Schramm, W. (1983). The unique perspective of communication: A retrospective view. *Journal of Communication, 33*(3), 6–17.

The Advent of Multiple-Process Theories of Communication

by Austin S. Babrow, Purdue University

This special issue of the *Journal of Communication* devoted to the "Future of the Field" reflects long-standing questions. Some have to do with disciplinary status. Is ours a scholarly discipline? If it is, has it sufficient substance to command the attention of others? Other questions have to do with debates about the paradigmatic status of the field. Should we strive for greater unity in disciplinary outlook? What are the rewards and costs of pluralism? Given answers to the preceding questions, how are we to represent the field to others, and what are they to make of our works? Ultimately, where should we be headed?

This essay describes a perspective that speaks to all of these questions. It assumes that a dialectic between the voices of theoretical pluralism and calls for broad covering perspectives in communication scholarship is desirable. Our disciplinary core—that which we want to claim as knowledge and to offer other disciplines—depends on the synthesis of efforts to unify and diversify the field. The present perspective represents one possible synthesis.

This paper's basic claim is that the field is moving beyond theorizing that treats communication as *a process* to what might be called *multiple-process theory*. Emerging views are sensitizing researchers to issues that have often been glossed over: (a) Communication involves multiple, substantively distinct processes; (b) these processes may be redundant, complementary, or contradictory; and (c) processes may mediate or moderate other processes.

This essay examines views of message-processing modes and effects, enlarges the scope of inquiry by considering a multiplicity of levels of analysis, and discusses a philosophical orientation compatible with multiple-process theory. It also considers future prospects and offers guidelines for theory development.

Austin S. Babrow is an associate professor in the Department of Communication at Purdue University, West Lafayette, IN. The author wishes to thank James P. Dillard and an anonymous reviewer for helpful comments on an earlier draft.

Communication, Cognition, Affect, and Motivation

Multiple-process theory is emerging in work on the interplay of communication, cognition, affect, and motivation that has begun to redress the excesses of cognitive science over the last two decades (see Dillard, 1992; Donohew, Sypher, & Higgins, 1988). Much of this work focuses on the multiform relations between cognition and affect, giving rise to what have been termed "warm" (Sorrentino & Higgins, 1986), "synergistic" (Eagly & Chaiken, 1992), or "cognitive-emotional fugue" (Dillard, 1993) theories. Dual-process models of persuasion are an important form of theorizing in this area.

Dual-process models are based on the important but long overlooked insight that responses to (persuasive) messages do not *only* take the form of controlled, capacity-intensive cognitive processes, or *only* the form of simple automatic processes involving little working memory. Message processing may be dominated by either form, or it may blend the two, depending on our ability and motivation to think about the substance of a message. Affective processes appear to (a) influence levels of motivation and ability to process in a thoughtful manner, (b) guide the retrieval of information from memory, and (c) provide cues to simple responses (see Babrow, 1991; Dillard, 1993; Eagly & Chaiken, 1992). Given the close interconnection of affective, motivational, thoughtful, and automatic psychological processes, the term *dual-process* models of persuasion is somewhat misleading, for it references only the distinction between highly mindful and less mindful dynamics. Alternatively, and taken as a whole, we might refer to these views as exemplars of multiple-process theory.

Despite the achievements of dual-process (and warm or synergistic) models, they neglect the significance of communication in the interplay of affect, cognition, and motivation systems. O'Keefe (1990) has noted that these models offer little direct insight into the substance of messages. Message meanings comprise dynamic mixes of words, manners of expression, implicature, and emotion (see Scheff, 1990). These complexities are barely realized in dual-process views, for they generally gloss over the dynamics of message interpretation. Consider one prominent example.

The *elaboration likelihood model* (ELM) of Petty and Cacioppo (1986) emphasizes the cognitive *elaboration* rather than the interpretation of message content. Elaboration denotes the working out of details from some *given* kernel of meaning. Consequently, ELM research has to date treated its two primary causes—motivation and ability to process a message—as if they were independent of the process of making sense of a message. One exception is Petty and Cacioppo's references to studies of message complexity or comprehensibility, where they recognize that thoroughly disordered messages undermine cognitive elaboration (1986, p. 76).

More sustained attention to the nature of messages and their interpretation is certain to illuminate fundamental interactions between message meanings and individuals' ability and motivation to process, as well as the directions such processing might take. In short, theories of (suasory) communication must unpack the ways that messages themselves condition ability and motivation to process, as well as the substance of that processing. Two other multiple-process theories offer relevant contributions.

Scheff's (1990) analysis of the *deference-emotion system* weaves together emotion, interpretation, discourse, and social structure. Briefly, he argues that the most fundamental social motive is the maintenance of secure social bonds. Pride and shame are crucial; they signal the state of the bond. In every social interaction, the deference one experiences determines one's sense of the state of the bond and associated feelings of pride or shame. Scheff tries to explain the ongoing interplay of discourse, motivation, emotion, and cognition within interpersonal and broader social relations.

A second relevant multiple-process perspective is based on the notion of *problematic integration* (Babrow, 1992). This model recognizes processes by which persons form and integrate probabilistic and valuative orientations (i.e., some state of the world is likely/unlikely and good/bad). Integration of these two basic orientations is thought to involve the interaction of probability and value judgments, as well as their incorporation into broader systems of knowledge, feelings, and motivations or intentions. But the perspective stresses those consequential circumstances where the integration of probability and evaluation is problematic (e.g., likely or certain sorrow, unlikely or impossible happiness, ambiguity, and ambivalence). Individual psychological processes, along with a variety of interpersonal and broader communicative processes (e.g., public responses to threats; practices instantiating ideologies of maturity, mental health, or particular political, religious, and economic orders), are thought to be constitutive of and therefore crucial to experiences with problematic integrations.

In short, studies of communication, affect, cognition, and motivation are giving rise to multiple-process theory. Rather than reviewing such theorizing in other substantive areas, this essay next discusses more general analytical and philosophical orientations that hold great promise.

Cross-Level Theory and Research

Efforts to cross analytical levels also promise to be a fertile source of multiple-process theory. For example, recent work has merged interpersonal, group, and mass communication processes (Babrow, 1990; Katz & Liebes, 1985; Lull, 1980); group membership theories have been applied to public opinion processes (Price, 1989); and cognitive mechanisms have been investigated in innumerable contexts.

Put most strongly, we might claim that cross-level analyses are necessary to understand human communication. Paisley (1984) implies a loosely similar position in his assertion that, by its very nature, the communication field gives attention to one category of behavior across many levels of analysis. While cross-level analysis might be essential to understanding human communication, and while Paisley's characterization may be true of the field as a whole, many specific communication theories and research practices skirt this important form of inquiry (Price, Ritchie, & Eulau, 1991b; Ritchie & Price, 1991). Moreover, research that crosses levels is not necessarily profitable (Nass & Reeves, 1991). Recent writings discuss several important considerations (see Alexander, Giesen, Münch, & Smelser, 1987; Price, Ritchie, & Eulau, 1991a).

Cross-level analysis offers its greatest potential contributions when it is conducted with an eye toward discontinuities as well as continuities across levels. Simple reduction of processes at one level to those at another risks important misunderstandings and missed insights. For example, only limited insights are available when various levels of communicative phenomena are reduced to nondistinctive cognitive processes. People undoubtedly "think" in dyads, organizations, and larger cultural, political, and socioeconomic orders; to understand the particularities of these phenomena requires attention to variations in the shape and mix of processes across levels. So too, simple subsumption of processes at a lower level into those at a higher level of analysis risks misunderstandings and missed insights.

An elegant and demanding approach to cross-level analysis illustrates many of the complexities of moving across levels: the search for theories specifying the relationships between variables at different levels of analysis. For example, Pan and McLeod (1991) propose the search for "cross-level auxiliary theories" of mass communication. These auxiliary theories might take the form of mathematical functions that translate linear relationships at one level (e.g., the "micro" or "individual") into their counterparts at another level (e.g., the "macro" or "aggregate").

Several trials confront those wishing to develop auxiliary theories of these sorts. The trials illustrate and signify the importance of multiple-process theory. For one, Pan and McLeod (1991, p. 166) point out that aggregation (i.e., statistical translation of "micro-level linear relationships into a macro-level linear relationship") involves specifying four associations: micro–micro, macro–macro (aggregates), and two cross-level relationships. For example, a full understanding of the effects of exposure to AIDS awareness messages on levels of HIV and AIDS in a population depends on understandings of processes linking (a) individuals' message exposure and risk behavior, (b) aggregate exposure with aggregate at-risk behavior rates (e.g., variations across communities), (c) individual with aggregate levels of exposure, and (d) individual with aggregate at-risk behavior rates.

Multiple-process thinking recognizes that variables, relationships, or processes at one level can interact with or moderate the form and functioning of variables, relationships, or processes at another level (see Pan & McLeod, 1991). Finally, as suggested above, multiple-process theory recognizes that within- and cross-level processes may be redundant, complementary, or contradictory in their nature and functioning. For example, psychological coping processes may be redundant with, complement, or contradict interpersonal responses to stressors (e.g., if interactants move through the grief process at different rates).

Multiple-process theorizing also arises when we surmount simplifying distinctions between levels of phenomena. As scholarship evolves, the distinction between macro- and micro-analysis erodes, as does the distinction among physiological, psychological, social, and cultural analyses (cf. Alexander, 1987; Chaffee & Berger, 1987). As we refine our appreciation of the continuum, we gain ever more insight into the range of processes participating in communication. Two examples illustrate this.

Babrow's (1992) attempt to frame levels of experience with problematic integration illustrates an underelaborated explicit view of the continuum. His framework implies numerous important and distinct processes that are glossed over by simplifying distinctions among individual, interpersonal, social processes. For instance, interactants' views of conversational norms governing talk about devout wishes or dire dreads (e.g., that such talk is risky or distasteful and hence best avoided, or that such talk is an important form of social support and connection) might be associated with particular utterances or patterns of messages within an exchange. Participants' views of conversational norms would also be associated with processes constituting their relationship to one another (e.g., particular private norms within the relationship, as opposed to general role-related norms), which might in turn be associated with processes that connect particular relationships to forces that structure relational types, which in turn implicate even broader arrangements of types of relationships characteristic of the social and historical context. While it might be possible to force these different levels of analysis into individual, interpersonal, and social categories, doing so obscures various phenomena. In short, there is much to be gained by exploring and refining the micro- to macro-analytic continuum.

Scheff's (1990) analysis of discourse, emotion, and social structure provides a more refined view of the continuum. Briefly, his work suggests that students of communication ought to think of their subject matter as interlocking structures and processes, any one of which implies a larger, more encompassing whole. Scheff's (1990) 10-step "part/whole ladder" illustrates this view of discourse: (a) At the most basic or concrete level are systems of words and gestures, (b) which are encompassed by sentences, (c) within exchanges, (d) in conversations, (e) within the relationship between the parties to the interaction (i.e., all of their conversations), (f) which exists in life histories of the parties, (g) and within all relationships of the same type, (h) which reflect the structures of the host society (i.e., all relationships), (i) which reflect the history of the host civilization, (j) manifesting the history and destiny of the human species. Each of these levels reflect distinct structures and processes, yet each interlocks with structures and processes at adjoining levels.

In sum, efforts to cross levels of analysis will be a fertile source of multiple-process theories in communication research. As these efforts bear fruit, the field may become more unified, and it may represent a more attractive source of information and inspiration to other fields (of course, whether this occurs will depend on numerous other processes: technological, political, economic, and the like).

Dialectical Perspectives

Dialectical perspectives on communication also embody multiple-process theory. Of course, dialectical analyses of one sort or another have been discussed for centuries (see Adler, 1952). As numerous dialectical thinkers have pointed out, however, several themes are consistent in these writings; these themes exemplify some of the most desirable potentialities of multiple-process theory.

The most elemental theme in dialectical thinking is that of opposition; "dialectic either begins or ends with some sort of intellectual conflict, or develops and resolves around such oppositions" (Adler, 1952, p. 350). Dialectical opposites are "mutually conditioning" (the occurrence, existence, or meaning of one pole is conditioned by its opposite) and at the same time "mutually excluding" (Marquit, 1981). For instance, sound presupposes but also excludes silence, and so too for amity and enmity, motion and stillness.

It is important to note that dialecticians have identified a number of different gradations of dialectical opposites (see Rawlins, 1989). Moreover, the idea of dialectical opposition does not imply that there is only one opposition to a given term or only one antithesis to a given thesis (Rychlak, 1976). One person may understand the antithesis of a given thesis in a way that diverges from another person's understanding. "This is because what is affirmed in any given thesis is always some meaning expressed in a complex of meaning extensions" (Rychlak, 1976, p. 15). In short, a fundamental project in dialectical theorizing is the identification of opposites in various forms and gradations, at times contradictory, and at others complementary. Multiple meanings, multiple oppositions, even around the same thesis, may be the rule rather than the exception.

Another key feature in dialectical thinking is totality. "Dialectical totality hinges on relatedness and contextuality. Discrete 'things' are inconceivable from a dialectical perspective; what comprises reality are relations and relations among relations" (Rawlins, 1989, p. 158). Hence, though dialectical analysis uncovers opposites, it transcends them by holding that dialectic relations are intrinsic. One pole (thesis) predicates an opposite (antithesis). Neither pole exists without the other; they coexist in their relation as a whole. The concept of totality also entails contextuality. Things have meaning only in relation to others, and those relations are conditioned by their participation in or relation to still other relationships forming even larger wholes. Some dialectical perspectives posit that "all things are connected to all other things in an infinity of interconnections" (Marquit, 1981, p. 309). Hence, another fundamental aim of dialectical analysis is to formulate what are inescapably partial descriptions of nested layers of interrelations.

Another feature of many dialectical perspectives is the notion of change (or process or motion). Change arises out of tensions between opposites; because reality is constituted by opposition, reality is inherently in motion and changing. Nonetheless, stability can occur when opposites reach a temporary balance. "That which in a conventional scientific light would appear as an ontological invariant, in a dialectical framework is a temporarily equilibrated form in transition" (Bopp & Weeks, 1984, p. 52). Therefore, another major aim in dialectical analysis is the study of change.

When researchers apply the notions of opposition, totality, and change to communication phenomena, all manner of distinct but interrelated processes are uncovered. A prominent example is Rawlins's (1992) work on interpersonal communication and friendship. Rawlins describes the operation of multiple antagonistic processes, the coexistence of which is constitutive of friendship: the dialectics of the private and the public, the ideal and the real, affection and instrumentality, acceptance and evaluation, dependence and independence, and expressiveness and protectiveness. When taken together as parts of a whole, some of the processes are at times redundant or complementary (e.g., expressiveness may be redundant with or may complement judgment or acceptance; so too for dependence and instrumentality). In sum, dialectical thinking leads to multiple-process theorizing. Processes imply opposing forces, and these relationships are associated with and encompassed by still broader relationships.

Complexities

Multiple-process perspectives recognize and attempt to explain the complexities of human communication and social experience. As these sorts of perspectives develop, they will foster and enliven discourse within the field and across disciplines. If communication phenomena are complexes of distinct though interrelated processes, and if specific theories and research practices in the field have until now tended to skirt these complexities, then the advent of multiple-process theory is potentially quite significant. The field will come of age—in the eyes of its members, allied disciplines, and academe as a whole—when it attends to the interplay of these many phenomena. Therefore, we ought to nurture the growth of multiple-process theory with great care.

References

Adler, M. J. (Ed.). (1952). *The great ideas: A synopticon of great books of the western world.* Chicago: Encyclopaedia Britannica.

Alexander, J. C. (1987). Action and its environments. In J. C. Alexander, B. Giesen, R. Münch, & N. J. Smelser (Eds.), *The micro–macro link* (pp. 289–318). Berkeley: University of California Press.

Alexander, J. C., Giesen, B., Münch, R., & Smelser, N. J. (1987). *The micro–macro link.* Berkeley: University of California Press.

Babrow, A. S. (1990). Audience motivation, viewing context, media content, and form: The interactional emergence of soap opera entertainment. *Communication Studies, 41,* 343–361.

Babrow, A. S. (1991). Tensions between health beliefs and desires: Implications for a health communication campaign to promote a smoking cessation program. *Health Communication, 3,* 93–112.

Babrow, A. S. (1992). Communication and problematic integration: Understanding diverging probability and value, ambiguity, ambivalence, and impossibility. *Communication Theory, 2,* 95–130.

Bopp, M. J., & Weeks, G. R. (1984). Dialectical metatheory in family therapy. *Family Process, 23,* 46–61.

Chaffee, S. H., & Berger, C. R. (1987). What communication scientists do. In C. R. Berger & S. H. Chaffee (Eds.), *Handbook of communication science* (pp. 99–122). Newbury Park, CA: Sage.

Dillard, J. P. (1993, May). *Rethinking the study of fear appeals.* Paper presented at the annual meeting of the International Communication Association, Washington, DC.

Donohew, L., Sypher, H. E., & Higgins, E. T. (1988). *Communication, social cognition, and affect.* Hillsdale, NJ: Lawrence Erlbaum Associates.

Eagly, A. H., & Chaiken, S. (1992). *The psychology of attitudes.* San Diego, CA: Harcourt, Brace, Jovanovich.

Katz, E., & Liebes, T. (1985). Mutual aid in the decoding of "Dallas." Preliminary notes from a cross-cultural study. In P. Drummond & R. Paterson (Eds.), *Television in transition: Papers from the first international television studies conferences* (pp. 187–198). London: British Film Institute.

Lull, J. (1980). The social uses of television. *Human Communication Research, 6,* 197–209.

Marquit, E. (1981). Contradictions in dialectics and formal logic. *Science and Society, 45,* 306–323.

Nass, C. I., & Reeves, B. (1991). Combining, distinguishing, and generating theories in communication: A domains analysis framework. *Communication Research, 18,* 240–261.

O'Keefe, D. J. (1990). *Persuasion: Theory and research.* Beverly Hills, CA: Sage.

Paisley, W. (1984). Communication in the communication sciences. In B. Dervin & M. J. Voigt (Eds.), *Progress in communication sciences* (Vol. 5, pp. 1–43). Norwood, NJ: Ablex.

Pan, Z., & McLeod, J. M. (1991). Multilevel analysis in mass communication research. *Communication Research, 18,* 140–173.

Petty, R. E., & Cacioppo, J. T. (1986). *Communication and persuasion: Central and peripheral routes to attitude change.* New York: Springer-Verlag.

Price, V. (1989). Social identification and public opinion: Effects of communicating group conflict. *Public Opinion Quarterly, 53,* 197–224.

Price, V., Ritchie, L. D., & Eulau, H. (Eds.). (1991a). Micro–macro issues in communication research [Special issue]. *Communication Research, 18(2).*

Price, V., Ritchie, L. D., & Eulau, H. (1991b). Cross-level challenges for communication research. *Communication Research, 18,* 262–271.

Rawlins, W. K. (1989). A dialectical analysis of the tensions, functions and strategic challenges of communication in young adult friendships. In J. A. Anderson (Ed.), *Communication yearbook 12* (pp. 157–189). Newbury Park, CA: Sage.

Rawlins, W. K. (1992). *Friendship matters: Communication, dialectics, and the life course.* New York: Aldine De Gruyter.

Ritchie, L. D., & Price, V. (1991). Of matters micro and macro: Special issues for communication research. *Communication Research, 18,* 133–139.

Rychlak, J. F. (1976). The multiple meanings of "dialectic." In J. F. Rychlak (Ed.), *Dialectic: Humanistic rationale for behavior and development* (pp. 1–17). Basel, Switzerland: Karger.

Scheff, T. J. (1990). *Microsociology: Discourse, emotion, and social structure.* Chicago: University of Chicago Press.

Sorrentino, R. M., & Higgins, E. T. (Eds.). (1986). *Handbook of motivation and cognition: Foundations of social behavior.* New York: Guilford Press.

Communication and the New World of Relationships

by Mary Anne Fitzpatrick, University of Wisconsin

The field of communication has come a long way concerning critical issues and research tasks since *Ferment in the Field* 10 years ago. Scholars have written numerous handbooks synthesizing the theory and research of the field (e.g., Berger & Chaffee, 1987; Knapp & Miller, 1985); the field has produced new journals specializing in communication (e.g., *Communication Theory, Health Communication, Language and Social Interaction*); and scholars have produced a four-volume *International Encyclopedia of Communications* (Barnouw, Gerbner, Schramm, Worth, & Gross, 1989). Together, these developments help to sustain not only the intellectual viability of the discipline but its public face as well.

Although the discipline has "come of age," this development does not mean that a concern for a universal paradigm for communication has been replaced by a comfortable acceptance of theoretical pluralism. In this essay, I would like to comment on the particular tensions within the area of interpersonal communication theory and research surrounding the study of social and personal relationships. It is my thesis that important issues in the study of social and personal relationships have been largely ignored as scholars pursue a spurious conflict between communication science and communication hermeneutics.

Research in Interpersonal Communication

In the 1950s, 1960s, and early 1970s, the study of interpersonal communication in this discipline was essentially the study of attitude change and persuasion. In the late 1970s, the focus shifted, allowing Berger (1977), in the first issue of *Communication Yearbook,* to see very different themes emerging. Arguably, the two most important shifts involved linking the study of interpersonal communication to the study of social and personal relationships as well as examining the communication process.

Mary Anne Fitzpatrick is a professor and the director of the Center for Communication Research at the University of Wisconsin. A version of this paper was presented in May 1992 as her presidential address to the 42nd annual meeting of the International Communication Association, Miami, FL.

The linking of interpersonal communication to relationships represented not only a theoretical move but also a response to the call for social relevance in research endeavors (Miller, 1983). By investigating the communication that occurs between lovers, business partners, or friends, scholars of interpersonal communication felt more capable of prescribing practical solutions to relational difficulties. A small group of scholars committed themselves to the observation and systematic analysis of the processes of interpersonal communication in initial interaction (Dindia, 1982) and in ongoing relationships (Fitzpatrick, 1988). Many interesting coding schemes were developed (e.g., Rogers & Farace, 1975), although the number of observational studies remained small.

Throughout the 1980s, the award-winning work in interpersonal communication was committed to the themes and topics first presented in 1977. The study of communication in social and personal relationships became the central focus of the subdiscipline of interpersonal communication. Despite the strengths of that early research, however, much of it was decontextualized and ahistorical. And, most of the work concerned itself with a limited number of relationships and a small number of communication behaviors. With the exception of a handful of observational researchers, the methods used to examine personal and social relationships were largely the same. Researchers often used laboratory experiments to test ideal theories. Even the work based on the rules paradigm, which argued for the importance of human agency rather than determinism, relied on laboratory experiments and statistical inferences as the way to test their conventional explanations for communication behaviors (Cronen, Pearce, & Snavely, 1979; Jackson, Jacobs, & Rossi, 1987).

I see us entering a new world not only of relationships but of research on relationships. Scholars in interpersonal communication need to be far more sensitive to the shifting cultural milieu and contexts in which relationships are formed and maintained. We often forget that neither friendship nor love has a fixed identity. What either means, and the practices both entail, vary historically and cross-culturally.

The Context of Personal and Social Relationships

Understanding communication in social and personal relationships requires attention to a number of important cultural and historical factors. Throughout the 20th century, advances in medicine and social hygiene radically changed the nature of sexual relations between males and females. Arguably more important than the fear of childbearing was the fact that most women, because of pervasive albeit minor gynecological problems, probably experienced great pain during intercourse. Modern advances in gynecological care have afforded women the opportunity for intercourse without physical pain (Shorter, 1980). The discovery of antibiotics that could cure the life-threatening sexually transmitted diseases, as well as the subsequent discovery of a pill that could decouple sex and procreation, facilitated the development of different types of relationships among heterosexual and homosexual pairs.

The rise of herpes, AIDS, and other sexually transmitted diseases may signal a change in American intimate culture. It is perhaps too early to consider the effect of these diseases on erotic behavior and its meaning. But it may prove to be the case that the 20 years between the widespread use of the birth control pill and the widespread understanding of AIDS and its means of transmission represented an unusual epoch in the history of the relations between intimates. Interestingly, it is during this period that we see the rise in studies of communication in social and personal relationships. Thus, we may come to challenge what we think we know about communication in social and personal relationships. Indeed, the future may find our work during this period the documentation of a very unusual time in the history of intimate human relationships.

In addition to considering the social response to historical and biological changes that affect relationships, scholars need to reconsider the ideological stance of their research on relationships as well as the ideological positions of their research participants. Simply put, ideology is the formation of a group's beliefs, values, and attitudes into a cohesive body of knowledge about what is good and what is not good. Space allows the discussion of only two illustrative ideologies that influence the study of communication in social and personal relationships.

The Ideology of Intimacy

In a classic piece, Parks (1982) exhorts us to beware of the ideology of intimacy as defining the research agenda of scholars of interpersonal communication. The ideology of intimacy sees the individual as involved in an unending quest for closeness and openness. Parks demonstrates that the ideology of intimacy encourages many scholars to ignore or devalue phenomena of fundamental importance. One is the importance of information control. Hiding is often as important as revealing. Perhaps more importantly, a focus on intimacy underplays the prevalence and utility of less intimate relationships in our lives. These weak ties to others may serve a number of functions. Such relationships appear to facilitate the diffusion of innovations, offer chances for social comparison, and promote large-scale cohesion and action (Parks, 1982).

My own work with couples over the past 20 years on communication in marriage supports many of Parks's (1982) insights. A major discriminating feature of modern marriage is the degree to which participants hold various ideological conceptions of relationships (Fitzpatrick, 1988). Indeed, all couples subscribe to the importance of good communication in marriage, but what constitutes "good" communication varies dramatically among couples. My colleagues and I have shown the persistence of a stable type of couple and an individual style that rejects open communication as the sine qua non of marriage and organizes married life around separate spheres (Fitzpatrick, 1988). What good communication means varies according to the ideology of married partners.

The Trinity Plus One

Many scholars add to the definition of ideology not only that it represents a system of values but that this system is organized to establish and maintain relations of domination. A central insight of this tradition is that the current economic structure systematically divides people into the trinity—class, race, and gender—and uses these divisions to limit access and opportunities. To these three, Kramarae (1992) would add age. It is these limitations that generate a "social environment within which collectivities of people build shared social realities based on their lived experience of coping, circumventing, and creating" (Meehan, 1988, p. 184).

The shocking disregard for race, class, gender, and age—so central to the interpretative and cultural scholars—makes the work of communication scientists of interpersonal communication less relevant to the field at large. Recently, we have made some strides in considering age (Giles, Coupland, & Wiemann, 1990), and at least have attempted to work on gender (Liska, 1992) in theoretically useful ways, but very few scholars of interpersonal communication focus on class and race.

These new insights warrant a continued focus on race and class as a factor in the development, maintenance, and decline of social and personal relationships. I suspect that the communication scientist will paint a markedly different picture of communication as it relates to race, class, gender, and age and their interrelationships from that proposed by the cultural theorists. But to ignore these central defining experiences of different social realities is to perpetuate the structures of domination.

Interpersonal communication needs an infusion of contextual factors. Many scholars of interpersonal communication erroneously believe, however, that in order to consider contextual and social facets of interpersonal communication, the theorist is forced to adopt an interpretive, hermeneutic (Leeds-Hurwitz, 1992), or even personal narrative style (Bochner & Ellis, 1992). For these theorists, to explain the meaning of human action in relationships in terms of purposes and reasons rules out any possibility of causal explanations. Furthermore, by rejecting a scientific basis for the study of relationships, social, interpretive, critical, and narrative approaches fail to specify how to judge among competing claims.

All scientists acknowledge that scientific knowledge involves the constructions of one's world, as any form of knowing does. But scientists make these constructions public so that they can be evaluated in terms of their ability to predict, explain, and give a sense of understanding about communication in relationships. We can have a social dimension to the study of communication in relationships without losing the scientific grounding of human communication study. It is possible to take very seriously the proposition that many actions are the product of human agency without seeing these actions as precluded by a causal science of communication.

Toward a Communication Science of Relationships

In this discipline, as soon as the empiricist account of human communication, based on the physical sciences, is shown to be inadequate, we immediately discard any systematic assessment of the phenomena under consideration in favor of a rhetorical, hermeneutic approach. We seem to boomerang from hard determinism to libertarianism where humans are considered such a special case that they must be understood apart from natural phenomena. Human choice is said to be determined exclusively by the person who makes it (Sappington, 1990).

The extreme form of communication science restricts us to explanation, prediction, and control of observable behavior. The most extreme form of hermeneutics restricts us to the explication of meaning in human communication. Most of the work on communication in relationships adopts either one of the other of these opposing positions. At base, both of these positions are empirical: sterile on one hand and relativistic on the other (Greenwood, 1991). Writers in both camps spend their time lighting matches to straw men and watching them burn.

Rather than rule out the unobservables, as positivists do, or rely on sensations, as hermeneuticists do, the realist account of communication in relationships argues that some theoretical terms refer to hypothetical entities, that some of these entities are candidates for existence, and that some candidates for existence *do* exist (Harre, 1985). At present, realism is the only position that does not make the success of science a mystery (Putnam, 1984). Realists are, of course, divided as to what constitutes success, how it is to be explained, and the role of realism in its explanation.

We need a *science* of communication in personal relationships. For thousands of years, relationships have been examined in narrative forms by poets and novelists. The recent turn to science for an understanding of social and personal relationships and the communication within those relationships was an important one for understanding these phenomena. Such a science of communication in relationships does not have to adopt the largely discredited positivist view of Hempel and his version of ideal theory. Nor does a science including prediction and control have to adopt a hard deterministic position in which human behavior is determined completely by forces outside the person.

Such a science can be centered on soft determinism and is not incompatible with free will (Sappington, 1990). People make conscious choices between different courses of action, and these choices affect them. The choices themselves, however, can be determined by other factors (Sappington, 1990). Importantly, the agent herself can choose not to make a choice. A communication science of relationships can either provide an empirical demonstration of the inaccuracy of what appears to be the most intuitively plausible reason for an action, or it can provide a theoretical account for how human agents are enabled to determine their own actions in relationships for the sake of reasons.

A realist philosophy of communication science allows us to maintain objectivity while at the same time stressing the social dimensions of mind and action (Greenwood, 1991). Such an approach to the study of communication in social and personal relationships must be anchored in what Greenwood (1991) considers the three important premises of empirical social science. In other words, a realist account is objective, advances causal explanations, and employs observational methods.

The study of communication in social and personal relationships needs to move beyond a positivist philosophy in order to consider relationships in contexts. A realist view of social and personal relationships allows us to examine human relationships and their symbolic exchanges not only in reference to causal determinants but also in terms of the explication of human action in reference to its purpose and reasons. A realist philosophy supports a variety of methodologies for developing objective, causal explanations for relationship processes. But the realist view allows us to approach this task systematically and objectively with public and clear justifications for our claims.

Epistemological Synthesis in the New World

It is only in taking each other seriously that we have a chance to develop important new communication theories. We have an intellectual trade deficit (Berger, 1991) because we look outside the discipline for theoretical insights. As a discipline, we are not capitalizing on the mutual relevance of each other's work. The shocking disdain we have for each other's intellectual contributions is holding us back. I am not arguing for any simple-minded hand holding across epistemological chasms. From time to time, however, each side could glance over the chasm. The best research on communication in interpersonal relationships in the next decade promises to be synthetic work, incorporating insights about relationships from a variety of different viewpoints.

In entering the new world of relationships, we cannot afford to jettison the important insights, ideas, and even techniques that we still need from both communication scientists and interpretative scholars. We have spent little time in capitalizing on the mutual relevance of each other's work. Many insights from interpretative and cultural studies have been systematically ignored by scholars in the empiricist tradition of interpersonal communication. And scientific studies of human communication have been slighted by scholars in the other traditions. Although not a panacea, a realist approach to the study of communication in relationships offers an alternative middle ground in the often spurious debate between empiricism and hermeneutics in the field of interpersonal communication in relationships. It may be the philosophical no-man's-land where both sides can meet to examine communication in relationships.

References

Barnouw, E., Gerbner, G., Schramm, W., Worth, T. L., & Gross, L. (1989). *International encyclopedia of communications*. New York: Oxford University Press.

Berger, C. (1977). Interpersonal communication theory and research: An overview. In B. D. Rubin (Ed.), *Communication yearbook 1* (pp. 217–228). New Brunswick, NJ: Transaction Books.

Berger, C. (1991). Communication theory and other curios. *Communication Monographs, 58*(1), 101–113.

Berger, C., & Chaffee, S. (Eds.). (1987). *Handbook of communication science*. Newbury Park, CA: Sage.

Bochner, A., & Ellis, C. (1992). Personal narrative as a social approach to interpersonal communication. *Communication Theory, 2*(2), 165–172.

Cronen, V., Pearce, B., & Snavely, L. (1979). A theory of rules structure and types of episodes and a study of perceived enmeshment in undesired repetitive patterns ("URPs"). In D. Nimmo (Ed.), *Communication yearbook 3* (pp. 225–240). New Brunswick, NJ: Transaction Books.

Dindia, K. (1982). Reciprocity of self disclosure. In M. Burgoon (Ed.), *Communication yearbook 6* (pp. 506–530). Beverly Hills, CA: Sage.

Fitzpatrick, M. A. (1988). *Between husbands and wives: Communication in marriage*. Newbury Park, CA: Sage.

Giles, H., Coupland, N., & Wiemann, J. M. (Eds.) (1990). *Communication, health and the elderly*. Manchester,UK: Manchester University Press.

Greenwood, J. D. (1991). *Relations and representations: An introduction to the philosophy of social psychological science*. London: Routledge.

Harre, R. (1985). *The philosophies of science* (2nd ed.). New York: Oxford University Press.

Jackson, S., Jacobs, S., & Rossi, A. M. (1987). Conversational relevance. In M. L. McLaughlin (Ed.), *Communication yearbook 10* (pp. 323–347). Newbury Park, CA: Sage.

Knapp, M., & Miller, G. R. (Eds.) (1985). *Handbook of interpersonal communication*. Beverly Hills, CA: Sage.

Kramarae, C. (1992). Gender and dominance. In S. Deetz (Ed.), *Communication yearbook 15* (pp. 469–472). Newbury Park, CA: Sage.

Leeds-Hurwitz, W. (1992). Forum introduction: Social approaches to interpersonal communication. *Communication Theory, 2*(2), 131–138.

Liska, J. (1992). Dominance-seeking language strategies. In S. Deetz (Ed.), *Communication yearbook 15* (pp. 427–456). Newbury Park, CA: Sage.

Meehan, E. (1988). Recentering television criticism. In J. Anderson (Ed.), *Communication yearbook 11* (pp. 183–193). Newbury Park, CA: Sage.

Miller, G. R. (1983). Taking stock of a discipline. *Journal of Communication, 33*(3), 31–41.

Parks, M. (1982). Ideology in interpersonal communication: Off the couch and into the world. In M. Burgoon (Ed.), *Communication yearbook 5* (pp. 79–108). New Brunswick, NJ: Transaction.

Putnam, H. (1984). What is realism? In J. Leplin (Ed.), *Scientific realism* (pp. 140–153). Berkeley: University of California Press.

Rogers, L. E., & Farace, V. (1975). Analysis of relational communication in dyads: New measurement procedures. *Human Communication Research, 1,* 222–239.

Sappington, A. A. (1990). Recent psychological approaches to the free will versus determinism issue. *Psychological Bulletin, 108*(1), 19–29.

Shorter, E. (1980). *A history of women's bodies.* New York: Basic.

Target Practice: A Batesonian "Field" Guide for Communication Studies

by Horace Newcomb, University of Texas at Austin

Three quotations from Gregory Bateson inform my comments.

> *It is a general assumption of this book that both genetic change and the process called "learning" (including the somatic changes induced by habit and environment) are stochastic processes. In each case there is, I believe, a stream of events that is random in certain aspects and in each case there is a nonrandom selective process which causes certain of the random selective components to "survive" longer than others. Without the random, there can be no new thing.* (1980, p. 163)

> *I begin from a discrimination I owe to Horst Mittelstaedt, who pointed out that there are two "sorts" of methods of perfecting an adaptive act. Let us suppose that the act is the shooting of a bird. In the first case this is to be done with a rifle. The marksman will look along the sights of his rifle and will note an error in its aim. He will correct the error, perhaps creating a new error which again he will correct, until he is satisfied. He will then press the trigger and shoot.*
>
> *What is significant is that the act of self-correction occurs within the single act of shooting. Mittelstaedt uses the term Feedback to characterize this whole genus of methods of perfecting an adaptive act.*
>
> *In contrast, consider the man who is shooting a flying bird with a shotgun or who uses a revolver held under the table where he cannot correct its aim. In such cases, what must happen is that any aggregate of information is taken in through sense organs; that upon this information, computation is completed; and that upon the (approximate) result of that computation, the gun is fired. . . . The man who would acquire skill with a shotgun or in the art of shooting pistols under the table must practice his art again and again, shooting at skeet or some dummy target. By long practice, he must adjust the setting of his nerves and muscles so that in the critical event, he will "automatically" give an optimum performance. This genus of methods Mittelstaedt calls Calibration.* (Ibid., pp. 215–216)

Horace Newcomb is a professor in the Department of Radio-Television-Film at the University of Texas at Austin.

> *There are subcycles of living and dying within the bigger, more enduring ecology. But what shall we say of the death of the larger system? Our biosphere? Perhaps under the eye of heaven, or Shiva, it doesn't matter. But it's the only one we know.* (Ibid., p. 230)

Bateson would perhaps quarrel mightily with my appropriation of the last quotation above to apply to the trivialities of disciplinary rather than ecological change. And even though I will shortly make a plea that the questions and issues surrounding the study of communication are indeed related to all sorts of survival, ecological among them, I will not claim to have anything approaching that "god's-eye view" of the world or of the field that he acknowledges. Perhaps Bateson himself achieved something of the sort in those late essays, but the rest of us had best recognize that only fools and saints pretend to such sweeping perspective. While I will not comment on the former, we all know there are precious few of the latter. It is for this reason that we need the more practical example of the second quotation, something we can apply to our consideration of a future.

With this in mind, I make a fairly small claim here. Disciplines generally produce rifles, rifle ammunition, and rifle shooters—men and women who take aim at small precise targets, measure their error, adjust, refine, and finally fire. It is very difficult, on the other hand, to attempt the other sort of learning, the calibration required to retrain one's entire body and mind; to cultivate and accumulate the swing of arm, eye, brain, toward the new; to feel the weight of the field shift and to see more clearly what we should aim for.

Yet as the first quotation above suggests, if we are to make any sort of progress toward survival—and I do not speak here of university departments, careers, or even disciplines—we must be ready to sense and recognize the new when it comes among us. We must be ready to follow as it shatters our received patterns of thinking, our research, our curricula.

If we could train for calibration rather than feedback, we might occasionally recognize the sway of entire bodies of knowledge, the repositioning of disciplines, and the sudden surge of interest among our students once again. It is tempting for some of us to suggest this is what happened with the introduction of humanities-based approaches to communication study from the late 1960s to the present. This is not, in my view, a matter of battles among paradigms, much less a matter of defeats and victories. Rather, I would suggest that the moves seemed to reinvigorate both the interloper and the interloped. They are best seen as matters of calibration, of significant shifts in "sights," and of adjustments made by all concerned. The result of such shifts is not necessarily an expansion or diminution of one or another method or mode of study. Rather, it is a matter of repositioning, an alteration of ratios of influence and analytical acuity, in some rare instances, of explanatory power.

It is equally important to recognize, however, that if the lay of the land, the field, was surveyed from a changed perspective, the newer approaches to communication study dragged with them their own, apparently indefatigable, rifle men and women. The field is now littered with more, as well as different, target shooters, with the rusty cases of our last rounds, discarded as we aim (theoretically, of course) ever more precisely at the targets pressed to our noses.

In the best case—in the long run, perhaps—such techniques, whether practiced by social scientists or humanists or the newer hybrids, produce important insight and significant knowledge. In the long run, we can see feedback as a means of leading to important change. But this, as I say, is the best case. Far more often feedback produces triviality, minor observation, the easily targeted.

Applied to research in our field or any other, the result is the clutter of repetitive and minor journal articles, indeed, the clutter of minor and repetitive journals. The well-armed research rifleperson can set a list of targeted journals on the nearest fence or bulletin board, ranked in descending order of significance, assuming those less important will be closer, more easily hit, with less work at adjusting after feedback. In fact, if the journal is needy enough, thin enough, there may be no feedback. The bullet too flawed for one publication will easily find another. Increasingly, this analogy should be applied to book publishing. Publishers' catalogues become supply depots. Bibliographies become bandoliers, filled with more precisely designed ammunition, and worn to professional meetings where we can ambush those less well informed.

Similarly, our curricula create more and more courses "aimed" at more and more narrowly defined targets. With all the journals, articles, and books available, with such extensive supplies, it would somehow be wrong not to create a specialized course on a smaller topic. Something must be significant if all that writing is moving closer and closer to the prey. We assume smaller, more focused courses result in knowledge. In fact, they may result only in "coverage." We pepper the field in hopes of chewing away at a target we have yet to define in any meaningful sense.

Pessimistically, I will suggest that the future of our field, of any academic field, tends toward such insignificant feedback. This is so because of the hindrance of institutional weight, the burdening corpse of received information—so different from the value of accumulated knowledge. We insist, with some justification, that students not only know, but be able to reproduce, every shot fired before them. Instead, I argue, we should be teaching students to calibrate. We no longer have time to waste on the reproduction of someone else's feedback. We should educate Batesons.

These problems result, I believe, from an interesting paradox. We have come to know our field in its parts, and have newly learned the significance of some of those parts. Senders have become institutions, artists, ideologies, social structures, mental categories. Messages have become texts. Receivers have become audiences, communities, interpreters, citizens. Such redefinitions barely scratch the surface of the changes we've seen.

At the same time, communication topics and issues are increasingly acknowledged as integrative and central. Communication is often recognized as nearly synonymous with culture. The distribution of power as applied to and organized by communication makes the subject equally synonymous with public, political life. On the one hand maintaining traditional wisdom and on the other sluicing toward nearly visible evolutionary alteration, old and new technologies are at the heart of human social experience. We have already viewed some form of social reorganization of the world on television. Other reorganizations, with other technologies, are in progress. The stakes are very high.

Too often these matters, when pursued from the rifle shooter's perspective, reduce academic life and work to a discussion of why the steaks are so expensive.

Training a new generation of scholars and teachers, retraining ourselves, is no simple matter of interdisciplinary work. We must aim for a curriculum, a mode of study and research that leads to means of recognizing truly significant questions.

Once, only once, when I was far too young to know better, I asked my father if he ever killed a man in the war. His answer—given, I later realized, with as much pain as patience—was that if so, it was without his knowledge. He, like all infantrymen of his day, had been instructed merely to fire toward enemy positions, to help "lay down a rifle barrage." Such a barrage, obviously, was intended to create, on a massive and more deadly scale, something like the result of a shotgun. Today's weapons of war, even in "small arms," are more deadly still. Now, a single soldier can create a rifle barrage with a single automatic rifle.

I make these observations because, in spite of Bateson's examples, I do not like the violent images they invoke. No more do I like the fact that his small example leads me to others, drawn from another sort of field. But these too are instructive, and I pursue them a bit further to make a final point.

I grew up hunting, and my father taught me to shoot both a shotgun (first, because it was safer) and a rifle (because it was more deadly, less predictable). Neither of us was a good shot. We ate what we killed (never much) not out of necessity, but to honor the memory of necessity. And because it tasted good.

More than utilitarian, however, our hunting served other purposes. I have no wish to romanticize here—there were never Faulknerian blood rituals, even when we hunted the woods Faulkner then still hunted. But our hunting was instructional in larger senses—aesthetic, rudely scientific. Perhaps, though I do not remember it so, the lessons were also epistemological. I do doubt this, however, just as I doubt that Bateson, if he ever fired a rifle or shotgun, had any epistemological epiphany while doing so. Most significantly, the questions I encountered while hunting were presented and addressed as moral issues. The lessons were about what was, is, at stake.

The same sorts of lessons must have been learned by others at other times in other ways—at ball games, in sewing and cooking, in what Norman Maclean would think of as "fishing." There are many ways of learning the important questions. But for me, the few moments spent in those "fields" with someone whose work usually went from before sunrise until after dark, were distinguished by their very lack of frequency—by their randomness—and therefore somehow far more significant.

I no longer hunt. I have fired neither a shotgun nor a rifle for many years. I find the violence repugnant as sport and take my meat from other sources. But I find Bateson's images far too central to dismiss simply because weapons are at their core. (Note, however, that I have not touched his bizarre remarks on the man who fires a pistol from beneath a table. Is it the case that the power of popular culture exceeds even our most committed imaginations?)

The point for the future of our field is that what we must teach and practice is twofold. We must teach in our classrooms and conduct research to teach one another on the assumption that our questions are first defined as significant moral and political problems. Then we must teach and practice in ways that enable us and our students to address and respond to those problems. This means more than offering another reading, another survey, another experiment to add to what we already know about injustice, terror, hunger, and fear. It means gathering knowledge and devising answers to questions. And I believe such goals are far different from those emerging from a focus on disciplines.

Disciplinary questions, method-driven, are too often trivial questions. We aim and miss, alter sights, aim and miss. Eventually we hit something, but at what expense (moral, social, political, intellectual)? Like target shooters in some commercial gallery, ear plugs in place, goggles shading out even the shooter beside us, we fire at our own puny targets, forever correcting for accuracy in small matters.

Instead, like the calibrator, our intent should be to see, to sense, the significant shift beyond us, feel other movements pulling us toward themselves. We must be ready to drop everything, lurch toward the significant, knowing that if we are willing to do that, we may recognize the field from a new vantage point.

The future of our field, of all academic fields, depends on learning to create as well as to respond to such change. The academy is under attack and rightly so. The attack is not justified because we have somehow neglected or refused to answer someone else's version of the right questions, but because we have too often failed to identify any significant questions or to justify the questions we pursue. Too often we have failed to explain how and why our work is important, perhaps because we cannot. We justify our pursuit of small matters by asserting that knowledge is built slowly, incrementally, that out of the detritus of our *vitae* emerges a kind of insight, a degree of truth. While this may be the case—and surely we must have both feedback and calibration—Bateson forces us to measure the cost. His significant questions, it seems to me—at least in later, philosophical work—dealt with why we could not learn better, produce more easily, and teach more effectively. Why, knowing as much as we do, must we be bound by old patterns?

I suspect the answer lies in our very knowledge of the evolutionary process he describes so eloquently. Shifting the great stochastic process with a bold new move—which should be our goal—might lead as easily into extinction as to the development of new species. Extinction is as painful in an academic institution as in the deep wetlands, though not nearly so important. Still, the response to this is that the more variations we put forward, the more likely we are to survive. But true variations, significant variations, are far different from the exploration of narrower and narrower paths forced by disciplinary and careerist opportunities.

For this reason, as much as any other, we are better off in our current state of disciplinary confusion about communication studies. So long as we are not shoved (do not shove ourselves) onto some isolated evolutionary ledge, our chances for significant invention are far better. So long as we are not easily pinned down into our evolutionary—or academic, institutional—corner, we have better opportunities to seek out and define the most important matters. Communication studies, for many of us, are central to social and cultural, political and moral matters of the gravest sort. We should turn our attention to them. So long as we are free to recalibrate, we may be the ones who shift fields rather than follow the shifts of others. We should be the random intervention that alters old courses, old choices, old patterns of thought.

I have no immediate strategy for such a direction, no program or curriculum plan to implement. I will assert, however, that our current agenda of research and teaching should be informed by the design and address of such questions. The immediate future of the field should center on efforts to halt the trivial and devote our meetings and publications to topics of significance, to discussion of what an extended and deepened purpose should look like. I do, however, also think that as a tactic we can recast one of the propositions put by our editors:

Communication scholarship lacks disciplinary status because it has no [accepted] core of knowledge—and is presently the better for it.

Reference

Bateson, G. (1980). *Mind and nature*. New York: Bantam.

Harmonization of Systems: The Third Stage of the Information Society

by Sandra Braman, University of Illinois

Postmodernism and information economics, mass communications and telecommunications, popular culture and managerial theory. Until now, they have been treated separately. Each offers a conception of the information society that is distinguished by disciplinary history and unit of analysis, yet each is like the others in addressing issues of social change stimulated by technological development and dependent upon the global information infrastructure, and in identifying dispersal of individuating subjects (whether biological, organizational, narrative, or political) and the loss of facticity as key characteristics of this era.

Analysts of postmodern culture identify this as the information age because information flows have replaced material and spiritual worlds as the basis of referentiality. Economists do so because information as both an intermediate and final product has come to dominate the economy, because the domain of the economy has expanded through the commodification of information, and because information flows have replaced the market as the key coordinating mechanism. Sociologists do so because information technologies are key to the restructuring of our social and political environments. For each, quantitative change has led to qualitative change.

The information society, however, did not emerge full-blown. Rather, three developmental stages can be identified: A first, beginning in the middle of the 19th century, was characterized by the *electrification* of communication. The second, beginning in the late middle 20th century, was characterized by *convergence* of technologies and by *awareness* of the centrality of information to society. The third stage, beginning with the 1990s, is characterized by *harmonization* of information systems with each other, with systems across national borders, and with other social systems. These changes in the communications environment are significant for the future of the field because they are shifts in the nature of our very subject matter itself, and require some new ways of thinking about it.

Sandra Braman is an assistant professor of communications at the University of Illinois, Urbana–Champaign.

Early Stages of the Information Society

The staging is suggestive; elements characterizing each period often first appear earlier and continue to exist later. Progress through the stages has neither been uniform geographically nor experienced in the same way by each class and each society.

Stage 1: Electrification of Communications

The information society was launched in the mid-19th century with the electrification of communications, technological developments that increased the speed and capacity of information flows and made possible the building of a fixed global information infrastructure.

These technologies gave birth to new types of organizations, such as the first international organization (the International Telecommunications Union) and national and international news agencies. They also made it possible for existing types of organizations to grow bigger and more extended geographically while retaining centralized control. And they made possible changes in the nature of organizations by stimulating the shift in control from owners to managers and from human decision makers to automation. This combined effect shifted both the locus of decision making and its nature (Chandler, 1977). By the end of the 19th century, information flows were central to the structuring of the global economy as well as both the subject and the tool of international relations (Blanchard, 1986; Headrick, 1990). The information society was global from the beginning.

Changes in organizational form contributed to commodification of information and with them information flows began to become both the engine of the market and its central coordinating mechanism (Antonelli, 1992). Commodification of information was further stimulated by news agency treatment of news as property. The wide clientele of news agencies encouraged the delocalization of facts (separation from the specific as they attain to the "general" or scientific) and their decontextualization (separation from a context, or story), tendencies that strip information of its noncommodity forms of value. The growth of advertising and multiplication of forms of mass culture contributed similarly.

The effect of these trends on narrative form was to fetishize facticity and extend the realms of discourse in which it was important. "Objective," or *New York Times*-style, journalism, in which isolated "facts" are offered, emerged and soon became the ideal. Surrealist and Dadaist artists pushed delocalization and decontextualization of facts to the extreme.

Two perspectives on the impact of these new information flows were extant during this period. Those involved in the design and building of the infrastructure, such as AT&T's first president Vail and his engineer Carty, were utopian in their visions, with dreams of global unity. Those concerned about the impact of the content of the information flows on society, however—whether sociologists looking at social "nervousness," or government officials dealing with propaganda—feared a dystopia instead.

Stage 2: Converging Technologies and Awareness of Information
The second stage of the information society began in the middle of the 20th century and was characterized by the convergence of computer and communication technologies and by awareness of the centrality of information to society.

By the 1960s, sufficient numbers of technologies were in use for their effects to begin to appear. Information flows became ever more ubiquitous and penetrated ever more deeply into society; vastly increased their capacity, speed, manipulability, and reach; and opened opportunities to design systems with fungibility, redundancy, interoperability, and intelligence. The world began a second round of building of the global infrastructure, and experimented with different ways of using the new technologies both to do traditional tasks more efficiently and to do new things. Since the information infrastructure determines organizational capacity (Fulk & Steinfeld, 1990), the use of new information technologies made possible the evolution of new organizational forms, most notably the transnational corporation.

It was during these decades that economists began to see information as the driving force of a new stage of societal development, following agricultural and industrial eras. Machlup explored the sectors of what he called the knowledge economy, and Porat developed a way of quantifying the information economy based on SIC (standard industrial classification) codes that formed the basis of statistical estimates of the information economy using neoclassical economics. Many, however, argued and argue that economics must be rethought in order to adequately deal with the intangibility of information (Babe, in press).

During the second stage of the information society, the notion of facticity came under attack. Within journalism, the struggle over its nature became at times vicious as "new" and "objective" journalists charged each other with lying. The stakes were high because the argument over the location and nature of facticity was a dispute over boundaries between radically different ways of perceiving and organizing the world, each with its own political implications (Braman, 1985). Within science, acceptance of the notion of paradigm change undermined existing bodies of fact. Heightened cross-cultural exposure—made possible through the use of new information technologies—widened awareness of the range of possible epistemologies. And at the popular level, there was an explicit rejection of the factual, manifested through chemical and religious experimentation and participation in or emulation of the life habits of other cultures.

This turbulence in narrative form led to what Geertz (1980) described as the blurring of genres. The yanking of information out of story lines, the widening of the geographic and cultural range from which information is received, and the juxtaposition of information flows from one medium with those received via another furthered the decontextualization and delocalization of information and made available to the general population the kinds of experiences earlier available only to artists.

The concept of the information society, coined by Umesao in 1962 in Japan, soon led to the notion of the process of informatization. These ideas were immediately brought to the attention of the mass audience, launched a well-funded research program, and became the center of Japanese national policy-making (Ito, 1991). In the United States, Bell's notion of the postindustrial society received widespread attention, though it was another decade before the concept of the information society came to North America, and Gottmann, not Bell, first used the term postindustrial (Hepworth, 1989). While in the United States there was some resonance between scholarly work and government planning, as with the wired city concept of the Johnson administration's Great Society, information was not a policy focus. In Brazil, on the other hand, the 1960s brought into power a regime focused on control of all types of information collection, processing, and flows in the service of state power. Gradually other governments acknowledged the impact of informatization. A number commissioned studies to serve as guides for policy-making. Notably, Sweden and France commissioned studies from the perspective of international impact.

Other ways of thinking about the effects of informatization emerged. Marshall McLuhan drew popular attention to the effects of the new technologies on individual perception and the social structure, reinforcing theoretically the popular multimedia experience of the 1960s. Calls for a New World Information Order—based on the insight that redistribution of global economic resources would not come into being without a redistribution of global information flows—demonstrated that a generalized consciousness of the centrality of information to society was already felt beyond the Western industrialized world. By the late 1980s, the example of newly industrialized countries (NICs) such as Singapore demonstrated that one route to economic success for developing nations was through participation in the information economy.

The Information Society in the 1990s

With the 1990s, we have entered the third stage of the information society, characterized by the harmonization of systems—communication systems with each other (e.g., linkage of broadcasting systems with telecommunications systems), national communication systems with those across borders (e.g., linkage of broadcasting systems internationally through shared ownership, production, and/or regional or global distribution), and communication systems with other social systems (e.g., the transformation of finance into a system of international information flows, or the dependence of just-in-time manufacturing, production, and distribution upon the telecommunications network). Harmonization has been dependent on the increased capacity, speed, and ubiquity of the global information infrastructure, on the mobility of its intelligence, and on its geodesic (nonhierarchical) nature (Huber, 1987).

The transnational corporation now dominates not only the global economic scene but, increasingly, the legal and political domains (see, e.g., Petersmann, 1991). They are diversely characterized by shifting combinations of centralization and decentralization, and by permeable boundaries as functions are variously externalized and internalized. These organizations are so highly interdependent that they have come to be described as *network firms*. The fundamental unit of economic analysis is no longer the firm nor the industry, but the project (long-term, multidimensional, contractual relationships); and the theory that is developing, *network economics*. Cooperation and coordination are now as significant as competition, and, because so much happens *within* organizations or networks, price is no longer a signal. Some scholars argue that the network model could provide the basis of all social analysis (Antonelli, 1992).

Harmonization of systems is pursued in a number of international decision-making arenas, with efforts in seemingly distinct realms echoing each other. The discussion about inclusion of trade in services under the General Agreements on Tariffs and Trade (GATT) (Braman, 1990b), for example, plays out in trade many of the same efforts at harmonization of systems promoted by arms control treaties within the defense arena (Braman, 1990a, 1991).

In this era, all three economic conceptualizations of the information society—information as the sector that dominates the economy, expansion of the economic domain through commodification of information, and replacement of market mechanisms with information flows—have come into their own. By any indicator, information as intermediate or final product now dominates the global economy. The commodification process has extended to that information which is most private and personal (including genetic), public (including government information gathered in "the public interest"), and global (as in remote sensing). And with the emergence of network economics, many market mechanisms have indeed been replaced with information flows.

In postmodern cultural experience, the fact is isolated not only from context and the specifics of the local, but from any reference to the material and spiritual worlds. The content of information flows now refers only to the content of other information flows, or hyperreality (Baudrillard, 1983). Thus the outcome of exaltation of fact has been to destroy it. The notion of facticity that has obsessed narrative form since Locke is dependent on the referential relationship between the sign and a material world separate from it. With the replacement of our sense of the real by the hyperreal, this referentiality is lost. Institutionalization of fact depended upon a consensus about methodologies for deriving and expressing fact. With the loss of this consensus, institutional claims to facticity lose credibility. This institutionalization also depended upon certain organizational forms, now dispersed into the network. Last, the concept of a fact included a sense of fixity. With the mutability of all forms of information, from genetic to visual, this confidence is also gone.

At the level of the individual, the loss of a narrative through which motives and forms for action can be derived is immobilizing; individuals become, as information processors, part of the harmonization of information systems with the very restricted and specialized function of consumption. At the level of the organization, dispersal of the locus of the corporate body decreases the ability of nation-states to act meaningfully.

Here the results of postmodern cultural enquiries and analysis of network economics come together. At the individual and societal levels of analysis, we see a decrease in the ability to act meaningfully and changes in the nature of effective action that significantly alter the domain of the political in ways that are largely yet not understood.

Concurrent with the loss of facticity, intellectual property, which establishes ownership in the *expression* of fact—and today, its processing—has become the strategically and tactically pivotal and economically most valuable form of property. This move delineates the new class divisions and emphasizes that in the information economy, relative position is more significantly determined by informational relationships to dynamic processes than ownership relationships to static capital and material resources.

Communications Research in the Harmonized Network Society

Our approaches to information policy, and to the purposes and methods of social science research, are inherited from a pre-information-society vision that no longer accurately portrays our actual environment.

In the pre-information-society environment, the "people" were understood to be a community, using communications in an effort to constitute their environment through engagement in participatory democracy. In Carey's terms (1989), communications was best characterized by the ritual model. During the first two stages of the information society, the community was transformed into an audience of consumers of commoditized information, of which Carey's transmission model of communication came to be the best characterization.

With emergence of the third stage of the information society, the people have been transformed again, into a network. We are still trying to understand what this means. Antonelli's (1992) work with network economics is one direction that offers some fruitful ways to think that are not only appealing conceptually but also useful for decision making.

The literatures on second-order cybernetics and chaos theory (Archer,

1988; Jantsch, 1989; Schieve & Allen, 1982) offer a complementary path. This work explores systems in healthy, continuous, and self-conscious change (morphogenesis) and emphasizes mutually beneficial self-amplifying causal processes. Coevolution of systems with other systems—including those in the natural world—is a central value. Social scientists working with these ideas are exploring ways to successfully navigate periods of extreme turbulence, such as those we are experiencing today. From this perspective, an autopoietic model of communication becomes the most characteristic—that is, a view of communication as the way in which the elements of a system creatively participate in the shaping of that system and its interactions with other systems. This would add a third to Carey's two categories.

Whatever approach is taken, key terms have new meanings. Organizations and networks should be studied as media; indeed, their effects as media may be far greater than those of the screen by which so many researchers are mesmerized. Content now importantly includes things like accounting systems, increasingly a powerful mode of exporting organizational form with the ultimate effect of harmonization of legal systems.

We need to explore the distinctive characteristics of the network as a unit of analysis. Microstructural and macrostructural forces are now central to our concerns, and links across levels need to be identified. With privatization of so many forms of previously public power, analysis of the effects of organizational communications becomes our most significant "press/state" discussion.

New questions arise: about the political effects of communication in the postmodern context, about appropriate decision-making tools for the range of values we need to address in this information environment, and about reconnecting the information society to the material world in time to save us from environmental destruction.

We have new concepts to explore, and phenomena and processes to analyze: the nature of coevolutionary and mutually beneficial processes (we know much more about those that are mutually damaging), morphogenesis, chaos, and autopoiesis.

Methodologically, the study of the harmonization effects of the third stage of the information society may be a particularly appropriate domain for the combination of various approaches in the study of specific problems, and for interdisciplinary work.

Finally it becomes clear that postmodernism and network economics, mass communications and telecommunications, popular culture and management theory are all talking about qualitative social change in the third stage of the information society. Since the effects and trends to which all these fields point resonate across levels of analysis—and since primary among these effects are fundamental shifts in the way we organize ourselves as a society—cross-fertilization among these literatures seems the most likely path to follow if we seek to emerge from this turbulent period with our goals enacted and our values intact.

References

Antonelli, C. (Ed.). (1992). *The economics of information networks*. Amsterdam: North-Holland.

Archer, M. S. (1988). Towards theoretical unification: Structure, culture and morphogenesis. In *Culture and agency: The place of culture in social theory* (pp. 247–307). New York: Cambridge University Press.

Babe, R. E. (Ed.). (in press). *Information and communication in economics*. Boston/Dordrecht/London: Kluwer Academic Publishers.

Baudrillard, J. (1983). *Simulations*. New York: Semiotext(es).

Blanchard, M. A. (1986). *Exporting the First Amendment*. New York: Longman.

Braman, S. (1985). The "facts" of El Salvador according to objective and new journalism. *Journal of Communication Inquiry, 9*(2), 75–96.

Braman, S. (1990a, December). The CSCE and information policy in the new Europe. Paper presented to Second Conference, Europe Speaks to Europe, Moscow, USSR.

Braman, S. (1990b). Trade and information policy. *Media, Culture and Society, 12,* 361–385.

Braman, S. (1991). Contradictions in Brilliant Eyes. *Gazette, 47*(3), 177–194.

Carey, J. W. (1989). *Communication as culture: Essays in media and society*. Boston: Unwin Hyman.

Chandler, A. D., Jr. (1977). *The visible hand: The managerial revolution in American business*. Cambridge, MA: Belknap Press.

Fulk, J., & Steinfeld, C. (Eds.). (1990). *Organizations and communication technology*. Newbury Park, CA: Sage.

Geertz, C. (1980). Blurred genres: The refiguration of social thought. *American Scholar, 49,* 165–179.

Headrick, D. R. (1990). *The invisible weapon: Telecommunications and international relations, 1851–1945*. New York: Oxford University Press.

Hepworth, M. (1989). *Geography of the information economy*. New York: Guilford Press.

Huber, P. (1987). *The geodesic network: 1987 report on competition in the telephone industry*. Washington, DC: U.S. Department of Justice, Antitrust Division.

Ito, Y. (1991). *Johoka* as a driving force of social change. *KEIO Review, 12,* 33–58.

Jantsch, E. (1989). *The self-organizing universe*. New York: Pergamon Press.

Petersmann, E. (1991). *Constitutional functions and constitutional problems of international economic law*. Fribourg, Switzerland: University Press.

Schieve, W. C., & Allen, P. M. (Eds.). (1982). *Self-organization and dissipative structures: Applications in the physical and social sciences*. Austin: University of Texas Press.

Beyond the Culture Wars: An Agenda for Research on Communication and Culture

by Dennis K. Davis, University of North Dakota, and James Jasinski,
University of Illinois

We live in an uncertain era, a time of enormous change when fundamental assumptions about the social world and our place in it are challenged everywhere. Long-dominant social values have lost their power to structure action. Long-suppressed ethnic cultures are gaining strength. International conflicts between ideological foes may be in decline, but conflicts over domestic cultural differences are escalating. Visions of a highly integrated, global New World Order are fading, and nightmares of civil disorder are springing up around the world.

Many explanations have been offered for the uncertainties of our times. These range from rapid technological change and the end of the Cold War to subversion by cultural elites. In our view, the cause is more profound. We are witnessing the gradual decline of modernism, a perspective that served as the foundation for Western culture for more than 4 centuries. The decline of modernism is profoundly troubling to many persons because most of our philosophical, religious, and political conceptions are grounded in it. In the space of a decade, we have moved from the Cold War to the Culture Wars, from the external threat of communism to the internal threat of postmodernism.

The field of communication has strong ties to modernism. Early communication theory was based on modernist assumptions. During the 1940s, quantitative communication research developed to describe, explain, and improve modern social institutions. This administrative research proved especially useful in planning and implementing promotional communication campaigns that could defend democracy while boosting sales for consumer goods and services. Highly profitable media industries were developed, guided by ratings data and formative research. But then came theory ferment and talk of a paradigm shift. For 2

Dennis K. Davis is a professor in the School of Communication at the University of North Dakota. James Jasinski is an assistant professor in the Department of Speech Communication at the University of Illinois.

decades, we have struggled to understand what is happening and where our field might be going.

We won't be able to understand what is happening until we view our field from a much broader perspective. Along with the other social sciences and humanities, we are being challenged to come to grips with a profound transformation of Western culture. In this essay, we will consider how the troubles affecting our field are linked to the decline of modernism. We will argue that the future of our field lies in exploring both microscopic and macroscopic social changes in a postmodern world. We begin by examining three fundamental assumptions of modernism and show how communication theory has been in the vanguard of the effort to challenge these assumptions. Then we discuss how communication research might address the many problems created by the decline of modernism and conclude by proposing an agenda for this research.

The Modern Era

The modern worldview persisted for 4 centuries because it was a very flexible perspective, one that has accommodated many social movements. Though strongly associated with the Protestant Reformation in the 16th century, it accommodated humanism, natural science, and classicism during the Renaissance. More recently, it was modified by the Industrial Revolution and the rise of capitalism.

Throughout its history, modernism has held to three fundamental assumptions:

Radical Individualism: The human individual is the fundamental and preeminent unit of society. Individuals are conceptualized as free and responsible agents who innately seek the True and the Good. Society exists primarily to assist individuals in their search by guaranteeing them certain basic rights. Social institutions, such as churches and public schools, are created by autonomous individuals to differentiate, socialize, and accommodate their needs.

Inevitable Progress Toward the True and the Good: Spiritual, social, and/or material progress toward the Good Society are/is inevitable only if artificial barriers to the individual search for the Truth about God and Nature can be removed (i.e., libertarianism). To find Truth, arbitrary constraints imposed by tradition or medieval social institutions must be overcome. The natural abilities (i.e., reason and conscience) of individuals must be freed so that they can find Truth.

The Preeminence of Modern Civilization: Modernism is assumed to be the highest form of culture. It evolved according to natural laws from, but is now qualitatively superior to, all forms of culture that preceded it. It will inevitably become the dominant world culture if it can be protected against subversion and trivialization by more primitive forms of culture.

The best accomplishments of modern culture deserve to be canonized and aggressively promoted throughout the world.

Challenging Modernism

Serious criticism of modernist assumptions has emerged over the past century at a time when modernism was achieving its greatest successes but also demonstrating profound limitations. Debates pitted critics who want to reform and restore truer forms of modernism against those who think all forms of modernism are doomed. For example, Habermas (1979) has advocated radical reform of what he terms the "modernist project" to realize the kernel of truth it contains; others argue that modernist assumptions are so fundamentally misleading or wrong that they must be abandoned. Some radical critics of modernism argue that it has become a conservative force that limits human freedom, retards progress, and impedes development of a humane basis for worldwide social order. The successes of modernism are questioned. Though material wealth has been created, a high price has been paid. Overly aggressive use of technology and the ruthless exploitation of human beings has decimated the physical and social landscape. Modernism is said to be at the root of world war, Cold War, genocide of minority groups, materialism, imperialism, colonialism, and global pollution. Modern social orders have succeeded *and failed* on a grand scale. Even in the wealthiest nations, the quality of human existence is eroding.

Social research, including communication research, has also begun to probe the limitations of modernist assumptions, its social institutions, and its social practices. This research rarely has been guided by postmodern notions. Rather it was intended to overcome practical problems. For example, in seeking a better understanding of the role of communication in modern life, research demonstrated the profound interdependence of human beings as they create culture and construct their social world. In psychology, the struggle of individuals toward independence and self-control has been shown to be fraught with inherent difficulties that can't be overcome without the support of others. Individuals have no inherent motivation or capacity to find essential Truth. However, they do have remarkably sophisticated abilities to perceive, communicate, experience, and remember a complex, often contradictory social world. Though none of the visions of the social world are eternally True, they can provide compelling and meaningful ways of structuring our daily lives and our relationships with others (Schutz, 1967).

Social science has gradually formulated conceptions and generalizations that seriously undermine the assumptions of modernism. Some of the most noteworthy of these notions are explained below.

First, the individual is a product of communication; human subjectivity, our capacity to experience and identify ourselves, emerges out of and is dependent upon ongoing patterns of internal and external communication practices (Epstein & Loos, 1989).

Second, the smallest social unit is best conceptualized as a community, not an individual. A community can be defined as a group of individuals who share a common culture that is capable of structuring a social world that members experience as meaningful and real (Berger & Luckmann, 1966; Davis & Puckett, 1991).

Third, the most basic and common function of communication is performance, not transmission of information. Performance occurs in rituals and other practices used within communities to coordinate experience and induce common perceptions of the social world. Performance practices serve to sustain culture through structuring of experience, not through transmission and learning of discrete bits of information (Carey, 1989). It promotes feeling with or involvement in, not knowledge of, the social world.

Fourth, two central problems of any society involve the structuring of individual experience on the one hand (stable and meaningful personal identity) and the large-scale coordination of action on the other (social order) (Giddens, 1979). Easy solutions to these problems don't work. For example, individuals can be coerced into adopting orderly practices, but these practices will not be experienced as meaningful. If too much coercion is present, formation of stable and meaningful identities will be jeopardized. On the other hand, highly idiosyncratic actions may be experienced as meaningful to individuals or small groups (e.g., urban gangs) but be quite disruptive or destructive for society as a whole.

Fifth, for each individual, the social world consists of many overlapping and interrelated realities (Berger & Luckmann, 1966). Each of these realities is learned and/or constructed during participation in a community. Each reality is routinely imposed upon fundamentally ambiguous environmental stimuli to construct a seamless experience of reality (Goffman, 1974). This effort routinely succeeds even when there are important gaps or inconsistencies between the various realities. Everyday life consists of moving between various realities and playing appropriate roles in each of them (Goffman, 1971). As we confront an increasingly multicultural world, our ability to create a seamless experience of reality is being severely tested. Troubles arise when gaps or inconsistencies become too great or when communities come into conflict.

The Communication Research Agenda

The present agenda for communication research is too often dominated by the search for solutions to the many practical problems created by new technology and the decline of modernism. Most research projects tend to be narrowly conceived. For example, efforts to probe and delineate the role of mass media in politics have come to focus on the inability of media to provide a panacea for the decline of Western democracy (Rucinski, 1991). Researchers are distressed to find that media haven't done much to strengthen democracy, and they have sought to understand why (Davis, 1990). Despite enormous growth in the accessibility of political information, public knowledge of politics remains low and is declining in some segments of the population. Media coverage of political campaigns has done little to stimulate political interest or participation (Wright, 1976). Should media be blamed for this situation? Should there be more public affairs programs tailored to the interests of specific audiences? Could news programs be structured differently to take into account the vast differences in viewer background knowledge (Davis & Robinson, 1989)? Would it help if media practitioners reconceptualized their job roles or their relationship to their audiences?

Narrow focus is also found in most other areas of communication research. Interpersonal and organizational communication research have investigated everyday communication practices with the goal of restructuring them in useful ways. If successful, this research could help people form stable personal identities and meaningful relationships with others. When applied to organizations, it could improve the quality of the work environment and boost efficiency.

While there is nothing inherently wrong with such research, it nevertheless fails to directly address the larger questions resulting from the decline of modernity. It is directed toward incremental, microscopic reform at a time when more significant change appears necessary. At best, this research will discover innovative ways of helping individuals to understand and cope with increasing fragmentation of their social world. It may help people endure meaningless social practices and accept institutions that foster these practices. However, it will not increase their overall understanding of their place in the social world or motivate them to seek innovative, qualitatively different solutions to their problems.

Focusing on Communication and Culture

Communication research should seek to contribute to the development of a constructive, postmodern perspective. It can do this by focusing simultaneously on research that investigates, first, the production of meaning within communities and, second, the negotiation of meanings between communities. The first body of research would systematically investigate how communication practices are used within various types of communities to foster personal identities which in turn structure perception, experience, and memory. Attention would be focused on the use of ritual in everyday life, including storytelling, everyday conversation, media use, and games. It may be especially heuristic to focus on meaning-production efforts in groups that are consciously seeking to forge new identities and/or empower members to engage in meaningful action

(Epstein & Loos, 1989). For example, therapy groups bring together persons who seek to overcome common problems by redefining themselves and their social world (Epstein & Loos, 1989). Social movements also foster similar efforts. We need to understand better how movements use ritual to shape the experience of their members.

In an ongoing project (Jasinski & Davis, 1990), we have conceptualized meaning production in terms of the generation of social capital. We have argued that four types of communication practices are used to create the capital. *Visualization practices* introduce, maintain, and disseminate political visions that identify and prioritize values and goals for individuals and for the society at large. *Signification practices* define public situations, actors, and actions; they literally create our experience of political reality by guiding and molding public perception of events. *Allocation practices* guide distribution and management of material resources, including the natural environment, public health, and the national economy. Finally, *legitimation practices* establish authority and the exercise of power by justifying and reinforcing the way that coercion is used to enforce practices.

Research on the negotiation of social capital would necessarily center on social institutions and the creation of public culture. Habermas (1989) has argued that one of the proud achievements of modernism was the development of the public sphere. In the medieval world, a small political and religious elite controlled all important social institutions. During the early modern era, the power of this elite was challenged by various communities (shopkeepers, Protestant sects, worker guilds, writers, academics, etc.) based in cities. These communities formed coalitions and carved out public spaces in which all persons from the constituent communities could interact regardless of their cultural background. In time, rituals, roles, and institutions were developed in these spaces. Sennett (1978) describes how coffee houses and theaters provided places in which a variety of new rituals were invented and various public roles were defined. In time, new institutions developed. For example, early newspapers informed people about events taking place in coffee houses.

Creation of the public sphere permitted otherwise powerless groups to unite in opposition to the arbitrary rituals and roles of kings and popes. Communities were freed to collaborate in creating and implementing modern visions (visualization) of what the social world should be and to establish new forms of authority (legitimation). Since these visions were forged by coalitions of communities, they tended to respect the cultural differences of constituent groups. Potentially troublesome differences were dealt with by defining them as private and therefore irrelevant to life in public (Sennett, 1978). However, most of these visions now appear to us as incredibly narrow and uninclusive. In many of them, women and ethnic minorities were relegated to low status. They were conceived of as emotionally and/or intellectually inferior to the men who dominated the coalition communities.

With the passage of time, creative and empowering activity within the public sphere declined. The institutions created by community coalitions became reified. Coffee houses disappeared and newspapers remained. The raucous, participatory theater of the Enlightenment gave way to mass entertainment supplied by a media industry and distributed to mass audiences. Most of the social institutions nurtured within the public sphere are now controlled by specialized, technocratic elites. Links to communities have been lost. In politics, public officials are elected on the basis of mass-media-based campaigns in which highly trained technocrats use polling data to target and manipulate voter groups. Public rituals and roles are no longer creatively evolved. They are routinely performed with ever decreasing commitment or sense of purpose. For example, the act of voting is rarely conceived and experienced as an assertion and affirmation of one's rights as a free person in a publicly controlled world. All too often, it is a routine act performed by alienated, powerless individuals who feel compelled to act in response to social pressure and media appeals.

It is through institutions that communities of all types should be able to confront and accommodate one another, to negotiate new forms of public culture. Communication research should suggest ways of transforming existing institutions so that they can serve this purpose. Research could identify and promote development of innovative communication practices, which can in turn serve to structure new rituals, roles, and ultimately new institutions. If we are to have a vital democracy, research is necessary to restructure public institutions so that they can become places where culturally diverse groups come together to elaborate public culture. For example, during the last presidential election campaign, a variety of new communication practices were used in an effort to reach and involve various groups of voters. Candidates appeared on specialized cable television channels and engaged in long discussions with program hosts and studio audiences. As public forums, these town meetings and talk show programs left much to be desired, but they may have been a useful alternative to routine news coverage and 30-second advertisements. Research could investigate the utility of such new practices and ideally recommend better alternatives.

Important and useful forms of institutional change are already underway. The civil rights movement and the feminist movement have challenged existing institutions and stopped some of their most destructive practices. However, most of these changes have not precipitated truly fundamental transformations. Communication research should examine the successes and failures of these movements. How is fundamental change being resisted by existing institutions? Can this resistance be overcome? For example, contemporary institutions often resolve minority or feminist complaints using old rules that are biased in favor of the groups that formulated them—even when these groups no longer exist or would now reject the rules. Existing journalistic practices lead to the production of news that trivializes or disrupts movements (Tuchman, 1978). We need research that illuminates such problems and can recommend alternatives.

It is highly unlikely that existing institutions will undergo useful, constructive change in response to challenges from social movements or ethnic minorities. Change will continue to be grudging and insufficient. No one will be satisfied with it. Charges of political correctness and reverse discrimination will be directed against increasingly adamant critics. Efficiency of institutions will decline and problems will persist. We need research that can produce a more complete understanding of institutions and the ways that they can be changed by negotiating new forms of social capital.

Our vision of institutional change retains one feature of modernism—its conception of public space and the public sphere as the proper arena for constructive social change. But we reject existing political institutions that to us represent at best 19th-century solutions to 21st-century problems. We see little point in tinkering with these institutions at a time when we have access to such a wide array of new communications media and such potential for cultural creativity within the myriad communities that inhabit our social order. Rather than fear the consequences of cultural diversity, we need to embrace it. However, this embrace must be grounded upon understanding, not the naive hope that we will somehow get along. Peace between communities doesn't just happen because the leaders of diverse communities reach rational decisions and declare a truce. It must be created through the development of institutions that enable social capital to be negotiated. Communication research should provide the insight needed to do this work.

References

Berger, P. L., & Luckmann, T. (1966). *The social construction of reality: A treatise in the sociology of knowledge*. Garden City, NJ: Doubleday.

Carey, J. W. (1989). *Communication as culture: Essays on media and society*. Winchester, MA: Unwin Hyman, Inc.

Davis, D. K. (1990). News and politics. In D. L. Swanson & D. Nimmo (Eds.), *New directions in political communication: A resource book* (pp. 147–184). Newbury Park, CA: Sage.

Davis, D. K., & Puckett, T. (1991). Mass entertainment and community: Toward a culture-centered paradigm for mass communication research. In S. Deetz (Ed.), *Communication yearbook 15* (pp. 3–34). Newbury Park, CA: Sage.

Davis, D. K., & Robinson, J. P. (1989). Newsflow and democratic society in an age of electronic media. In G. Comstock (Ed.), *Public communication and behavior, Vol. 3* (pp. 59–102). New York: Academic Press.

Epstein, E. S., & Loos, V. E. (1989). Some irreverent thoughts on the limits of family therapy: Toward a language-based explanation of human systems. *Journal of Family Psychology, 2,* 405–421.

Giddens, A. (1979). *Central problems in social theory*. London: MacMillan Press.

Goffman, E. (1971). *Relations in public: Microstudies of the public order.* New York: Basic Books.

Goffman, E. (1974). *Frame analysis: An essay on the organization of experience.* New York: Harper & Row.

Habermas, J. (1979). *Communication and the evolution of society.* London: Heinemann.

Habermas, J. (1989). *The structural transformation of the public sphere.* Cambridge, MA: MIT Press.

Jasinski J., & Davis, D. (1990). *Political communication and politics: A theory of public culture.* Paper presented to the Political Communication Division of the American Political Science Association, San Francisco.

Rucinski, D. (1991). The centrality of reciprocity to communication and democracy. *Critical Studies in Mass Communication, 8,* 184–194.

Schutz, A. (1967). *The phenomenology of the social world.* Evanston, IL: Northwestern University Press.

Sennett, R. (1978). *The fall of public man: On the social psychology of capitalism.* New York: Vintage Books.

Tuchman, G. (1978). *Making news: A study in the construction of reality.* New York: Free Press.

Wright, J. D. (1976). *The dissent of the governed: Alienation and democracy in America.* New York: Academic Press.

The Hierarchy of Institutional Values in the Communication Discipline

by Jennifer L. Monahan, University of Georgia, and Lori Collins-Jarvis, Rutgers University

While numerous scholars have examined controversies within the communication discipline, few have looked at the underlying value differences that sustain these controversies. On this 10th anniversary of *Ferment in the Field,* we explore the fundamental values that guided our discipline in the past and propose how these values will undoubtedly shape the future.

A value is an "enduring belief that a specific mode of conduct or end-state of existence is personally or socially preferable to an opposite mode of conduct or end-state of existence" (Rokeach, 1973, p. 5). According to Rokeach, individuals first acquire values in an isolated, all-or-nothing fashion. As individuals mature, however, they increasingly encounter social situations where their values come into competition, forcing them to weigh one value against the other and decide which is most important in a particular situation. Eventually, individuals learn to integrate their values into an organized system in which each value is ranked in order of importance relative to all other values.

As the discipline of communication matures, it too must move through the process of value integration. While we have examined the values that guide our discipline, we have not approached this mature level of value integration. Most scholarly discourse has been limited to two general forms of argument. Some scholars champion one particular discipline-level value, advocating that the future of the field depends upon our whole-hearted adoption of this value as a fundamental (and often, exclusive) guiding principle (e.g., Berger, 1991; Rakow, 1992). Other scholars recommend that future growth depends upon the reconciliation of two diametrically opposed values, arguing that each value, in its own way, makes an essential contribution to our discipline (e.g., Purcell, 1992; Rogers & Reardon, 1988). Rather than generating a progressive discussion

Jennifer L. Monahan is an assistant professor in the Department of Speech Communication at the University of Georgia. Lori Collins-Jarvis is an assistant professor in the Department of Communication at Rutgers University.

Copyright © 1993 *Journal of Communication* 43(3), Summer. 0021-9916/93/$5.00

culminating in a greater collective understanding, these approaches have, instead, led to a seemingly endless repetition of arguments.

This essay proposes that past discussions of discipline-level values have been limited by the tendency to consider values in isolation, rather than as part of hierarchically organized systems. A systemic approach to value examination reminds us that all discipline-related values are desirable, but that the degree to which any one value can be attained depends upon its importance (to a particular individual, in a given situation) relative to all other values.

In reviewing past debates about the discipline, five overarching values seem to us to emerge as central to our disciplinary disputes:

- *Connectedness* (that we possess a shared core of knowledge)
- *Creativity* (our ability to generate original theory)
- *Distinctiveness* (that we do not replicate other disciplines)
- *Pluralism* (that we encourage theoretical diversity)
- *Social relevance* (that our endeavors improve the human condition).

This essay demonstrates how the relative importance of each value changes depending upon prevailing social, economic, and intellectual conditions. This analysis is based on the assumption that values are social constructions, influenced by the departments in which we teach or conduct research, the colleges and universities by which those departments are governed, and the academic community as a whole.

To demonstrate the utility of a value system perspective, we first explore these values in a familiar context, the discipline in the 1980s.

Discipline-Level Values in the 1980s

All five institutional values noted above were prized by different subgroups within the field during the 1980s. However, certain values were clearly on the rise, while other values appeared to be on the wane. The following value hierarchy dominated the discipline at that time: (1) pluralism, (2) social relevance, (3) distinctiveness, (4) creativity, (5) connectedness (ranking values from most to least salient).

An analysis of the original *Ferment in the Field* reveals that pluralism, or theoretical and methodological diversity, was the most salient academic value discussed. The call for increased theoretical and methodological diversity within communication science was first a reaction against connectedness—a value that had formerly dominated the discipline (Gitlin, 1978). The expansion of communication science between the 1950s and the 1980s may explain the shift in value emphasis from connectedness to pluralism. In the 1950s and 1960s, the predominance of the behavioral science perspective served as a necessary unifying (and legitimating) mechanism for the fledgling academic discipline of communication science (see Rogers, in press). By the 1980s, however, the discipline that had once been populated almost exclusively by (mostly American) behavioral scientists now included an eclectic mix of scholars representing diverse theoretical and methodological perspectives (e.g., critical, cultural, literary, naturalistic, etc.). Confronted with the limitations of a discipline where a unified but narrowly defined approach was highly valued, these new scholars sought to define the field in more pluralistic terms (Katz, 1983; Smythe & Van Dinh, 1983).

Economic factors also supported this pluralistic trend. The 1980s were primarily a period of economic stability or growth for universities in general and communication departments in particular (Rogers, in press). Like most institutions that experience economic growth, communication departments expanded and diversified. As departments became more differentiated (or, in many cases, fragmented into separate, more specialized entities), they came to include a more diverse mix of scholars (Mader, Rosenfield, & Mader, 1985).

Social relevance emerged as the second most salient value in the 1980s, due in part to its association with pluralism. Many scholars who championed theoretical and methodological diversity also argued that the discipline should be increasingly devoted to improving the human condition. Critical scholars, for example, challenged the predominant administrative research approach, arguing that more theory and research should be undertaken to spur social change and emancipate dominated social/political/economic groups, not reinforce the status quo (Smythe & Van Dinh, 1983). Even members of the discipline representing the more established, behavioral science approach agreed that 30 years of empirical research had resulted in few findings of social consequence (Katz, 1983).

Like social relevance, distinctiveness has long been a salient value in communication science. As Bochner and Eisenberg (1985) contend, questions such as "What characterizes the field of communication?" "What do we do that other academic disciplines do not?" and "What makes us unique in the world?" have dogged the field since its inception. Few people who attend communication conferences or read communication journals can fail to note the field's ongoing identity crisis.

Distinctiveness and pluralism are intimately related. During the 1980s, the discipline's boundaries expanded and its internal connectivity decreased: There were more kinds of communication researchers and less shared knowledge between subfields (Wiemann, Hawkins, & Pingree, 1988). The ramifications of these trends for distinctiveness seem clear. If making the field more open to new perspectives was salient, obviously the importance of precise definitions of who we are was less so.

Another outcome of the growth in pluralism and social relevance was a curtailed emphasis on creativity (narrowly defined in this essay as the ability to generate original theory). Some might find it curious that creativity was ranked so low as fourth in the value hierarchy, since creativity is, after all, the *ideal* academic value, the one held most sacred. Yet a close look at the epistemological battles of the 1980s reveals that creativity was not a central point of contention. For example, most articles in the initial issue of *Ferment in the Field* did not address creativity or addressed it as a justification for a value shift (i.e., we are not creative because we are not pluralistic), but not as an end in itself.

Finally, the 1980s were a time of embracing new perspectives, not integrating them with the old. Supported by stable economic resources, the fragmentation of the discipline intensified. Scholars paid little attention to how these fragments fit into a functional whole. Thus, the value of connectedness (i.e., the degree to which members of the discipline have a shared core of knowledge) was the least salient of the five overarching values. The discipline was less concerned with strengthening existing connections and more concerned with opening doors to new perspectives.

In summary, during the 1980s, economic expansion and dissatisfaction with the status quo resulted in an increase in the field's boundaries and a decrease in its internal connectivity. The utility of a value hierarchy is apparent when considering the relationships between two values within this system: pluralism and connectedness. While the values of pluralism and connectedness are not *necessarily* in opposition, it is difficult to embrace new theoretical and methodological perspectives while simultaneously forming connections between these perspectives and preexisting ones. Hence, while pluralism and connectedness may both be worthwhile values, the reality is that as one of these values takes on increased significance, it often does so at the expense of the other.

The 1990s and Beyond

Economic expansion and dissatisfaction with the status quo helped open the door for pluralism and social relevance in the 1980s. This section illustrates how value structures are likely to shift in the 1990s, depending upon which mechanism (economics or status quo) is emphasized.

Reactions to Existing Value Structures

The current status of the discipline includes scholars from a multiplicity of theoretical perspectives, using disparate methodological means to achieve diverse intellectual ends (Barnett & Danowski, 1992). This diversity may leave some scholars adrift, feeling that they do not really understand what communication is, what communication scholars do, and more to the point, what it is that communication scholars do *not* study. A natural reaction to pluralism is a yearning for connections. Now that the discipline has opened its doors to many perspectives, some members may want to close the doors and answer the question "Who are we?"

Reactions to pluralism may take two forms. Barnett and Danowski (1992), for example, argue that the field should be cut in two, separating the humanists from the social scientists. Burgoon (1989) also calls for a "divorce," primarily between speech communication and communication sciences. Others, however, call for increased connectivity and increased ties among the discipline's various subparts. Berger (1992), for example, suggests that scholars should define a "relatively small set of overarching questions" around which to organize their efforts.

Whether the argument concerns dividing the discipline or forging new links, *connectedness* is a value currently generating considerable debate. Related to the desire for connections is the reemergence of distinctiveness as a highly salient issue. The field is now a loosely held together amalgamation of various perspectives, located in various types of university departments (e.g., journalism, speech, radio-TV-film, mass communication). Consequently, it is now more difficult to know who we are than perhaps at any point in the past.

A renewed emphasis on creativity is also apparent. The resurgence of creativity debates (e.g., a recent "Chautauqua" in *Communication Monographs,* 1992) may be a response to the emphasis on social relevance in the 1980s. Research that is relevant only to some particular context or social phenomenon (i.e., in keeping with the value of social relevance) is often at odds with the the norms associated with creating useful theory, such as keeping research generalizable to a diversity of situations and perspectives.

On the other hand, the renewed emphasis on creativity may be an outcome of increased connectedness. Forming new connections between various subgroups can lead to new ways of considering old problems (Rogers & Reardon, 1988). Among doctoral students we see a burgeoning interest in integrating seemingly incompatible perspectives to create new ways of understanding communication (e.g., uniting critical theory with empirical methods).

If connectivity and distinctiveness become more salient, social relevance and pluralism may take on less significance. As noted above, it is difficult for a discipline to embrace new perspectives as it attempts to understand and form connections among preexisting ones.

In reaction to the social relevance and pluralistic perspective of the 1980s, new voices are crying out for severing the ties that bind, or alternatively pleading for increasing the bridges and pathways between interest groups. Either way, there is likely to be a renewed interest in defining the discipline for its own members and for the outside world.

Reactions to Economic Pressures

Value shifts are influenced not only by a discipline's response to a previously established value system, but also by economic realities. Distinctiveness and social relevance are two values that are particularly sensitive to economic conditions.

While most academic institutions in the United States experienced economic growth or stability in the 1980s, the 1990s have been characterized by contracting academic resources. Distinctiveness becomes most salient during economic downswings, when resources contract and universities eliminate departments. Communication departments are particularly vulnerable to elimination, since their relative youth and multidisciplinary nature often lead to the perception that such departments lack distinctiveness. If university officials perceive that communication departments cannot offer something different from other academic disciplines, what then is the justification for their continued existence (Mader et al., 1985)?

The ability of communication scientists to perform socially relevant teaching and research may also serve as a justification for their economic existence. Thompson (1992) recently articulated a shift in priorities for federal and state funding agencies away from basic to applied research. To attract sufficient funding in this new era, communication scientists may need to emphasize research that is directly applicable to present social problems. For example, research that capitalizes on communication scientists' unique ability to understand the forces that shape our social world (e.g., information technologies) may become an increasingly important revenue source.

To remain economically solvent, communication departments that previously splintered into independent entities may be forced to unify their resources. Connectivity thus becomes a salient value to communication scientists who must jointly explore their commonalities in order to survive.

In summary, the discipline's response to economic forces and its reactions to the previously established value system will, in large part, dictate the values to which scholars devote much of their attention, as well as the values they are forced to consider as lesser priorities. Economic conditions and dissatisfaction with the status quo may result in a renewed emphasis on distinctiveness and connectedness and a declining interest in pluralism. Thus, in the future we expect to find that distinctiveness will once again dominate our value debates.

The relative importance of other values depends upon which factor dominates. For example, social relevance might remain an important value (in response to economic scarcity) or it may become less attractive (in reaction to a previously established value system). Consider how these factors might also effect creativity. If a backlash against social relevance predominates, creativity should continue to generate much debate. Creativity may not be a salient value, however, for scholars preoccupied with economic survival.

The Ideal Value Hierarchy for the Future

This essay began by proposing that past discussions of discipline-level values have been limited by the tendency to consider values in isolation, rather than as part of hierarchically organized systems. We end with a proposed value hierarchy for the future.

From our vantage point, an ideal value hierarchy is exemplified by the values of the Chicago School of sociology in the 1930s. In the Chicago

School, scholars from multiple perspectives worked together to generate an extraordinary amount of integrated, original communication theory. Through intense observations of their surroundings, scholars such as Robert Park sought to understand how communication influenced (and was influenced by) relevant social problems. However, the ultimate goal of such research was not simply to understand any one particular social phenomenon, but to generate original theoretical conceptions that are as valid today as they were 60 years ago. In our vision of the future, then, communication scholars should be guided by an interaction between the core values of creativity and social relevance.

A fundamental difference between the Chicago School and communication scientists today centers on the value of connectedness. The Chicago scholars were divided by their multiple perspectives but united by their common social interests, whereas communication scholars today share neither common perspectives nor interests. Perhaps then, Berger's (1991) call for an overarching set of questions to organize the discipline's efforts around should be heeded.

Finally, consider the discipline's long-standing concern with distinctiveness. The Chicago model suggests that our discipline's obsession with this value is misplaced. None of the voluminous writings about the Chicago School (e.g., Delia, 1987; Rogers, in press) suggest that distinctiveness was ever an issue for debate. Yet, guided by the values of creativity and social relevance, they generated a truly distinctive approach to communications (Bulmer, 1984).

As the discipline moves toward the 21st century, we need to recreate the excitement and interest in communication processes evident in a bygone era. We thus call for the return of creativity and social relevance as our guiding values.

References

Barnett, G. A., & Danowski, J. A. (1992). The structure of communication: A network analysis of the International Communication Association. *Human Communication Research, 19*(2), 264–285.

Berger, C. R. (1991). Communication theories and other curios. *Communication Monographs, 58,* 101–113.

Berger, C. R. (1992). Curiouser and curiouser curios. *Communication Monographs, 59,* 101–107.

Bochner, A. P., & Eisenberg, E. M. (1985). Legitimizing speech communication: An examination of coherence and cohesion in the development of the discipline. In T. H. Benson (Ed.), *Speech communication in the 20th century* (pp. 299–322). Carbondale: Southern Illinois University Press.

Bulmer, M. (1984). *The Chicago School of sociology: Institutionalization, diversity, and the rise of sociological research.* Chicago: University of Chicago Press.

Burgoon, M. (1989). Instruction about communication: On divorcing Dame Speech. *Communication Education, 38,* 303–308.

Delia, J. (1987). Communication research: A history. In C. R. Berger & S. Chaffee (Eds.), *Handbook of communication science*. Beverly Hills, CA: Sage.

Gitlin, T. (1978). Media sociology: The dominant paradigm. *Theory and Society, 2,* 205–253.

Katz, E. (1983). The return of the humanities and sociology. *Journal of Communication, 33*(3), 51–52.

Mader, T. F., Rosenfield, L. W., & Mader, D. C. (1985). The rise and fall of departments. In T. H. Benson (Ed.), *Speech communication in the 20th century* (pp. 322–340). Carbondale: Southern Illinois Press.

Purcell, W. H. (1992). Are there so few communication theories? *Communication Monographs, 59,* 94–97.

Rakow, L. F. (1992). Some good news–bad news about a culture-centered paradigm. In S. Deetz (Ed.), *Communication yearbook 15* (pp. 47–57). Beverly Hills, CA: Sage.

Rogers, E. M. (in press). *The history of communication*. New York: Free Press.

Rogers, E. M., & Reardon, K. K. (1988). Interpersonal versus mass media communication: A false dichotomy. *Human Communication Research, 15,* 284–303.

Rokeach, M. (1973). *The nature of human values*. New York: The Free Press.

Smythe, D. W., & Van Dinh, T. (1983). On critical and administrative research: A new critical analysis. *Journal of Communication, 33*(3), 117–127.

Thompson, D. (1992, November 23). Science's big shift. *Time,* pp. 34–35.

Wiemann, J. M., Hawkins, R. P., & Pingree, S. (1988) Fragmentation in the field—and the movement toward integration in communication science. *Human Communication Research, 15*(2), 304–310.

Argument for a Durkheimian Theory of the Communicative

by Eric W. Rothenbuhler, University of Iowa

My hope and expectation for the future of the field is that it be wildly pluralistic, broader and deeper, and more diverse than it is now. Such pluralism must be based on a set of individual voices. So I offer one vision of a future communication study, in the hope that it will be only one of the programs that make up the field, only one flavor in our stew.

Communication theory must come to terms with the full range of the communicative as an abstracted element of social life. Communication studies may have been founded out of concern for the frustrations and achievements of people as they set out *to communicate*. But the most general category of the subject matter of our field is *the communicative,* all that goes with the capacities of communication. This abstraction of a general category depends on an analytic framework, and together they define the disciplinary logic of our field. The communicative can be set alongside the biological, psychological, economic, sociological, political, cultural, historical, aesthetic, and others as more or less distinct frameworks for the analysis of a given subject matter. Just as there can be political or psychological analyses of the problem "to communicate," so can there be communicative analyses of psychology and politics.

Once we abstract the communicative and begin to explore for the limits of its generality, we turn our theories of communication into general social theories, and two things happen: Theories of communication become theories that are communicative in concepts and logics rather than by dint of their reference to a concrete subject matter, and our theories address, or can be used to address, the whole of the social world. Along the way we will inevitably encounter previous travelers. It will be our respon-

Eric W. Rothenbuhler is an associate professor in the Department of Communication Studies, University of Iowa. The author thanks Thom McCain and Jane Frazer, codirectors of the Center for Advanced Study in Telecommunication, where he was scholar in residence January through March 1992. During that period he was able to reread a substantial piece of Durkheim's work and update his bibliography of the secondary literature. He thanks also Sam Becker who helped edit this essay; Greg Shepherd and John Peters, with whom intellectual conversations are a frequent pleasure and a primary influence on his work; and Mark Levy, Michael Gurevitch, and the *Journal* staff.

sibility to come to terms with the writings of those who are remembered for asking questions of general social theory before we began to.

So in addition to pushing communication theory forward toward the abstraction of the communicative as a ubiquitous feature of social life and the development and generalization of a theory of the communicative as a framework for the analysis of social life, I advocate what might appear a kind of "pushing backward" of the field: to explore more deeply the possibilities of our intellectual inheritance from scholars we have not usually considered part of our canon.

There are many from whom we can draw. But given the large potential of his work to contribute to ours, it is particularly unfortunate that communication scholars have neglected Émile Durkheim.

Durkheim and Communication Studies

Durkheim's conception of sociology was not limited to the boundaries of our current sociology departments. It was a general social science conceived as the study of the form and function of social facts, wherever they appear. Social facts were defined as all those ways of feeling, thinking, and acting that are diffused throughout a society and function independently of the individual. This independence makes them conditions external to individual will. Social facts fall along a continuum from infrastructures through institutions to movements of moods and ideas that Durkheim called "social currents." Collective representations are ideal rather than material social facts (language, for example, is an infrastructure in the realm of ideal social facts). Durkheim and his students took their conception of sociology into a great diversity of fields, including what we would be more inclined to think of as anthropology, economics, linguistics, social history, and, as I am proposing here, communication.

The roles of collective representations in the action of individuals, and in the larger social processes that are of concern in communication studies, are part of Durkheimian general sociology. Indeed Durkheim's study of religion, Mauss's of magic and gift, Halbwachs's of memory, Simiand's of money, as well as many other studies of many other topics by members of Durkheim's school, were essentially studies of the forms and functions of communicative systems.

But in communication studies today Durkheim is hardly anywhere to be seen. Schudson (1986) is correct to point out that there is neo-Durkheimian work done in media studies, but it is mostly implicitly so. Durkheim is little discussed. The *International Encyclopedia of Communications* includes an entry for Durkheim. But in the short space available all the authors can do is assert that Durkheim's "concept of social reality assigns a central role to communication" (King & Cuzzort, 1989, p. 66).

A review of the *Social Sciences Citation Index* for the years 1966 to 1991 yields only 19 citations of Durkheim in communication journals, excluding my articles. Of these, none is a prolonged engagement of Durkheim's theoretical system. Though a few are careful applications of one or another of his ideas (the most popular appears to be the social functions of deviancy), most are only passing references, often citing him for historical interest only. The *Handbook of Rhetorical and Communication Theory* (Arnold & Bowers, 1984) does not include a single index or bibliography entry for Durkheim; the *Handbook of Communication Science* (Berger & Chaffee, 1987) includes only three references, all just passing examples of historical precedent. The only popular communication textbook I know of that mentions him is DeFleur and Ball-Rokeach's (1989) *Processes and Effects of Mass Communication*.

The work of Jim Carey provides a telling example of the extent to which Durkheimian theory is undervalued in communication studies. One of the most sensible and sensitive students of communication, Carey (1988) proposes our project be the use of communication

> *as a site . . . on which to engage the general question of social theory: How is it, through all sorts of change and diversity, through all sorts of conflicts and contradictions, that the miracle of social life is pulled off, that societies manage to produce and reproduce themselves? . . . [For] whatever the details of the production and reproduction of social life, it is through communication, through the intergraded relations of symbols and social structure, that societies, or at least those with which we are most familiar, are created, maintained, and transformed.* (p. 110)

Admittedly, Carey is antidisciplinary while Durkheim and I are both promoters of disciplinary thinking. But substantively, this is a strikingly Durkheimian project. Not only is the general question—how do societies reproduce themselves—one more thoroughly associated with Durkheim than any other member of the founding generation of sociologists, but Carey's answer to the question—through communication—is the answer Durkheim himself gave (though not in so many words). But Carey, who is one of the few communication scholars to cite Durkheim, usually distances himself, denying that his project is Durkheimian (but *contra,* see Carey, 1988, pp. 19, 23). Rather, Durkheim is associated with functionalism construed in only its vulgar forms, to which Carey's project is portrayed as an alternative. That portrayal is valid, but only because the functionalism considered is a kind of individualistic social psychology deriving from Radcliffe-Brown and Malinowski and not Durkheim's theory (cf. Carey, 1986; Rothenbuhler, 1987).

It is clear that Durkheim's legacy is absent in communication. Hinkle (1960) argues that American sociologists' acceptance of Durkheim was prepared by certain theoretical and methodological exigencies. Most prominent was the acceptance, from cultural anthropology, of a concept of culture as independent of the individual, and an understanding, from the pragmatists, of the importance of the symbol in social conduct. Both of these theoretical exigencies are also important in communication studies. What could communication studies be without having at its center a conception of symbolic activity? As regards culture, nothing has so revolutionized and revitalized communication studies in the last 20 years as the coming of cultural studies. The conception of culture as an independent force in social process and order has radically changed media studies, revitalized rhetorical studies, and is having an impact even in the areas of interpersonal, small group, and organizational communication studies.

If conceptions of the role of symbolization in social interaction and of culture as an independent force in social life are what prepare the ground for the growth of Durkheimian theory, then surely the field of communication studies is ready. Indeed, one wonders why it has not yet developed. As Hinkle (1960) pointed out more than 30 years ago: "Durkheim has become a central figure in a number of specialized fields. It seems likely that his impact could be observed in such specialties as the sociologies of . . . [among many others] communications" (p. 289).

Durkheimian Contributions to Communication Theory

Recent social theory places yet more emphasis on independent cultural orders and symbolization processes, bringing a concept of communication into discussions of social order, process, and the person. But Durkheim did it first, and we can learn from his work.

Durkheimian social order is a pattern across the actions, thoughts, and feelings of individuals. The social transcends the border of the individual body in symbolic form and in the form of languages, logics, competencies, value and belief systems, ideologies, shared sentiments, morals, mores, and so on. The constituents of order exist as much inside the individual as outside. But the person is not wholly socially constituted. Persons are also individual beings with their own bodies, characters, and personal activities. Similarly, the world is not wholly socially constituted and certainly not smoothly so; it includes a wealth of unique positions, experiences, and individuating activities. Social processes of continuity and change, then, are outcomes of tensions, adjustments, displacements, rhythms, and shifts in the balance and nature of sociating and individuating phenomena.

From the Durkheimian perspective two things become obvious about communication that are not treated so in our literature. First, communication is an activity that must always be socially structured. I do not mean this in the small sense of communication being interpersonal, but in the large sense of its only being *possible* through systems of language, logic, and presumption that are not in any sense the property of individuals. But simultaneously, to the extent it is authored on the one hand and interpreted on the other, communication is always also an individuating activity (see Rothenbuhler, 1987; Shepherd & Rothenbuhler, 1991). One important task of the scholar is to understand the balance of social structuring and individuating forces in any given case. The Durkheimian perspective, thus, is more subtle, flexible, and realistic than either the emphasis on the personal in so much of communication studies, or the emphasis on social control in so much of the rest.

The second point that the Durkheimian perspective helps us to recognize as obvious is that in being a socially structured but individuating symbolic activity, communication as we usually think of it—talking, listening, writing, reading, recording, filming, broadcasting, viewing—is only one part of a larger category. Durkheim may have thought of it as collective representations, but we can see that that category is the communicative. Wherever activities or artifacts have symbolic values that articulate individuals into positions vis-à-vis each other or their collectivities, the communicative is present. We need to be studying this whole field and we need a theory to guide us in the work. Durkheim's will do for starters; eventually we will alter it until it becomes our own.

Durkheimian communication theory would be both a new way to think about our intellectual roots and a new way to think of our future in communication studies. In either form it has great potential benefit to the field. By seeing that our current work has important intellectual continuities with Durkheim and the beginnings of modern social science, new problematics become apparent that will increase the scope of our concern. This insight can lead us to explore the roles of communication everywhere in social life and not restrict ourselves arbitrarily to the traditional problematics of the field.

But more than this empirical problem of exploring the role of communication wherever it occurs, perceiving ourselves in the Durkheimian tradition facilitates a shift to more abstract levels of analysis and to generalizing our frame of interpretation. A Durkheimian communication theory, ultimately, would conceive society as a communication system, history as a symbolic relation of present and past, politics as a contest of expressions, and economics as an elaborate system of symbolic exchanges. The future of communication theory, as plotted by a Durkheimian perspective, would be a successive elaboration of every corner of social life as communicatively founded. We could conceive of no more ambitious future.

References

Arnold, C. C., & Bowers, J. W. (Eds.). (1984). *Handbook of rhetorical and communication theory*. Boston: Allyn and Bacon.

Berger, C. R., & Chaffee, S. H. (Eds.). (1987). *Handbook of communication science*. Newbury Park, CA: Sage.

Carey, J. (1986). Overcoming resistance to cultural studies. *Mass Communication Review Yearbook, 5,* 27–40.

Carey, J. (1988). *Communication as culture: Essays on media and society*. Boston: Unwin Hyman.

DeFleur, M. L., & Ball-Rokeach, S. J. (1989). *Theories of mass communication* (5th ed.). New York: Longman.

Hinkle, R. C. (1960). Durkheim in American sociology. In K. H. Wolff (Ed.), *Émile Durkheim, 1858–1917: A collection of essays, with translations and a bibliography* (pp. 267–295). Columbus: Ohio State University Press.

King, E. W., & Cuzzort, R. P. (1989). Durkheim, Émile (1858–1917). In *International encyclopedia of communications,* Vol. 2 (pp. 66–68). New York: Oxford University Press.

Rothenbuhler, E. W. (1987). Neofunctionalism for mass communication theory. *Mass Communication Review Yearbook, 6,* 66–89.

Schudson, M. (1986). The menu of media research. In S. J. Ball-Rokeach & M. G. Cantor (Eds.), *Media, audience, and social structure* (pp. 43–48). Newbury Park, CA: Sage.

Shepherd, G. J., & Rothenbuhler, E. W. (1991). A synthetic perspective on goals and discourse. In K. Tracy (Ed.), *Understanding face-to-face interaction: Issues linking goals and discourse* (pp. 189–203). Hillsdale, NJ: Lawrence Erlbaum Associates.

Implications of Public Relations for Other Domains of Communication

by James E. Grunig, University of Maryland

In the 1960s and 1970s, research on public relations consisted mostly of biographies of leading practitioners, case studies of public relations practice, and some highly applied studies—such as research on the factors leading to the acceptance of news releases or the proportion of content in the news media that comes from public relations sources. In addition, public relations educators and practitioners considered much of communication research to be relevant to their problems—although they did little of this research themselves. A recent bibliometric study by Pasadeos and Renfro (1992), however, showed that public relations authors increasingly are citing each other and that they are citing public relations scholars increasingly more than practitioners.

In the last decade, public relations researchers have made remarkable progress, and their work has reached the point where a general theory of public relations is in sight. Scholars from other communication disciplines now could learn much from public relations—especially about publics; symmetrical communication; the management of communication; the effect of organizational structure, environment, culture, and power on communication behaviors; and the impact of gender and diversity on the practice of professional communication. Public relations research has progressed through three levels of problems. The *micro* (individual) level refers to the planning and evaluation of individual public relations programs. The *meso* (group) level refers to how public relations departments are organized and managed. The *macro* (environmental) level refers to explanations of public relations behavior and the relationship of public relations to organizational effectiveness.

The Micro Level: Individual Public Relations Programs

The micro level was the logical place for researchers to start because this level deals with the applied concerns of public relations practitioners—the

James E. Grunig is a professor in the College of Journalism at the University of Maryland.

Copyright © 1993 *Journal of Communication* 43(3), Summer. 0021-9916/93/$5.00

planning and evaluation of communication programs. The micro level was also an easy place to borrow theories from other domains of communication science.

The Effects of Public Relations

Public relations educators and researchers have used or contributed to the research on persuasion or the effects of the media because they believed originally that these were the most important effects of public relations programs. Most of the researchers did not think of their research as public relations research, however, preferring instead to think of themselves as more generic communication researchers. The communication researchers whose work came closest to public relations were those studying the effectiveness of public information campaigns (e.g., Rice & Atkin, 1989; Salmon, 1989).

Many public relations researchers have broken away from the idea that messages, campaigns, or the mass media must persuade (change attitudes or behaviors) to be effective. They have conceptualized public relations as a process of symmetrical dialogue, often basing their research on Chaffee and McLeod's (1968) concept of coorientation (e.g., Broom, 1977; Culbertson, 1989; Pearson, 1989)—which isolated several cognitive effects of public relations as well as effects on attitudes and behaviors.

Segmentation of Publics

As public relations scholars began to study the coorientational relationships of organizations and publics, they began to break from the assumption that public relations is mass communication—and began to define the term *public* in public relations. *Publics* arise in response to the consequences that an organization has on people as it pursues its mission—in communities, governments, employees, shareholders, consumers, and activist groups.

Publics organize around problems and seek out organizations that create the problems. They make issues of the problems by seeking information, seeking redress of grievances, pressuring the organization, or seeking government regulation. As publics move from being latent to active, organizations have little choice other than to communicate with them, whereas organizations can choose to ignore markets if they wish (J. Grunig & Hunt, 1984, chap. 7).

Dewey (1927) and Blumer (1946) developed classic theories of publics, and Cobb and Elder (1972) described types of publics in a theory of public opinion. Communication researchers, however, have done little to define and explain the behavior of publics. Price (1992) has called attention to this oversight, although he seems unaware of the research tradition on publics in public relations. Strategic management researchers have used the related concept of *stakeholders,* which Freeman (1984) defined as "any individual or group who can affect or is affected by the actions, decisions, policies, practices, or goals of the organization" (p. 25).

All of these elements from theories of mass communication, public opinion, and strategic management are relevant to a theory of publics, but none of them constitute such a theory alone. I have incorporated—or explained—most of these concepts in a situational theory of publics (for reviews, see J. Grunig, 1989; J. Grunig & Repper, 1992). The situational theory works from the assumption that a population can be segmented into publics by the extent to which they passively or actively communicate about a problem and the extent to which they actively behave in a way that supports or constrains the organization's pursuit of its mission. Publics are more likely to be active when the people who constitute them perceive that what an organization does *involves them* (*level of involvement*), that the consequences of what an organization does is a *problem* (*problem recognition*), and that they are *not constrained* from doing something about the problem (*constraint recognition*).

Active publics are important targets for public relations programs because they are most likely to be aware of and concerned with what the organization is doing. In addition, if an organization does not communicate with active publics and attempt to manage conflict, those publics can become activist groups that can limit the ability of an organization to accomplish its goals—either directly through protest, boycott, or strike or indirectly through government regulation.

I have used the situational theory of publics in research on communication and development in Third World countries (e.g., J. Grunig, 1969; J. Grunig, 1971; L. Grunig & J. Grunig, 1990) and to integrate uses and gratifications theories (J. Grunig, 1979). The theory also has much to offer mass communication scholars who are interested in media audiences, public opinion, and uses and gratifications.

Strategic Management of Public Relations

The situational theory of publics has been most valuable when it is linked to theories of strategic management developed by management scholars and organizational sociologists. Organizations use strategic management to relate their missions to their environments. Public relations contributes to overall strategic management by building relationships with publics that support the mission of the organization or that can divert it from its mission. Organizations plan public relations programs strategically, therefore, when they identify the publics that are most likely to limit or enhance their ability to pursue the mission of the organization and design communication programs that help the organization manage its interdependence with these strategic publics.

In contrast to this strategic approach, most organizations carry out the same public relations programs year after year without stopping to determine whether they continue to communicate with the most important publics. Dozier and L. Grunig (1992) have pointed out that at some point in their history, most organizations probably develop their public relations programs strategically—that is, the presence of a strategic public probably provides the motivation for initiating public relations programs. As time passes, however, organizations forget the initial reason for the programs and continue communication programs for publics that no longer are strategic. Public relations then becomes routine and ineffective because it does little to help organizations adapt to dynamic environments.

J. Grunig and Repper (1992) have developed a theory that links strategic management of the public relations function to issues management and to the overall strategic management of the organization. That theory is a component of a general theory of excellence in public relations that guides a multiyear study funded by the Research Foundation of the International Association of Business Communicators (IABC). The IABC research team, which I head, has concluded that involvement in strategic management is one of 17 characteristics of excellent public relations departments. The IABC research team has studied 300 organizations in depth in the United States, Canada, and the United Kingdom to determine the extent to which each of the 17 characteristics of excellence are related to organizational effectiveness. Initial results (J. Grunig, L. Grunig, Dozier, Ehling, Repper, & White, 1991) confirm strongly that public relations is most likely to make organizations more effective when it is part of strategic management.

Practitioners in few other communication domains think of communication strategically. Most could benefit, however, by doing so. Managers of mass media organizations, for example, should look at their role as providing the strategic information needed by key stakeholders in a community or political system.

The Meso (Managerial) Level

Conceptualizing public relations as a management function allows scholars to link the microlevel theories of communication planning and evaluation to the meso level of an organization—the level of a group or department. At the meso level, public relations researchers left the familiar confines of communication theory and used theories of organizations and management to help construct original theories of public relations. Communication scholars, even those studying organizational communication, had not looked at the consequences of organizational structures and roles on the communication activities of organizations until public relations researchers began to do so (e.g., J. Grunig, 1976, 1992a).

At the meso level, researchers have asked how the communication function must be managed in an organization for public relations to be excellent—that is, to contribute to organizational effectiveness. The IABC research team has identified eight meso-level characteristics of excellent public relations (Grunig, 1992b). Three are highlighted here because they have been developed from among the most extensive programs of research on public relations.

Models of Public Relations

I began a program of research in 1976 to identify typical ways that organizations practice public relations and to explain why they practice it differently (J. Grunig, 1976)—typical patterns I now call *models* of public relations. J. Grunig and Hunt (1984) first identified four models in the history of public relations. The *press agentry* model describes public relations programs whose purpose is getting favorable publicity for an organization in any way possible. The *public information* model sees public relations only as the dissemination of information—although the information is more truthful than that produced by press agents. Both of these models are one-way: Neither uses research and strategic planning. In addition, their purpose is asymmetrical—to change the behavior of publics but not of the organization. The other two models are more sophisticated and professional. The *two-way asymmetrical* model uses research to develop messages that are likely to persuade strategic publics to behave as the organization wants. The fourth model, the *two-way symmetrical,* features public relations based on research that uses communication to manage conflict with strategic publics.

Extensive research (J. Grunig & L. Grunig, 1989, 1992) shows that organizations do practice public relations as these models describe, that different types of organizations typically practice different models (e.g., government agencies the public information model, corporations the two-way asymmetrical model), and that the two-way symmetrical model is the model most likely to make an organization effective. Theorists (J. Grunig & L. Grunig, 1992; Murphy, 1991) and researchers (J. Grunig et al., 1991) have found that effective organizations often mix the two-way models. Organizations have mixed motives and must balance attempts to persuade publics with attempts to negotiate with them. Generally, however, excellent public relations practitioners move their organizations toward the symmetrical end of a continuum with the two-way asymmetrical model on one end and the two-way symmetrical model on the other.

Finally, critical essays on the ethics of public relations (Pearson, 1989) have made cogent arguments that the symmetrical model is more ethical than the other models because it does not require an organization to know whether it is right or wrong on an issue. Rather, symmetrical public relations opens the question of right or wrong to negotiation.

Similar theories of symmetrical communication exist elsewhere in the communication discipline—such as in rhetoric, interpersonal communication, conflict resolution, organizational communication, communication and development, and to some extent in mass communication. However, research on the relative effectiveness of symmetrical communication, on its ethics, and its relative effectiveness is most extensive in public relations, and the other domains could profit from studying that research.

Gender Differences in Public Relations

Extensive research has shown that female practitioners fill the managerial role in public relations less often than males—primarily but not exclusively because of discrimination against women in management (Hon, L. Grunig, & Dozier, 1992; Wright, L. Grunig, Springston, & Toth, 1991). The public relations profession now has a majority of females practicing in the United States, and over two-thirds of the students studying public relations there now are women. Women, therefore, enhance the potential of the public relations department because—having formally studied public relations—they are more likely to have the knowledge to practice public relations strategically and symmetrically. Keeping them out of the managerial role thus limits the excellence of most public relations departments.

Although much of the research on women in public relations began with differences in roles, a large and growing community of scholars has begun to use feminist theory to criticize public relations (e.g., Creedon, 1991; L. Grunig, 1989; Rakow, 1989; Toth, 1988). They argue, in essence, that current public relations practice is based on male models of competition, hierarchy, and dominance. Public relations would be more ethical, responsible, and effective, they add, if instead it were based on female models of cooperation and equity.

Researchers in other communication domains also have studied gender and communication. Because of the severity of gender problems in public relations and the extensiveness of the research there, these scholars would learn much from that research.

The Macro Level: What Makes Excellent Public Relations Possible

At the macro level, researchers have looked at the conditions in and around organizations that explain why some organizations practice public relations in an excellent way and others do not. At this level, again, public relations researchers have used variables rarely used elsewhere in the discipline of communication. At first, researchers thought that *organizational structure* would explain how organizations practice public relations: They believed that decentralized structures in which employees at all levels participate in decisions should facilitate excellent public relations (J. Grunig, 1976). In addition, research demonstrated that symmetrical systems of internal communication are part of these decentralized structures (see J. Grunig, 1992a, for a review).

Structure explained some of the variance in the practice of public relations but left most of it unexplained. Thus, researchers next studied the extent to which the presence of complex, turbulent environments stimulates excellence in public relations (Schneider [aka L. Grunig], 1985; J. Grunig, 1984). After research showed that the environment explained only a small portion of public relations behavior, researchers turned to the power-control perspective in organizational sociology (e.g., Robbins, 1990). That perspective maintains that organizations behave as they do because the people with the most power in the organization—the dominant coalition—choose to do so. Initial results from the IABC study (J. Grunig et al., 1991) have confirmed the explanatory value of power for public relations. Public relations cannot be effective unless the senior public relations person has power to affect organizational decisions, either formally or informally.

Finally, research suggests that both the characteristics of excellent public relations and the power of the senior public relations person are affected strongly by the culture of the organization (Sriramesh, J. Grunig & Buffington, 1992) and that organizational culture, in turn, is affected by the culture of a country or region of a country (Sriramesh & White, 1992). Excellent public relations departments exist more often in a *participative* culture than an *authoritarian* culture, as do the other macrolevel characteristics that bring about excellent public relations departments.

Effects of Public Relations: An Integrative Theory

The last element of a general theory of public relations, then, is an integrating theory that explains the value of excellent public relations for an organization. Research based on middle-range theories from the micro and meso levels confirms that communication programs based on strategic, managerial, symmetrical, and diversity principles most often achieve their microlevel communication objectives. Achieving these micro communication objectives helps organizations achieve their missions and goals at the macro level. The literature on organizational effectiveness (e.g., Robbins, 1990) shows that effective organizations choose appropriate goals and then achieve them. When public relations helps the organization build relationships, it saves the organization money by reducing the costs of litigation, regulation, legislation, pressure campaigns, boycotts, or lost revenue that result from bad relationships with publics— publics that often become activist groups. Good public relations also helps make money by cultivating relationships with donors, consumers, shareholders, and legislators.

This theoretical connection between excellent public relations and organizational effectiveness therefore shows why organizations benefit from public relations. With that integrating principle, we can explain why middle-range theories of strategic management, symmetrical communication, gender diversity, organizational cultures, power, structures, and environment are integral parts of excellent public relations.

This general theory represents, I believe, an outstanding accomplishment of public relations researchers—a grand theory rarely found in communication and one that other scholars in the discipline should emulate.

References

Blumer, H. (1946). The mass, the public, and public opinion. In B. Berelson and M. Janowitz (Eds.), *Reader in public opinion and communication* (2nd ed., 1966, pp. 43–50). New York: Free Press.

Broom, G. M. (1977). Coorientational measurement of public issues. *Public Relations Review, 3*(4), 110–119.

Chaffee, S. H., & McLeod, J. M. (1968). Sensitization in panel design: A coorientational experiment. *Journalism Quarterly, 54,* 661–669.

Cobb, R. W., & Elder, C. D. (1972). *Participation in American politics: The dynamics of agenda building.* Baltimore, MD: Johns Hopkins University Press.

Creedon, P. J. (1991). Public relations and "women's work": Toward a feminist analysis of public relations roles. *Public Relations Research Annual, 3,* 67–84.

Culbertson, H. M. (1989). Breadth of perspective: An important concept for public relations. *Public Relations Research Annual, 1,* 3–25.

Dewey, J. (1927). *The public and its problems.* Chicago: Swallow.

Dozier, D. M., & Grunig, L. A. (1992). The organization of the public relations function. In J. E. Grunig (Ed.), *Excellence in public relations and communication management* (pp. 395–418). Hillsdale, NJ: Lawrence Erlbaum Associates.

Freeman, F. E. (1984). *Strategic management: A stakeholder approach.* Boston: Pitman.

Grunig, J. E. (1969). Information and decision making in economic development. *Journalism Quarterly, 46,* 565–575.

Grunig, J. E. (1971). Communication and the economic decision-making processes of Colombian peasants. *Economic Development and Cultural Change, 19,* 580–597.

Grunig, J. E. (1976). Organizations and publics relations: Testing a communication theory. *Journalism Monographs, No. 46.*

Grunig, J. E. (1979). The budgets, level of involvement, and use of the mass media. *Journalism Quarterly, 56,* 248–261.

Grunig, J. E. (1984). Organizations, environments, and models of public relations. *Public Relations Research & Education, 1*(1), 6–29.

Grunig, J. E. (1989). Publics, audiences, and market segments: Segmentation principles for campaigns. In C. T. Salmon (Ed.), *Information campaigns: Balancing social values and social change* (pp. 199–228). Newbury Park, CA: Sage.

Grunig, J. E. (1992a). Symmetrical systems of internal communication. In J. E. Grunig (Ed.), *Excellence in public relations and communication management* (pp. 531–576). Hillsdale, NJ: Lawrence Erlbaum Associates.

Grunig, J. E. (1992b). Communication, public relations, and effective organizations: An overview of the book. In J. E. Grunig (Ed.), *Excellence in public relations and communication management* (pp. 1–31). Hillsdale, NJ: Lawrence Erlbaum Associates.

Grunig, J. E., & Grunig, L. A. (1989). Toward a theory of the public relations behavior of organizations: Review of a program of research. *Public Relations Research Annual, 1,* 27–66.

Grunig, J. E., & Grunig, L. A. (1992). Models of public relations and communication. In J. E. Grunig (Ed.). *Excellence in public relations and communication management* (pp. 285–326). Hillsdale, NJ: Lawrence Erlbaum Associates.

Grunig, J. E., Grunig, L. A., Dozier, D. M., Ehling, W. P., Repper, F., & White, J. (1991, September). *Excellence in public relations and communication management: Initial data analysis.* San Francisco, CA: IABC Research Foundation.

Grunig, J. E., & Hunt, T. (1984). *Managing public relations.* Fort Worth, TX: Holt, Rinehart and Winston.

Grunig, J. E., & Repper, F. (1992). Strategic management, publics, and issues. In J. E. Grunig (Ed.). *Excellence in public relations and communication management* (pp. 117–158). Hillsdale, NJ: Lawrence Erlbaum Associates.

Grunig, L. A. (1989, June). *Applications of feminist scholarship to public relations: Displacing the male models.* Paper presented to the National Women's Studies Association, Towson, MD.

Grunig, L. A., & Grunig, J. E. (1990). Strategies for communicating on innovative management with receptive individuals in development organizations. In M. Mtewa (Ed.), *International science and technology: Philosophy, theory and policy* (pp. 118–131). New York: St. Martin's.

Hon, L. C., Grunig, L. A., & Dozier, D. M. (1992). Women in public relations: Problems and opportunities. In J. E. Grunig (Ed.). *Excellence in public relations and communication management* (pp. 419–438). Hillsdale, NJ: Lawrence Erlbaum Associates.

Murphy, P. (1991). The limits of symmetry: A game theory approach to symmetric and asymmetric public relations. *Public Relations Research Annual, 3,* 115–132.

Pasadeos, Y., & Renfro, B. (1992). A bibliometric analysis of public relations research. *Journal of Public Relations Research, 4,* 167–187.

Pearson, R. (1989). *A theory of public relations ethics.* Unpublished doctoral dissertation, Ohio University, Athens, OH.

Price, V. (1992). *Public opinion.* Newbury Park, CA: Sage.

Rakow, L. F. (1989). From the feminization of public relations to the promise of feminism. In E. L. Toth & C. G. Cline (Eds.), *Beyond the velvet ghetto* (pp. 287–298). San Francisco, CA: International Association of Business Communicators.

Rice, R. E., & Atkin, C. K. (1989). *Public communication campaigns* (2nd ed.). Newbury Park, CA: Sage.

Robbins, S. P. (1990). *Organization theory: The structure and design of organizations.* Englewood Cliffs, NJ: Prentice-Hall.

Salmon, C. T. (Ed.) (1989). *Information campaigns: Balancing social values and social change.* Newbury Park, CA: Sage.

Schneider [aka Grunig], L. A. (1985). *Organizational structure, environmental niches, and public relations: The Hage-Hull typology of organizations as predictor of communication behavior.* Unpublished doctoral dissertation, University of Maryland, College Park, MD.

Sriramesh, K., Grunig, J. E., & Buffington, J. (1992). Corporate culture and public relations. In J. E. Grunig (Ed.), *Excellence in public relations and communication management* (pp. 577–598). Hillsdale, NJ: Lawrence Erlbaum Associates.

Sriramesh, K., & White, J. (1992). Societal culture and public relations. In J. E. Grunig (Ed.). *Excellence in public relations and communication management* (pp. 597–614). Hillsdale, NJ: Lawrence Erlbaum Associates.

Toth, E. L. (1988). Making peace with gender issues in public relations. *Public Relations Review, 14*(3), 36–47.

Wright, D. K., Grunig, L. A., Springston, J. K., & Toth, E. L. (1991). *Under the glass ceiling: An analysis of gender issues in American public relations.* New York: Public Relations Society of America Foundation.

Making a Difference in the Real World

by Robert K. Avery, University of Utah, and William F. Eadie, California State University, Northridge

Periodically, some of the field's senior scholars wonder about our status in both the academy and the real world (e.g., Becker, 1992; Berger, 1991; Berger & Chaffee, 1988; Burleson, 1992; Delia, 1979; Miller, 1981; Miller & Sunnafrank, 1984; Redding, 1985; Wartella, in press). Even the previous *Ferment in the Field* devoted several articles to this issue (Blumler, 1983; Miller, 1983; Rogers and Chaffee, 1983; Thayer, 1983; Tunstall, 1983). While a chronological reading of these essays shows that as a discipline our self-esteem has improved, the bottom line of even the most recent of these articles is that we are not yet influential "enough" as a field.

When the topic came up at the 1987 Salt Lake City meeting of the Western Speech Communication Association, psychologist Irwin Altman, at the time the academic vice president at the University of Utah, told the conventioneers not to worry. The same questions had been laid at the doorstep of psychology some years earlier. In response, a spokesman for the psychologists said that the most his discipline could say about its significance was that there was no evidence they had done any harm.

In retrospect, we view the indictment that "we haven't done any harm" as being a serious one, even when stated in good humor. The truth is, we think, that psychology *has* made an enormous positive impact on the collective mental health of contemporary society. Thousands of subscribers from all walks of life regularly read *Psychology Today,* applying the research findings and professional experiences reported therein to daily living. If a poll were taken of professional people over the age of 30, it is likely that a significant portion of those surveyed would say that psychology has influenced their lives positively. If the same poll were to substitute the word "communication" for "psychology," we would argue that a lot of the respondents would ask, "What's that?"

The discipline has long claimed deep historical roots and a foundation based on praxis (see Craig, 1990). However, those arguments do not seem to have impressed our colleagues in other disciplines terribly much. Nor

Robert K. Avery is a professor of communication at the University of Utah and founding editor of *Critical Studies in Mass Communication.* William F. Eadie is professor of speech communication at California State University, Northridge, and the first SCA editor of the *Journal of Applied Communication Research.*

have they impressed the public at large. Eadie's godson, a student at a selective liberal arts college, recently selected communication as his major. His father, an old friend and alumnus of that same college, was clearly skeptical of his son's choice, primarily because there had been no communication department at that college when he attended, but secondarily because he was unsure about what his son would learn from such study. Neither the public nor those who fund the various aspects of higher education are yet convinced of our value. There would be a public outcry if a financially troubled university would attempt to eliminate its philosophy department, even if that department served few majors and had a mediocre faculty at best. On the other hand, public protest could be minimal if that same university attempted to eliminate a communication program that served large numbers of majors and had a good, if not excellent, faculty.

Talking About Communication

We believe that the discipline has not progressed for two reasons: First, we talk much more to each other than to those outside our field; and second, we have not clearly defined what we are about to ourselves, our colleagues, our students, and the general public.

While it is relatively easy to see that we enjoy talking to each other much more than to our academic colleagues or the public (the fact that 10 years after *Ferment in the Field* we are engaged in yet another intradisciplinary debate of this kind supports that assertion), a bit of analysis of some publications should help make the point more completely. The authors have both edited scholarly journals: Avery was the founding editor of *Critical Studies in Mass Communication* (*CSMC*) and Eadie was the first editor of the *Journal of Applied Communication Research* (*JACR*) once it was acquired by the Speech Communication Association (SCA). These journals have substantially different missions, yet in both cases there were hopes that the journals would have an impact on the public at large, if only indirectly.

Avery remembers well the day he presented a proposal to Alan Wurtzel at ABC for a grant that would help launch *CSMC*. Wurtzel was highly supportive and said that the ABC funding would be forthcoming. But Avery felt that in good conscience he could not leave Wurtzel's office without letting him know what the content of *CSMC* would undoubtedly be. Avery remembers telling Wurtzel, "Alan, you need to know that this journal is going to be highly critical of American broadcasting as a hegemonic force in modern capitalist society. Writers *will* take ABC and the other networks to task."

Wurtzel smiled and then laughingly said, "Bob, you don't really expect anything written in your journal to make any difference here. Don't worry about it."

Examining the last several years of *CSMC,* it is clear that while quality is high, applicability is a mixed bag. Maybe a quarter of the articles published have potential applicability, mainly because their authors have focused on social or personal issues. The largest number of those articles can be categorized as focusing on multiculturalism, while perceptions of television news runs a close second. *CSMC* has published titles such as "Out of Work and On the Air: Television News of Unemployment," "Chernobyl: The Packaging of Transnational Ecological Disaster," "Foucault and Dr. Ruth," and "Covering the Homeless: The Joyce Brown Story." But indeed, it has also published "Work Songs, Hegemony, and Illusions of the Self," and "Hegemonic Masculinity in *thirtysomething*." To its credit, *CSMC* was featured as a key journal in the area of cultural studies in a *Chronicle of Higher Education* article, and it is available for sale to the general public at Borders Bookstores. Whether or not the articles are immediately usable, the journal has made an impact.

JACR's mission is specifically oriented to publishing usable research, and in the first two years of its life as an SCA journal it has published articles on a variety of topics, with the largest number being on communication aspects of sexual harassment, communication about environmental risks, and treatment of communication apprehension. However, many of these articles have only indirect application at best, and are more interesting to colleagues in the field than to the public at large. A special section on sexual harassment focused on analyses of sexual harassment stories from persons *within* the field of communication, and article titles have included "Coping With Writing Apprehension," "Translation as Problematic Discourse in Organizations," and "An Expectancy Theory Explanation of the Effectiveness of Political Attack Television Spots: A Case Study." Eadie has made a practice of writing press releases about research appearing in *JACR*. His strategy has been to write short "fillers" for use in local newspapers. So far, however, the only publication he is aware of that publishes these is *Spectra,* SCA's internal newsletter.

When SCA acquired *JACR* there were a variety of ideas for what it would become, just as there have been a variety of notions about what constitutes "applied communication research" (cf. Eadie, 1990). Some wanted a publication along the lines of what might be called *Communication Today,* but others remembered what happened to a publication along those same lines called *Today's Speech*.

Reaching the Public

In 1953, Professor Robert T. Oliver convinced the Speech Association of the Eastern States to begin publishing a journal aimed at teachers and the general public. Oliver, who had worked with Toastmasters International to develop a self-help public speaking program, was convinced that the field needed to reach beyond its boundaries. Initial articles in *Today's*

Speech were short and clearly aimed at the general reader. Early titles included, "How to Prepare a Talk," "Are You Saying It Right?" and "Getting the Most Out of Those Meetings." (Interestingly, there are any number of magazines being produced for business today that feature articles with similar titles.) Ten years into publication, the articles were still short and general. Titles included, "Conversational Bores," "Who Says You Should Be a Better Speaker?" as well as short rhetorical studies such as "George Romney: From the Mission to the Mansion." Over the next 10 years, however, *Today's Speech* began to publish longer and more "scholarly" articles such as, "Agitation over Aggiornamento: William Buckley vs. John XXIII," and "Kenneth Burke's Concept of Motives in Rhetorical Theory." By 1975, the journal had changed its name to *Communication Quarterly* and was entirely written for, and aimed at, academics in the field of communication.

Eadie's analysis was that *Today's Speech* had never really found an audience outside the field and eventually was not deemed credible by its primary audience of university professors. Consequently, he set about to make *JACR* credible to the field before attempting to make it so to any other audience. He published the best of what he received and invited very little; he also deliberately published commentaries aimed at the field itself. In short, he tried to produce a journal that the field would find applied, rather than one of use to the general public.

The diversity of what *JACR* has published is a clue to the lack of definition within the discipline. This point is not a new one: Burleson's (1992) explanation—that the reason the field has so few theories is that we do not value *communication itself* but rather assume its presence—is a telling one. Becker (1991) has told how in the early days of our doctoral programs courses in "phonetics, acoustics, and the anatomy of the vocal organs" (p. 1) were included in every course of study as a way to gain academic respectability, whether or not such courses had anything to do with the candidate's research interests. More recently, Becker (1992) has bemoaned the "cafeteria approaches" taken by many undergraduate institutions in order to serve increasing numbers of majors and has suggested that we find ways to insure that undergraduates are all exposed to some unifying view of the field. Perhaps we have become so accepting of diversity, in terms of methodologies, philosophical orientations, and topics of study, that we have lost track of what it is we are studying.

It is also a difficult task for the media to think of the communication discipline when preparing stories where our research could contribute insights. During the 1992 political campaign, *Time* magazine ran a cover story on lying in politics. Quoted were psychologists, sociologists, and political scientists; communication scholars were notably absent. Eadie, in his role as *JACR* editor, wrote a letter to the editor adding information to the piece based on research from the communication discipline. *Time* published it as the lead letter on that story, but all references, except those to Eadie's own research, were edited out before publication.

We are not totally ignored by the media as a field, but most of us do not work very hard to gain media attention. The scholars of presidential rhetoric have done the best job, and Kathleen Hall Jamieson has led the way by producing outstanding scholarship and by being available to discuss her insights cogently on the air or in interviews. Other topics of research in communication that have attracted media attention have included safe sex talk, courtroom communication processes, and the effects of television on children. Yet, much of what we do remains hidden, and coming out of hiding is not an easy task.

One answer may be to adapt to a larger scale a credibility strategy used by many communication departments. Communication faculty have often overcome academic snobbery by deliberately providing service—especially serving on university committees where our colleagues can see that we handle ourselves well and have good ideas. The same principle can be applied to gaining media exposure; if we are willing to speak up more publicly and from positions of authority, we will gain more coverage than we have now. If we wait to be "discovered," we probably won't be.

We are a devoted lot; we'd like our work to make a difference. To do so, we need to stop talking so much to ourselves and start talking clearly and cogently about our work to others. That way, we'll avoid passing from this life and having the best thing that could be said about our work is that we "didn't do any harm."

References

Becker, S. L. (1991, March). Don Quixotes in the academy: Are we tilting at windmills? Paper presented at the conference on Applied Communication in the 21st Century, Tampa, FL.

Becker, S. L. (1992). Celebrating spirit, commitment, and excellence in communication. *Southern Communication Journal, 57,* 318–322.

Berger, C. R. (1991). Communication theory and other curios. *Communication Monographs, 58,* 101–115.

Berger, C. R., & Chaffee, S. H. (1988). On bridging the communication gap. *Human Communication Research, 15,* 311–318.

Blumler, J. G. (1983). Communication and democracy: The crisis beyond and the ferment within. *Journal of Communication, 33*(3), 166–173.

Burleson, B. R. (1992). Taking communication seriously. *Communication Monographs, 59,* 79–86.

Craig, R. T. (1990). The speech tradition. *Communication Monographs, 57,* 309–314.

Delia, J. G. (1979). The future of graduate education in speech communication. *Communication Education, 28,* 271–281.

Eadie, W. F. (1990, November). Being applied: Communication research comes of age. *Journal of Applied Communication Research,* Special Issue, 1–6.

Miller, G. R. (1981). "Tis the season to be jolly": A yuletide 1980 assessment of communication research. *Human Communication Research, 7,* 371–377.

Miller, G. R. (1983). Taking stock of a discipline. *Journal of Communication,*

Miller, G. R., & Sunnafrank, M. J. (1984). Theoretical dimensions of applied c
research. *Quarterly Journal of Speech, 70,* 255–263.

Redding, W. C. (1985). Stumbling toward identity: The emergence of organizat
munication as a field of study. In R. D. McPhee & P. K. Tompkins (Eds.), *Org*
communication: Traditional themes and new directions (pp. 15–54). Beverly
Sage.

Rogers, E. M., & Chaffee, S. H. (1983). Communication as an academic discipline
logue. *Journal of Communication, 33*(3), 18–30.

Thayer, L. (1983). On "doing" research and "explaining" things. *Journal of Commu*
tion, 33(3), 80–91.

Tunstall, J. (1983). The trouble with U.S. communication research. *Journal of Comn*
tion, 33(3), 92–95.

Wartella, E. (in press). Communication research on children and public policy. In P.
(Ed.), *Beyond agendas: New directions in communication research* (proceedings c
Wichita Symposium, Wichita, KS, September 1992).

A Policy Research Paradigm for the News Media and Democracy

by W. Lance Bennett, University of Washington

Should communication scholarship stop trying to influence the practice of journalism and focus more on socially relevant research? The answer I offer to this provocative question is an emphatic *NO*. It is hard to imagine a more opportune time than now to promote dialogue between academics and journalists about the democratic roles and social responsibilities of both journalism and communications scholarship. Even drawing a distinction between policy analysis and "socially relevant research" seems ill advised.

Policy formulation and socially relevant research become inseparable if the problem is formulated as *discovering what conditions join people, politicians, and the press in open, critical public debates about the uses of power.* This essay proposes a framework to join journalists and communication scholars in dialogue about the conditions that best promote the public dialogue of democracy itself.

There are reasons to suspect that both journalists and academics may be receptive to thinking anew about communication in democracy. On the journalistic side, the world in which journalists operate is undergoing tremendous change. Begin with the upheaval in news organizations: The ownership, the technology of production, and the modes of distribution and sale of what we call news are transforming the production of information in astounding ways (Bogart, 1991). The implications of news being a commercial product have never been more apparent, whether viewed globally, as world news wholesalers affecting the supply of daily information, or viewed locally as market forces driving decisions about the content and format of newspapers and broadcast organizations (Bagdikian, 1987; Wallis & Baran, 1990). At the same time, the interface between journalists and sources is increasingly clouded by public relations and news management technologies that journalists are often unable to combat even when they are aware that the news is compromised as a result. In recent decades, even the publics who consume the news have grown increasingly savvy or cynical, depending on one's point of view, often blaming the messenger for unwelcome features of the message itself.

Lance Bennett is a professor of political science at the University of Washington. The support of the Social Science Research Council is gratefully acknowledged.

All of these conditions have made journalists increasingly open to discussion of issues involving their public credibility and their professional and political responsibilities. At the same time, there has been a tremendous growth of forums, foundation-sponsored research, and conferences dedicated to exploring the proper responses of journalists to changing political and economic pressures. Not only are journalists talking about redefining their professional roles, but some observers have spotted early signs of changes in political reporting (Hallin, 1992).

There are important ways in which communications scholars can contribute to this policy dialogue. This does not mean that there are any easy answers about how public debates should be organized, or how much of the burden of orchestrating them should fall on journalists. However, there are encouraging signs that a major obstacle to credible scholarly input is slowly being overcome. In the past, scholars have had trouble, despite some noble attempts, in getting a grip on a broad, empirically grounded conception of mass mediated politics as a dynamic process. Policy analysis and recommendations to journalists, therefore, either came from very thin empirical slices of much more complicated realities, or from ideological positions that were debatable from the outset. Journalists may be willing to engage in considerable self-criticism, but they often join professional ranks against criticism from academics who are either too narrowly scientific or too broadly ideological. Such scholarly input may say little to those who have to meet daily deadlines within the unforgiving constraints of market-oriented management, inattentive but judgmental publics, and unreasonably idealistic media critics.

The possibilities for more useful scholarly input are enhanced simply because researchers are beginning to put together a better story about how the news is constructed and why it matters. Research is converging around some general agreements about how the news is produced, how it is consumed, and the conditions that enrich and impoverish the political dialogues in the news media forum. With a better empirical grasp of an admittedly complex process, there is some hope for building a new paradigm to study, compare, and constructively criticize policy debates in the news. The argument about how to build this paradigm is developed in three sections: (a) a sketch of what a news policy paradigm might look like, (b) a brief review of research indicating important areas of (pre-paradigmatic) agreement about the dynamics of press–government–public interactions, and (c) a look at some steps toward a future field of study built around the research and policy implications of these emerging perspectives.

A News Policy Paradigm

In the past, the study of mass political communication has been fragmented by abstracting narrow-but-researchable topics out of their more complex surroundings. Blumler (1985) has referred to "an American tendency virtually to equate communication science with microscopic, individual level investigation . . . ." Even when the individual is not the unit of analysis, paradigms have developed around popular, if fragmentary, research topics, further isolating the parts from the whole. Thus, it became difficult for researchers studying the effects of news information on individuals to share their insights with scholars interested in the effects of corporate ownership on the production of information in the first place. As noted by many scholars, the field of mass communications showed little interest or ability to build bridges or hold dialogues across paradigms as different as behavioralism and neo-Marxism. There are some welcome signs that this may be changing (Dervin, Grossberg, O'Keefe, & Wartella, 1989).

The best hope for a paradigm shift to bridge some of these gaps is to recognize that many previously isolated research topics bear in important ways on different aspects of larger communication processes. This important point has been made by others calling for more holistic research frames, most notably Gurevitch and Blumler (1977) and Blumler and Gurevitch (1982). However, the failure of any particular holistic paradigm to sweep the field suggests that putting the pieces of larger pictures together is not likely to be done by any of the existing canned holistic approaches. No matter what their ideological or epistemological stripes, systems theory, structurationism, neo-Marxist, or neo-Gramscian approaches all turn out to be as divisive in their own ways as the microparadigms like behavioralism that they seek to replace.

Here is a more modest proposal for building broader analytical perspectives adequate to the task of studying the real world: simply carve out broad problem areas, and put together the research pieces necessary to say something important about them. It is unlikely that scholars actively doing interesting research are all going to run to the same paradigm, no matter how useful it might be. However, they might be persuaded that they are all studying important parts of the same problem, and ought to work toward more coordinated efforts to define research questions, conceptual vocabularies, and the like. If broader schools of thought emerge from these attempts at ground-up paradigm building, so much the better. If not, at least we may make some progress on understanding complex mass communications processes.

For example, suppose we are interested in comparing and theorizing about different patterns of policy debate that occur in the news coverage of different political situations in which issues or conflicts engage policymakers and publics in political dialogue. We might propose a series of case studies of different kinds of issue-situations (foreign, domestic, technical, moral, issues with interest groups attached, issues with little organized public support, etc.). It is not hard to imagine a research design built around a cluster of studies aimed at tracking the news coverage of these developing political situations over time. Different studies would look at content samples of news coverage, along with organizational analyses of both journalistic and government decisions.

In fact, such an approach has been under way in a group study of the policy debate following Iraq's invasion of Kuwait and the subsequent Gulf War (reported in Bennett & Paletz, in press). Scholars involved in the project, some of whose findings are mentioned in this article, have been interested in questions like these: Whose voices and views were heard in the news how much of the time? How much of the news hole was devoted to the situation for how long a period? Was marketing research on audience reactions fed back into story assignment decisions? How did public relations techniques affect the developing story? What happened when the story died down for lack of new activity from government officials or other actors centrally involved? Were new voices and views brought into the news by news organizations themselves, and if so, how? How did public opinion form in response to these patterns of coverage? How did the above characteristics of the news debate affect the policy decisions of elites in the situation? Did the news record leave markers for going back later to address the effectiveness of the policy—for example, establishing criteria for judging success or failure, or identifying particular officials as accountable? How did journalists assess their own performances afterwards, and did those assessments indicate particular normative themes?

These kinds of questions suggest the possibility of opening up larger research frames to study more complex communication processes. The focus here is on something we might call a *policy debate record,* embedded within and, with regard to public opinion and policy decisions, responsive to the news coverage of a situation. With the introduction of more sophisticated news data bases and archives, the most daunting empirical aspects of such research designs are made relatively manageable. More importantly, thinking about broad communications processes as the units of analysis invites the pooling of different (paradigmatic) understandings about how press, governments, and publics interact through the news.

This news debate framework is broad enough to give us new purchase on normative questions about information and democracy, and it also permits an empirical foundation for those normative concerns through researchable questions of the sort asked above. These research possibilities allow scholars to cover the range of politically important questions about the production and consumption of information in democracy. This kind of empirical flexibility removes the weak empirical and blanket ideological criticisms of press coverage that may have, with reason, been ignored by journalists in the past. Cases as different as Vietnam, Tiananmen Square, and the Gulf War can be compared, and differences explained within the same theoretical framework. I will return at the end of the discussion to consider how such research may reshape debates about press policy in general and democratic journalism norms in particular.

Merging Research Traditions

As noted above, studying broader policy debates in the news requires drawing on different research traditions that have focused on smaller parts of the political communication picture. For example, early work on media effects established important understandings of how people respond to information, including the important insight that specific messages often have little effect in complex and noisy communication environments. However, more recent experimental approaches have controlled for the noise and sharpened our understanding of the kinds of informational cues that people respond to. It is clear that political signals from the media affect not only what people think about (Iyenger & Kinder, 1987) but important aspects of how they think as well (Iyengar, 1992).

Thinking in political process terms raises the next question of how these messages are produced. Here we borrow from another stream of research on press–government relations and the creation of news itself. Early studies alerted us to the symbiotic relations among reporters and officials (Cohen, 1963; Gurevitch & Blumler, 1977; Tuchman, 1978) and to the ways in which government officials dominate news stories (Sigal, 1973). Later scholars pointed out that open and sustained policy debates were more likely to occur when consensus among elites broke down (Hallin, 1986). This has led to a somewhat more formalized claim that it is not just the broad historical breakdown of consensus (although that may be going on) but very specific patterns of elite debates in specific situations that affect many of the other qualities of policy debates in the news. When prominent politicians or members of key institutions oppose one another, the media gates open for more sustained debates in which more diverse groups and ideas are likely to be heard. Even the editorial positions of prestige newspapers like *The New York Times* (i.e., papers that want to be players in national policy debates) are indexed to the swings of power in Washington (Bennett, 1990).

What drives the indexing that guides much of the script in political debates? Journalists use implicit norms about representation and democracy to guide their coverage, generally presuming that democracy is working unless proven otherwise—meaning that officials represent the people, and the job of the press is to report to the people what their representatives are doing. Once powerful officials take a viewpoint seriously, it becomes relevant to look to society for other expressions of similar views, and to public opinion for reactions. This norm of *presumed democracy* is reinforced by beat reporting, which keeps the usual official sources on tap for quick reference in developing news scripts (Cook, 1991). The idea that journalists use implicit norms about how to cover a story and how to play sources is not new. This insight, too, is borrowed from another research tradition following the work of Tuchman (1978), Gans (1979), and others.

The information flows within these news debates can also be shaped by economic forces involving audience reactions and marketing considerations (which also brings into the picture the research tradition based on the corporate ownership paradigm). We might hypothesize that corporate owners and managers of news organizations are politically neutral as a class to the presumed democracy norm used by journalists at street level, since that norm tends to limit policy debates in the news to the political terms acceptable to government officials. There is little evidence of organized (i.e., class) opposition between corporate elites and political elites. However, other corporate norms involving profits and the sale of the news as a commodity may intrude into the content of policy debates in certain circumstances.

The direction or bias of economic influence may not always be obvious, as, for example, when coverage of Tiananmen Square and the Chinese democracy movement became a huge story that applied considerable pressure to policymakers to take actions at odds with current administration policies. In the case of Tiananmen, there was relatively little prior Washington debate. A dramatic news event came out of the blue and captured the imagination of vast news markets. The development of the story and its political themes reflected that relatively rare case of an interaction between journalists and audiences that was not heavily mediated by Washington politics (Pratt, 1989; Shanor, 1989).

In cases like Tiananmen Square, audience responses appear to have driven the story more than relatively weak elite debates. By contrast, in the case of the Gulf War, audience responses seem to have shaped the story around cultural themes of patriotism and community much in keeping with the prevailing elite consensus at the outbreak of the war. Hallin and Gitlin (1991) note the importance of market reactions in local television decisions about how to play the war on the home front. The Gulf War appears to be a case in which the dominant journalistic norms meshed poorly with situational factors, producing a limited policy debate about an important and fateful government action (Entman & Page, 1991).

If we add up the legacy of these various traditions (media effects, press–government symbiosis, the sociology of journalistic norms, corporate ownership and economic constraints on the media), we get a fairly broad and powerful conception of how media debate records are constructed and how they affect people. But we are not finished yet. The framework for studying news debates also requires attention to public opinion formations in response to evolving news characteristics.

Not only do journalists structure news information around elite cues, but public opinion forms in response to elite cues in the news as well. In studies of major foreign policy situations since Vietnam, Zaller (1991) finds that members of the so-called informed public are likely to be most influenced by the themes of elite debates. In addition, it appears that opinion rallies in support of presidents in foreign policy crises are likely

to rise as long as elite opposition remains minimal, and to fall when opposition becomes strong (Brody, 1991). Thus, public opinion appears to
be structured both on the input side (as it is reported in the news) and the
output side (as it is shaped by the news) by the dynamics of debates inside the Washington establishment.

This does not mean that news debates are seamless webs of elite control. To the contrary, as situations like Tiananmen indicate, factors inside
news organizations and strong public reactions in society can combine to
shape coverage independently. For example, the economics of market responses can be important, particularly when a developing story is script-
ed around safe and sacred cultural themes like democracy and freedom.
Moreover, the presumed democracy theme is not always off-base. In
many situations it is reasonable to assume that democracy works fairly
well: Elite debates can be volatile and nonconsensual on many issues,
and the emergence of strong blocs of public opinion and interest-group
organization produce notable effects on policy choices and elite positions, as indicated by prolonged debates on abortion, civil rights, health
care politics, and many other important policy areas. The problem comes
in using the presumed democracy norm as a primary guide in all situations, meaning that it continues to shape news debates even in cases
where normal democratic processes are not working well. Since these
cases are generally obvious only in retrospect (i.e., after disasters happen), it might make sense to expand the journalistic guidance system to
include norms that anticipate some probability of democratic failure, and
hedge all stories systematically against this possibility. This will be discussed further in the next section.

There is a final element of the news analysis frame being developed
here: the impact on policy itself. When the dynamics of debates do not
move quickly toward consensus and easy policy resolution (and media
attention remains strong), elites are likely to be affected by the news debate itself, making the news an important two-way agent in the democratic dialogue. The questions of whether news debates contain evidence of
changes in policy options, elite positions, public approval of officials, or
evaluative criteria for thinking about policy outcomes, complete the
frame of analysis around the news policy debate.

The Future of a News Policy Paradigm

What emerges from this merger of different research traditions is a broader picture of democratic communications that is open to empirical differences and theoretical comparisons across political situations. The hope is
to build new theories to explain how the same media system can produce
very different patterns of political coverage with very different effects on
public opinion and government policies. The broad analytical approach
outlined above offers more satisfying answers to several important ques-

tions: why the press misses some stories and takes a leading role in the development of others; why the public is featured prominently in open policy debates on some issues and marginalized in restricted public debates on others; and why some public policy debates seem to come much closer to others in producing understandable options, wise choices, and reasonable accountability.

Answering these empirical questions is a step toward thinking about a new normative agenda for press–government relations. Evidence from a number of cases already suggests that when the combination of press norms currently in use does not produce sustained and open policy debates in the news, journalists are among the first to recognize that they missed the stories. Loud episodes of press self-criticism have been heard after stories as different as the savings and loan collapse and the Gulf War. Might it not be wise to consider looking squarely at the reporting norms and practices that were employed (just as forthrightly) in these missed stories as in the journalistic triumphs?

This is not a call to jettison existing norms. As noted above, the problem with the presumed democracy norm is not so much that it is wrong, but that it is not right all the time. Thinking in this way may create an opening for dialogue about how to surround existing norms with additional guidance systems for reporting political stories. Without such a dialogue, the danger is that self-criticism may lead to the spurious conclusion that what happened in missed stories was that the norms were not applied properly, when in fact, the problem is that they were applied all too (routinely) well.

The normative question thus becomes how to supplement existing, reasonable normative standards with other guidance systems to create a more uniformly high standard of political debates in the news. A set of supplementary norms and practices may emerge from debates about things like:
- changing the beat system,
- running more "news analysis" pieces, and blurring the distinction between analysis and news,
- expanding the trend of covering more independent policy agendas constructed by news organizations themselves,
- recognizing when potentially important stories are dying because of the lack of elite debate and finding other ways of sustaining coverage of these situations,
- establishing accountability markers in the news for future evaluation of policies,
- incorporating public opinion into policy debates in more systematic, self-conscious ways,
- evaluating the legitimacy of sources and experts beyond common sense measures of reputation and the not always reliable assumption that the democratic system is working and self-policing.

How can we promote such normative dialogues? We can begin by recognizing that we are capable of sophisticated empirical research on the news and its role in democracy. First, demonstrate that comparisons among complex cases can identify the general conditions that affect the quality of political debates in the news. Next, recognize that these understandings are useful only if we think about normative questions differently. Research has already made clear that journalists do not often miss stories because they fail to apply their own standards properly, but because those standards simply do not always achieve the desired results.

Assuming a common interest in the democratic enterprise, the right kind of research can stimulate a different kind of normative debate. As with any good dialogue, all parties should change, meaning that scholars will develop new ideas about productive research questions and theoretical formulations. This is how paradigms grow. If the news policy paradigm aims at enhancing the possibilities for democratic communication, communications research may move beyond its old sticking points and, in the process, gain a measure of social recognition based on offering useful recommendations to journalists, politicians, and news consumers alike.

References

Bagdikian, B. (1987). *The media monopoly* (2nd ed.). Boston: Beacon Press.

Bennett, W. L. (1990). Toward a theory of press-state relations in the United States. *Journal of Communication, 40*(2), 103–125.

Bennett, W. L., & Paletz, D. (Eds.). (in press). *Just deserts: The news media and U.S. foreign policy in the Gulf War.* Chicago: University of Chicago Press.

Blumler, J. G. (1985). European-American differences in communication research. In E. M. Rogers & F. Balle (Eds.), *The media revolution in Western Europe* (pp. 185–199). Norwood, NJ: Ablex.

Blumler, J. G., & Gurevitch, M. (1982). The political effects of mass communication. In M. Gurevitch, T. Bennett, J. Curran, & J. Woollacott (Eds.), *Culture, society and the media* (pp. 236–267). London: Methuen.

Bogart, L. (1991). The American media system and its commercial culture. *Media Studies Journal, 5*(4), 13–33.

Brody, R. A. (1991, September). *Crisis, war and public opinion: The media and support for the president in two phases of the confrontation in the Persian Gulf.* Paper presented at the Social Science Research Council Workshop on the Media and Foreign Policy, University of Washington, Seattle.

Cohen, B. C. (1963). *The press and foreign policy,* Princeton, NJ: Princeton University Press.

Cook, T. E. (1991, September). *Domesticating a crisis: Washington newsbeats, human interest stories, and international news in the Persian Gulf War.* Paper presented at the Social Science Research Council Workshop on the Media and Foreign Policy, University of Washington, Seattle.

Dervin, B., Grossberg, L., O'Keefe, B., & Wartella, E. (1989). *Rethinking communication. Vol. 1: Paradigm issues.* Newbury Park, CA: Sage.

Entman, R. M., & Page, B. I. (1991, September). *The news before the storm: The limits to media autonomy in covering foreign policy.* Paper presented at the Social Science Research Council Workshop on the Media and Foreign Policy, University of Washington, Seattle.

Gans, H. (1979). *Deciding what's news.* New York: Pantheon.

Gurevitch, M., & Blumler, J. G. (1977). Mass media and political institutions: The systems approach. In G. Gerbner (Ed.), *Mass media policies in changing structures* (pp. 251–268). New York: Wiley.

Hallin, D. C. (1986). *The uncensored war: The media and Vietnam.* Berkeley: University of California Press.

Hallin, D. C. (1992). The passing of the "high modernism" of American journalism. *Journal of Communication, 42*(3), 14–25.

Hallin, D. C., & Gitlin, T. (1991, September). *Cultural themes and news about war.* Paper presented at the Social Science Research Council Workshop on the Media and Foreign Policy, University of Washington, Seattle.

Iyenger, S. (1992). *Is anyone responsible? How television frames political issues.* Chicago: University of Chicago Press.

Iyenger, S., & Kinder, D. R. (1987). *News that matters: Television and American public opinion.* Chicago: University of Chicago Press.

Pratt, L. (1989). The circuitry of protest. *Gannett Center Journal, 3*(4), 105–115.

Shanor, D. R. (1989). The "Hundred Flowers" of Tiananmen. *Gannett Center Journal, 3*(4), 128–136.

Sigal, L. (1973). *Reporters and officials,* Lexington, MA: D.C. Heath.

Tuchman, G. (1978). *Making news: A study in the construction of reality.* New York: Free Press.

Wallis, R., & Baran, S. (1990). *The known world of broadcast news: International news and the electronic media.* London: Routledge.

Zaller, J. (1991, September). *Political awareness and susceptibility to elite influence on foreign policy issues.* Paper presented at the Social Science Research Council Workshop on the Media and Foreign Policy, University of Washington, Seattle.

The Centrality of Media Economics

by Douglas Gomery, University of Maryland

We need to chart a direction where communications scholars regularly analyze the mass media as industries. The production, distribution, and presentation of television, radio, newspapers, and magazines require substantial investment and generate vast profits. And mass media businesses do routinely take in and spend vast sums of money. The mass media in the United States are fundamentally economic institutions. However one evaluates their place in the United States (and I shall stick to this country for my discussion and examples), the mass media begin (and continue to exist) only as economic institutions.

A decade ago in *Ferment in the Field,* Jeremy Tunstall argued that "the central mistake was to base a discipline on a combination of practical journalism and social psychology" (Tunstall, 1983, p. 92). Tunstall called for studies of both the economics of the Hollywood motion picture industry and the economics of TV program production; he pleaded for close examinations of technical change in the newspaper business. Scholars have responded to these pleas, yet media economics hardly stands at the core of communications study. It should.

Media economics should move into the center of communications study by offering more powerful and flexible methods by which to address core concerns. Marxist critical studies and free-market empiricism lack appeal because they ask us to analyze a subject when we already "know" a predetermined answer. From the left, critics of the mass media assume an all-encompassing conspiracy of media monopolies. Yet a cursory examination of the contemporary radio and magazine industries undercuts such a monolithic image. By contrast, conservative free-market advocates assume that efficient operation represents the paramount and only goal for any enterprise, even ones so vital to democracy and quality of life as mass communication and mass entertainment. Studying the economics of mass communication as though one were simply trying to make toaster companies run leaner and meaner is far too narrow a perspective (Besen, Krattermaker, Metzger, & Woodbury, 1984; Schiller, 1986).

No research in public communication can hope to make its mark in the world of public policy unless it addresses questions of economic influence and effect. Critics from all sides have long found problems with the

Douglas Gomery is a professor in the College of Journalism at the University of Maryland.

media and asserted a plethora of corrective regulations by which to improve industry operation and content production. However, we should not restrict analysis of media economics to today's problem industries only. Instead, the whole range of media industries and institutions needs to be analyzed regularly to establish a base from which to understand and evaluate the workings of the mass media, today and tomorrow.

This new emphasis on media economics needs to have at its core the study of changing conditions of quality—referred to as the study of performance. That is why I favor a model for media economic analysis that not only examines questions of who owns the media (economic structure) and media conduct, but also looks closely at performance. Based on the pioneering work found in industrial organization literature of economics, the media economist first should establish and define the basic conditions of an industry, then seek to establish its major players (structure), then define the behavior dictated by this structure (conduct), and finally evaluate the core questions of industry performance (Gomery, 1989; Scherer & Ross, 1990).

Simply listing who owns the media is not enough. One needs to hypothesize and understand how a particular form of industrial structure leads to certain corporate conduct. Recognizing that single-firm industries (e.g., monopolies) are ruthless or that competitive industries lead to greater choice and more far-reaching expression offers only the first part of a well-rounded analysis. We need a system for media economic analysis of the linkage among structure, conduct, and performance that leads to discussions of the need for public policy reformulations. Examining performance of media industries ought to be the ultimate step in media economics analysis.

It is at the level of performance analysis that all communication scholars can and should take an interest in media economics. It is here where we need to foster a connection between media economics and the long-time concerns of our field, whether this deals with questions of how best to promote diversity or how best to foster freedom of speech and discussion. If we can link the study of the economics of ownership and corporate behavior to the communication qualities we desire, communication scholars can begin to make recommendations for policy change that the players in real-world public policy discussion will take seriously (McQuail, 1992).

Our field has resisted the study of industry performance. Social science is supposed to be objective and positive, and not deal with normative issues. But as issues of media freedom and audience choice, of proper news objectivity and depth, continue to swirl, we should stop being afraid to combine empirical research and normative concerns.

What defines good media performance has long assumed an outcome of competitive pressures and a plethora of voices, a flourishing marketplace of ideas. But this purely competitive (to use the economics term) ideal is rarely met in our modern world. We need to combine media economic analysis and normative analysis to see how we might deal with the plethora of perceived public policy concerns about the mass media. Economic analysis can best help us make more informed choices of appropriate government action and assess the range of policy influences and effects (Busterna, 1988; Entman, 1985).

Economic Analysis

Before we are able to grapple with the difficult questions of performance, we need to closely examine the economics of market structure and conduct. First, the analysis of economic structure seeks to establish the number of buyers and sellers in a market, to identify barriers to entry for potential new competitors, to isolate the effects of horizontal and vertical ownership patterns, and to study the consequences of conglomerate control (Gomery, 1989).

It may seem like we are approaching a world of *one media industry,* yet in economic reality there exists a defined collection of media businesses competing for customers. It is best not to take on a single whole of media economics, but to restrict examinations to an industry-by-industry basis and then sum up and look at the total picture.

Analysis of economic conduct can begin by noting that revenues for media businesses fall into two distinct classes. On one hand, there is direct payment for books, popular music, movies, and pay television, businesses that sell their wares directly to the public. On the other hand, there is the world of indirect payment, characterized by advertisers "buying" audiences; over-the-air television, radio, magazines, and newspapers, for example, rely on advertising dollars to create the bulk of their revenues. These media may have a small initial charge (e.g., the subscription price of a newspaper), but advertising fees generate the bulk of the revenues (Lacy, 1984; Picard, 1989).

The important difference here for the study of industrial conduct is that with direct payment, customers are able to telegraph their preferences directly. For advertising-supported media, the client is the advertiser, not the viewer or listener or reader. Advertisers seek out media that can best help sell products or services; advertisers desire placement in media that can persuade customers who can be convinced to change their buying behavior and have the means to execute new purchases.

Given this duality of revenue generation, the industrial organization economic model postulates that the structure of a media industry determines the particular characteristics of its economic behavior. Once a certain market structure is established, the media economist then looks for certain techniques of price setting and program production, for certain types of marketing and promotion. In short, a certain category of market structure leads to specific corporate conduct.

Market structure and conduct in the media world fall into three distinct categories: monopoly, oligopoly, and monopolistic competition. We need to examine these three forms closely.

Monopoly

In a media industrial monopoly, a single firm dominates. The basic cable television franchise and the single-community daily newspaper provide two examples of media monopoly. To take advantage of this power and exploit economies of larger scale operation, cable television and newspaper corporations collect their monopolies under one institutional umbrella. In the cable television world we call them multiple system operators or MSOs; the industry leaders include TeleCommunications, Inc. and Time Warner. Newspaper owners form groups, headed by Gannett, Times Mirror, and Knight Ridder.

A monopoly fosters economic behavior that many cynically summarize as: "We don't care, we don't have to, we're the . . . ." This is all too familiar to any cable television subscriber. In any local jurisdiction there is typically only one cable television company. If one does not like the lone cable television corporation's offerings and prices, then the choices are to not subscribe or to move. The monopoly cable company has little incentive to keep prices down or to offer high-quality service. And, not surprisingly, in October 1992, in the heat of a presidential race, the U.S. Congress passed cable reregulation over the veto of President George Bush.

Oligopoly

In an oligopoly a handful of firms dominate. The most heralded example of this is the longtime three (now four) TV networks. Other oligopolies can be found in the six major music record labels and the six commanding Hollywood major studios that offer up not only most movies we see but also most prime-time TV programs (Owen & Wildman, 1992; Vogel, 1990).

To maintain their positions of power in recent decades, media oligopolies have diversified. For example, Capital Cities/ABC not only owns and operates a famous television network, but also a score of successful television and radio stations, the sports cable television network ESPN, a string of newspapers, and a collection of highly profitable magazines. That way Capital Cities/ABC is not dependent on the business cycle of a single operation. Unprofitable subsidiaries can be reconstructed and repositioned with funds generated from other profitable ongoing businesses. This enables an oligopoly to offer a high barrier to entry; potential rivals lack this conglomerate protection.

An oligopoly sees its small number of firms operate in reaction to each other. The metaphor is a poker game with five or six players. Each player knows a great deal about what the other is up to, but does not possess perfect knowledge. Take the case of the four dominant over-the-air TV networks. When NBC offers a new comedy at a particular time on a particular day, its rivals—ABC, CBS, and Fox—counterprogram. This leads to some experimentation, although all too often it means only a numbing generic sameness where like programs (e.g., comedies, dramas, or soap operas) face off against each other.

Economists have a great deal of trouble modeling oligopolistic behavior. The outcomes of oligopolistic corporate interplay depend on how many firms there are, how big they are in relationship to each other, past corporate histories, and sometimes the whims of individual owners. Analyzing a purely competitive situation is easier. A firm then can only operate myopically in its best interest. Formulating corporate action in response to hundreds of other rivals is too costly and makes no sense in a world of profit maximization.

Competition

Monopolistic competition denotes a marketplace where there are many sellers, but for any specific product or service there are but a few competing differentiable products. Today's radio and magazine industries are monopolistically competitive. For example, although there are thousands of magazines, they can be grouped by identifiable genres, from hobbyist quarterlies to scandal sheets to serious monthlies. Within a single genre only a small number of publications compete for a reader's purchase and attention. The same competition can be said to be at work—in large media markets—for radio broadcasting, with their range of familiar formats, from "album-oriented rock" to "country" to "all-news" (Husni, 1989; Owen, 1976).

Monopolistic competition, with its large number of firms making differentiated products, offers customers more choice at reasonable prices. But within a single format or genre category, magazines and radio stations do cluster and counterprogram. For example, fans of professional wrestling or news analysis or country music quickly become quite cognizant of their magazines or their radio stations. There are but a handful of players within each genre or format. And there are barriers to entry for new competitors. Radio stations, for example, need an FCC license to broadcast. In medium- and small-sized cities, some marginally attractive options never are tendered. Many a medium-size community has no all-news or classical-format radio station simply because there are too few licenses and company leaders go with more widely popular (and more profitable) formats. This is not a purely competitive situation where all companies produce similar products, and so monopolistically competitive situations produce lots of choice. It's a situation that leaves some customers (e.g., classical music fans) unsatisfied.

Performance

Analysis of economic structure and conduct initiates and logically leads to analysis of performance. Indeed what media scholars and critics care about most are economic linkages to media performance. Remedies are proposed when proper industry performance is not met. We need to select performance criteria that are precise and operational. How well has a media industry functioned when compared to some ideal standard? If there is market failure, then is there a regulatory remedy? (Litman & Bridges, 1986; McQuail, 1992).

For example, a generation ago the Federal Communication Commission (FCC) instituted the so-called Prime-Time Access Rule. The then three dominant over-the-air TV networks were deemed to have too much power and to be operating not in the public interest. To mitigate oligopoly power and lead to "better behavior," the FCC required that "local" programs be presented during one half hour of prime time. In short, the television station licensee affiliated with a network was required to behave a certain way in exchange for the valuable franchise he or she held. But this policy was ill-conceived and backfired; continual runs of "Wheel of Fortune" or "Family Feud," while selected by local stations, hardly make for what most observers judge as "quality local programming."

Prior to heralding a new regulation or filing an antitrust suit, policymakers struggle with locating criteria by which to judge economic performance. What do we want out of a media industry? Let me suggest six media performance norms that encompass most judgments and take them up in order of ease of use. That is, the first criteria considered are easier to deal with than those further down the list (McQuail, 1992).

First, media industries should not waste resources; they should be as efficient as possible. Monopolists waste resources to maintain their position of power. But what about control by a few firms? Many argue that this is just as bad; that these industries regularly cooperate through powerful trade associations and thus hardly represent lean and mean business operations. Watchers of the network TV business are familiar with the excesses and waste aptly labeled the "Cadillac culture" operations of the 1960s and 1970s.

Second, media industries ought to bring new technologies to the marketplace as quickly as possible. It has long been known that monopolies and collusive oligopolies resist the innovation of new technologies in order to protect their highly profitable status quo positions. Through the 1980s and into the 1990s, for example, there have been constant struggles over proper cable television regulation to ensure that customers are not faced with gouging price increases and deteriorating service.

Third, media industries ought to distribute their products and services to rich, poor, and all of those in between. The mass media, because of scale economies and underwriting by advertisers, have long been justifiably praised because they are so cheap to acquire. For example, broadcast television has long been widely available to even the poorest members of society. But access is becoming more and more restrictive as a larger share of the mass media go to direct payment. Cable TV charges run up into the hundreds and even thousands of dollars a year. If television is an important link in our democracy, how will our process of government change if a third of the population does not have access to cable television?

Fourth, the mass media ought to facilitate free speech and political discussion. A democracy needs freedom of expression to make it go, and the mass media ought to be open enough to promote debate of all points of view. For example, do the new reality-based TV shows—tabloid television—help the process of public debate? Should we citizens care and press for regulation?

Fifth, the mass media ought to facilitate public order. In times of war, violence, and crime, how should we regulate the media (if at all) to ensure differences? This is a growing area of concern as the media easily jump across national (and local) boundaries.

Finally, the media ought to protect and maintain cultural quality, and offer to play some role in education. Many regularly ask: Can advertising-generated-revenue companies not simply dish up more sensationalism?

Policy

The history of the media in the United States suggests that we will always have to choose among media industry monopolies, oligopolies, or situations of monopolistic competition. We will always have to choose among less than perfectly desirable market structures and conduct. Idealists speculate on a world of one media industry; realists know that in modern capitalism corporations will operate differently within industry market structures that are not purely competitive.

Once we understand media industry structure and conduct, we will need to be clear about what performance criteria we wish to prioritize and work from there. The media economist can play a central role by evaluating proposals for regulation and analyzing their effects on structure, conduct, and performance. But the media economist should not seek to impose performance criteria. Media scholars can help specify what appropriate criteria might be. Communication scholars should seek to influence the public policies toward public communication rather than simply leave the field to lawyers. Only with media economics at the core of our field can we hope to effectively influence public policy.

We should take this task of media economics, as I have broadly defined it, very seriously. Future societies will depend upon complex electronic information networks offering information and communication. Distribution will be globalization. We need to help judge proper performance and (if necessary) tender corrective public policy actions.

Should we offer strong countervailing governmental power? Or should we let the free market tap the energies (and rewards) of technical innovation? Or should we, as we do at the present, seek to optimize the mixture of a regulated and unregulated media world?

In the future we will have to worry about many, many channels, with greater ease of access *and* growing expense of use. For the media of the future we will have to worry more and more about bridging information and media gaps; about securing political involvement; about maintaining creativity, independence, and diversity; about social solidarity and minority rights; and about cultural autonomy and identity. Media economics will be able to help us sort though these thorny public policy questions. This is why media economics ought to move to the core of our field as we move into the new media environment.

We need to rationally adopt public policies that facilitate all citizens having access to many channels and sources, offering different content, alternative voices. We will need to know how the common structures of message production and distribution and presentation affect audiences. Thus, in the future we need to examine, on an industry-by-industry basis, the structure, conduct, and performance of changing media economics.

References

Besen, S. M., Krattermaker, T. G., Metzger, A. R., Jr., & Woodbury, J. R. (1984). *Misregulating television*. Chicago: University of Chicago Press.

Busterna, J. C. (1988). Antitrust in the 1980s: An analysis of 45 newspaper actions. *Newspaper Research Journal, 9*(1), 37–48.

Entman, R. M. (1985). Newspaper competition and First Amendment ideals: Does monopoly matter? *Journal of Communication, 35*(3), 147–165.

Gomery, D. (1989). Media economics: Terms of analysis. *Critical Studies in Mass Communication, 6*(1), 43–60.

Husni, S. A. (1989). Influences on the survival of new consumer magazines. *Journal of Media Economics, 1*(1), 39–50.

Lacy, S. (1984). Competition among metropolitan daily, small daily, and weekly newspapers. *Journalism Quarterly, 61*(3), 640–644.

Litman, B., & Bridges, J. (1986). An economic analysis of daily newspaper performance. *Newspaper Research Journal, 7*(1), 9–26.

McQuail, D. (1992). *Media performance: Mass communication and the public interest*. Newbury Park, CA: Sage.

Owen, B. M. (1976). Regulating diversity: The case of radio formats. *Journal of Broadcasting, 21*(3), 305–319.

Owen, B. M., & Wildman, S. S. (1992). *Video economics*. Cambridge, MA: Harvard University Press.

Picard, R. G. (1989). *Media economics: Concepts and issues*. Newbury Park, CA: Sage.

Scherer, F. M., & Ross, D. (1990). *Industrial market structure and economic performance* (3rd ed.). Boston: Houghton Mifflin.

Schiller, H. I. (1986). *Information and crisis economy*. Norwood, NJ: Ablex.

Tunstall, J. (1983). The trouble with U.S. communication research. *Journal of Communication, 33*(3), 92–95.

Vogel, H. L. (1990). *Entertainment industry analysis: A guide for financial analysis* (2nd ed.). New York: Cambridge University Press.

Reconnecting Communications Studies With Communications Policy

by Eli Noam, Columbia University

Ten years ago, in summarizing *Ferment in the Field,* the predecessor to this volume, George Gerbner (1983) noted the centrality of communications in modern society and agreed with the observation that "if Marx were alive today, his principal work would be entitled *Communications* rather than *Capital*" (p. 358). Marx's scholarly writing on capital influenced politics and policy. But could the same be said for writings about today's central economic activity, the communication of information, since *Ferment in the Field* appeared?

The answer is no. Communications studies played only a minor role in the enormous changes in the public treatment of the communications system. During the past decade, individualized and mass electronic media were transformed from national monopolies and oligopolies to new structures that may, in time, resemble print media and their distribution (Noam 1992a, 1992b). At the time communications was on the table of national policy, when new institutional arrangements were being established, the field of communications did not communicate well to governmental decision makers, whether in Washington, Brussels, or other capitals. There were some exceptions, including Gerbner's own work on violence, or studies on advertisements aimed at children, because social science could bring specialized expertise and tools to these questions. But, while communications research has often been quite openly and legitimately political, as in the discussion over the New Information Order (Schiller, 1983), communications scholars absented themselves from actual policy—perhaps questioning, with Lasswell, whether "the concern for enlightened policy [can] survive close involvement in the process itself" (in Melody & Mansell, 1983, p. 109).

Consequently, communications scholarship has been without a real-

Eli M. Noam is a professor of finance and economics at Columbia University Graduate School of Business. He is also the director of the Columbia Institute for Tele-Information, and has served as Commissioner for the New York State Public Service Commission. The author would like to thank John Carey, Barry Cole, Everette Dennis, William Drake, Martin Elton, Herbert Gans, Richard Kramer, Milton Mueller, Aine NíShuilléabháin, Michael Noll, and Robert Pepper for their helpful comments. Most literature referenced in this essay is from the "Ferment in the Field" special issue of *Journal of Communication.*

world role, in contrast to some other fields, such as environmental studies, which overcame the structural impediments that limit academia's influence and participation in the public arena. Policymakers often ignore social science research (Hamelink, 1983), but scholars also underestimate their own weight. Ideas may not win, but they matter. While convenient ideas may get amplified.more than those that threaten, and while the realistic set of policy options may be narrow (Haight, 1983, p. 231), the policy process is also a voracious consumer of ideas. They are used to illuminate, legitimate, and do battle. The test of an academic field is not and should not be its instrumentality. Communications studies have made important contributions to the understanding of the media that surround us, the processes that contribute to their outputs, and their cultural, political, and economic significance. They have addressed broader issues of class, gender, history, control, language, audiences, the human interface, and content, to mention a few. Even so, when a discipline that is by now fairly substantial in terms of numbers and maturity is largely absent in the shaping of society's treatment of the very subject of its study, one must take note.

Communications Scholarship's Tendency to Defend the Status Quo
In communications' transition from monopoly, communications scholars have often protected the existing order. The old policy arrangements had some undeniable social merit as well as power and benefits to disburse to their participants. In most countries, communications were a public service oriented to the public welfare. But the reality has been more complex. In point-to-point telecommunications—never a popular research subject, despite their pervasiveness in personal and organizational information exchange—long-standing monopolies had become bloated and slow. Technological decisions tended to be captured by domestic supplier industries. Even so, the change to a more open network environment was accompanied by scholarly assertions of impending social doom, few of which were retracted when the predicted calamities failed to materialize.

In television, too, the reality of the traditional public monopoly broadcast system that existed in many countries fell far short of the idealized expectation of quality programming. The pervasive politicization of the powerful public institutions was not given much research attention. Nor was there much study of the negative impact on national and regional cultures and on artistic independence resulting from a system in which a single national public broadcast monopoly served as the gatekeeper and chief financier of the film and video creativity of an entire society. Despite a vast body of political science research, it was often assumed that such an institution would act for the public, without regard to its self-interest or that of its political patrons. Such assertions were even made on behalf of state broadcasting in the predemocratic regimes of Eastern Europe (Szecskö, 1983).

Have other academic disciplines been more involved in the transformation of communications? Technologists provided some of the tools that enabled change. They (and business-school strategy researchers) often played a booster role that looked overoptimistically at the potential of technical progress. But their role outside technology policy has been comparatively small. Political scientists and historians have had an astonishingly low profile considering the magnitude of change and its long-term implications for the political and social system. Legal academics have played some role by framing antimonopoly arguments, extending free speech principles to electronic media, and crafting new communications statutes. Among social scientists, economists have probably been the most influential, providing the general free-market case which helped to destabilize the "natural" monopoly system. Economists and lawyers were also active—often in the employ of interest groups—in the implementation of change. But once the argument for removing entry barriers had been accepted, they contributed little to a vision of the future.

In the academic pecking order, theory is more prestigious than empiricism or policy (Miller, 1983). It is produced for academic receptors, in ever narrowing networks of specialists communicating in jargons. Yet theory must refer to a fast-changing reality, especially if it has political implications and if it is to guide applied research. Ten years ago Pool argued that beliefs were no substitute for empiricism: "Avoid measurement, add moral commitment" (1983, p. 260). Similarly, Melody and Mansell (1983) pointed to an interrelation: "In the policy debate, the theoretical and methodological trappings of research are directly confronted by reality and the test of relevance" (p. 113). But parts of the field have remained inhospitable to empiricism, despite its significant contributions in earlier days. With inadequate incentives inside academia, the empirical and policy base of communications research was further weakened by a brain drain of those with a strong fact base into private consulting, think tanks, and nonacademic dissemination.

Issues for the Future

Given these problems, communications studies are not well poised to deal with some of the issues of the new information order that is emerging or to have an impact on it. Will communications research become increasingly sophisticated methodologically yet less publicly relevant? Or will the next generation of researchers prove to be up-to-date and involved? This requires the identification today of tomorrow's issues, their transmission to the next generation of scholars, and their presentation to the public. What are some of these trends?

Beyond the Nation-State
Under the old information order, territorially organized electronic communications networks were based, technologically, on the need for a net-

work architecture that minimized transmission distances; politically, on the desire of the state for control over communications; economically, on incumbent organizations' desire for profitable protection; and socially, on the shared reference of national culture. But in the future, with the cost of transmission dropping and distance-insensitive, both telecommunications and mass media networks will become globally organized. This will have important effects. For example, "electronic democracy" tends to be viewed as the use of communications media for political participation within established political units. Yet, this is merely one step in the creation of "virtual" communities. Group formation is based on economic and social interaction. Communications media will not create a global village, but instead help organize the world as a series of electronic neighborhoods transcending national frontiers.

Historically, the nation-state was at tension with cross-border allegiances. The new environment weakens national cohesion in favor of both an internationalism and particularism. It is difficult for a state to extend its powers beyond traditional frontiers, but it is easy for network groups to do so. Through communication—the process through which a shared culture is created—they establish themselves as new cultural units, affecting both cultural fragmentation nationally and postmodernist homogenization internationally (Carey, 1983, 1993). They have to set individual contributions to cover their cost, and in the process create their own de facto tax and redistribution mechanisms. They have to mediate the conflicting interests of their members, determine major investments, set standards, decide whom to admit and whom to expel. As group networks becomes more important and complex, control over their management becomes fought over. Elections may take place. Constitutions, bylaws, and regulations are passed. Arbitration mechanisms are established. Financial assessment of members takes place. Networks thus become political entities and quasi-jurisdictions.

Beyond Regulation

The replacement of communications monopolies by a partly competing, partly collaborating, interconnected, and nonhierarchical *network of networks* will fundamentally change the face of media industries as we know them. To provide integration of the various discrete networks, specialized *systems integrators,* which will become in time the central institutions of communications, are emerging and replacing many of the roles of today's telephone companies, broadcasters, and cable operators. They will put together individualized *personal* networks, and interconnect them with each other in what may be described as a *system of systems.*

Such a structure will be radically different from the present media system, and it invites academic analysis. For example, what would be the role of public control? Could some overall and beneficial equilibrium emerge out of decentralized suboptimizing actions? What traditional goals of public policy are left unresolved and what new ones would need to be addressed? Will they be overshadowed by trade concerns? (Drake & Nicholaïdis, 1992). How do partially regulated environments function? These are important questions for theory and policy. Communications scholars must continue to write about the dangers of marketplace transformation, such as information poverty (Mosco, 1983), but it is just as important to think about what a new support mechanism should look like in the future system of systems. Similarly, concern with a property-rights-based information order and with the invasion of privacy spheres is no substitute for analyzing remedies.

Similarly, the interrelation of the various electronic communications networks must be thought through carefully now that they are beginning to happen. Access rules define the rights of various media and thereby the participatory rights of their users. They are nothing less than a constitutional framework for the communications infrastructure. Leaving such fundamental communications issues to the technical specialists of various disciplines would be like leaving war to the generals (Comstock, 1983).

Beyond National Culture

In the past, the scarcity of electromagnetic spectrum permitted only a tiny number of television channels, resulting in program content bridging many viewing interests to aggregate large audiences. The outputs of a medium are defined by its structure. In what ways then will the change in the media structure alter production, news, programs, and distribution? These are areas that were underresearched 10 years ago (Blumler, 1983; Gans, 1983; Tunstall, 1983), and they still are. The broadening of transmission bandwidth beyond traditional limited television leads to a measurable widening of program options and viewer differentiation, both in the high- and the low-culture ends of the program spectrum (Noam, 1992b). This process will take several decades, but it is on its way. Future media based on electronically accessible video libraries will further drastically affect program differentiation, viewer control, and program provision from alternative sources. Viewers will also end up paying much more for their television viewing in the new environment, raising distributional concerns. On the other hand, the production of programs will be encouraged and cultural activity increased.

Beyond Information Scarcity

Recent decades have seen giant strides in the distribution of information and in its production. The weak link in the information chain is the increasingly inadequate absorption capacity of individuals and organizations. Computer technology does not help much—unless underlying information is quantitative and structured, and questions are well-defined. These conditions are rarely met in real life. The mismatch of inflows and absorptive capacity raises questions for communications research. One example is a likely change in the way information gets presented. While the traditional print alphabets are geared to slow and narrowband communications channels, in the future multiple information tracks will be provided in a parallel fashion so as to widen information access to an individual. This change is unlikely to be content neutral, and literacy, culture, and creativity will be affected.

Another approach to enhance the ability to absorb information is to automatize its screening so that less clutter reaches the individual. This will not be neutral, either. In order to make content and meaning more intelligible to machines, many forms of communication will be subject to some standardization of format, syntax, and style. Thus, echoing Innis (1950), the written language itself is likely to be changing with technology, and with it how we think, interact, and conduct politics.

What Is to Be Done?

Communications scholarship has not kept pace with the concrete questions of public treatment of media, even though its subject of study, the communication of information, has achieved centrality in society and the economy (Garnham, 1983). Much of the field has been insular, disconnected, and often invisible. The issue is not the absence of political power by academia. It is one thing to be weak, and quite another to be left behind. Obviously, ideology and media shape each other and are shaped by economic and political conditions. But being mesmerized by the potential of communications media or by the power of their owners is no substitute for thinking along and ahead, providing the world with visions, details, and ways of protecting traditional concerns in the new communications environment.

For the field of communications studies to blossom it must not react to its centrifugalism by narrowing its focus; to the contrary, it must expand. First, it must broaden into adjoining media. In the past, communications studies have concentrated on mass media, paying little attention to point-to-point and computer communications. Yet the blurring of boundaries separating electronic media and the creation of multimedia technologies, group networks, and interactive personal communications render many distinctions obsolete. This can hardly be stated too strongly. Traditional journals, associations, and curricula must recognize that the other forms of electronic communications are an integral part of their subject.

Second, communications studies must broaden beyond the bounds of pure academia. Communications scholars must both address and occasionally venture into a real world, whether in production, government, media business, or public-interest advocacy, to name a few. While one must be determined to avoid excessive closeness, research and teaching will benefit overall from such experience (Schramm, 1983).

Third, even within the academic realm, communications studies must overcome insularity. The field will hopefully maintain and strengthen its own disciplinary multiculturalism, be it by historians of communications, philosophers, sociologists, interpreters of culture, to name a few. Yet, despite communications studies being broad in concept, there is an absence of strong links and even some hostility to some disciplines not at the center, such as technology, operations research, political science, law, and economics. Ten years ago, Mattelart called for a "decompartmentaliz[ation of] the problems of information" (1983, p. 65). This challenge to reclaim the multidisciplinary approach as the comparative advantage of the field is even more critical today, in both research and curricula.

Fourth, communications scholars might mute their ideological conflicts (Lang & Lang, 1983), in which critical scholars castigate others for serving the status quo (Grandi, 1983) and for ignoring the political processing of technological change, while their opponents dismiss them in turn as reactive and lacking a positive agenda. There is nothing wrong with a vigorous dialectical process of contending ideas, but in communications studies it seems to dissipate disproportionate energy.

And fifth, communications studies must reestablish a strong empirical and applied base within the field, so that theory, methodology, empiricism, and policy will reinforce each other again.

Without such efforts, communications studies will not be able to identify the future of communications or illuminate society's understanding of it. Nor will it be able to delineate its own field. If the chasm between an academic field and its subject matter of study becomes too wide, a self-correcting mechanism takes over. The rapidly moving world of communications media, technology, and infrastructure will force communications studies to change focus, directly or through the next generation of students in the field, and this process of change will no doubt transform the field as we know it.

References

Blumler, J. C. (1983). Communication and democracy: The crisis beyond and the ferment within. *Journal of Communication, 33*(3), 166–173.

Carey, J. W. (1983). The origins of the radical discourse on cultural studies in the United States. *Journal of Communication, 33*(3), 311–313.

Carey, J. W. (1993). Everything that rises must diverge. In P. Gaunt (Ed.), *Beyond agendas: New directions in communications research*. Westport, CT: Greenwood Press.

Comstock, G. (1983). The legacy of the past. *Journal of Communication, 33*(3), 42–50.

Drake, W. J., & Nicholaïdis, K. (1992). Ideas, interests, and institutionalization: "Trade in services" and the Uruguay Round. *International Organization, 46*(1), 37–100.

Gans, H. J. (1983). News media, news policy, and democracy: Research for the future. *Journal of Communication, 33*(3), 174–184.

Garnham, N. (1983). Toward a theory of cultural materialism. *Journal of Communication, 33*(3), 314–329.

Gerbner, G. (1983). The importance of being critical—in one's own fashion. *Journal of Communication, 33*(3), 355–362.

Grandi, R. (1983). The limitations of the sociological approach: Alternatives from Italian communications research. *Journal of Communication, 33*(3), 53–58.

Haight, T. R. (1983). The critical researcher's dilemma. *Journal of Communication, 33*(3), 226–236.

Hamelink, C. J. (1983). Emancipation or domestication: Toward a utopian science of communication. *Journal of Communication, 33*(3), 74–79.

Innis, H. A. (1950). *Empire and communications.* Oxford, England: Clarendon Press.

Lang, K., & Lang, G. E. (1983). The "new" rhetoric of mass communication research: A longer view. *Journal of Communication, 33*(3), 128–140.

Mattelart, A. (1983). Technology, culture, and communication: Research and policy priorities in France. *Journal of Communication, 33*(3), 59–73.

Melody, W. H., & Mansell, R. E. (1983). The debate over critical vs. administrative research: Circularity or challenge. *Journal of Communication, 33*(3), 103–116.

Miller, G. R. (1983). Taking stock of a discipline. *Journal of Communication, 33*(3), 31–41.

Mosco, V. (1983). Critical research and the role of labor. *Journal of Communication, 33*(3), 237–248.

Noam, E. (1992a). *Telecommunications in Europe.* New York: Oxford University Press.

Noam, E. (1992b). *Television in Europe.* New York: Oxford University Press.

Pool, I. D. S. (1983). What ferment?: A challenge for empirical research. *Journal of Communication, 33*(3), 258–261.

Schiller, H. I. (1983). Critical research in the information age. *Journal of Communication, 33*(3), 249–257.

Schramm, W. (1983). The unique perspective of communication: A retrospective view. *Journal of Communication, 33*(3), 6–17.

Szecskö, T. (1983). Communication research and policy in Hungary: Partners in planning. *Journal of Communication, 33*(3), 96–102.

Tunstall, J. (1983). The trouble with U.S. communication research. *Journal of Communication, 33*(3), 92–95.

The Traditions of Communication Research and Their Implications for Telecommunications Study

by Willard D. Rowland, Jr., University of Colorado, Boulder

There is a strong contemporary movement to develop and find a place in the academy for programs or even formal departments known as "telecommunications." This tendency is in part a natural outgrowth of an increasing academic awareness of the importance of communications technologies in modern society and a desire to create more intellectual space for their study and interpretation. The tendency is likewise driven by the professional training imperatives of the university—to explore the opportunities for developing degree programs in fields of applied practice that will continue to be important aspects of economic and cultural life.

To many in communication research and teaching, this new arena is seen as an extension, perhaps just another subfield, of the interdisciplinary mix of interests that have over the decades fostered an increasingly central place in the academy for the various programs associated under the communication rubric—journalism, speech, media studies, broadcasting, advertising, film, and mass communication. In that light, telecommunications is understood as part of the liberal arts nexus, particularly in the social sciences and humanities, that have informed all aspects of communication study and to which the field itself has made significant contributions.

Little recognized, however, has been that aspect of the development of telecommunications that has emerged quite independently of the arts and sciences heritage, a strain that is lodged much more thoroughly in the applied professional realms of such other fields as engineering and business. Because such programs tend to be more technical and management oriented, they attract a strongly applied, vocational student interest and command relatively rich sources of support from the principal relevant industries (e.g., telephone and computer) and from those state and federal

Willard D. Rowland, Jr. is dean of the School of Journalism and Mass Communication at the University of Colorado, Boulder. An earlier version of this paper was presented at the National Telecommunications Forum Symposium on Telecommunications Education at the Sugarloaf Conference Center, Temple University, Philadelphia, PA, February 8, 1992.

government agencies concerned about fostering improved conditions for a postindustrial, information-based economy. With such backing and appeal, these programs are likely to become increasingly seductive to university administrations and to offer considerable challenge to communication study just as it has begun to articulate a rich intellectual environment for research in new communications technologies.

The purpose of this paper is in part to point to this challenge and to suggest the need for those in communication to take cognizance of it. Another purpose is to examine the traditions of communication inquiry and to suggest what values they bring to the field of telecommunications that are essential in the continuing debate within the academy about its proper intellectual domain and institutional locus.

The principal argument is that, due to its heavy technological emphasis, telecommunications will tend toward instrumental, applied forms of research, adopting an administrative orientation that is grounded in an atheoretical futurism, that focuses principally on problems of technical capacity and promise, and that approaches policy questions largely in economic and market terms. The corollary argument is that there rest in certain aspects of communication research alternative traditions that take telecommunications to be socially and culturally problematic and that point it toward a more critical policy discourse. These are the patterns of inquiry that see telecommunications as part of the social history of communications technology, as illustrative of all the issues of meaning, ideology, power, and culture characteristic of posteffects communication research.

The Legacy of Communication Research

To help develop an understanding of the importance of these questions in telecommunications, it will be useful to recall certain special features of the origins and legacy of communication research. As part of this analysis, it is possible to show how much of what is thought of as telecommunications study had already come into existence within communication, with important intellectual implications that are downplayed or ignored in the more technocratic approaches. Those origins and associations suggest much about an alternative set of expectations for the future of research in telecommunications technology and policy.

Communication education dates back to the early part of the 20th century, where it began to grow out of two quite distinct emphases in the academy: journalism and speech. Both of these fields developed in conjunction with the receptivity in the turn-of-the-century university to new constellations of professional degree programs (Brubaker & Rudy, 1976, pp. 198–218; Kohlstedt, 1988, pp. 46–47). They reflected an increasing academic interest in the communication aspects of the new industrial order.

Journalism instruction emphasized the craft of writing and reporting. However, there were several aspects of its initial curricula that
emerged relatively early and that helped stamp media education with a
certain flavor that always made it much more than narrow training in
newspapering skills. For one thing, because of concerns about the legal
and economic contexts in which American journalism operated, those
programs began to develop an interdisciplinary character that included
instruction in media law and in the financial and advertising aspects of
the news. Simultaneously, because of the debates about the significance
and impact of the press, journalism education began to look to other disciplines, particularly in the social sciences, for guidance about the social
role of the media. As that process of inquiry developed, it increasingly
dovetailed with the debates within the profession itself about its own
standards and codes of practice (Commission on Freedom of the Press,
1947; Siebert, Peterson, & Schramm, 1956).

Simultaneously, speech programs were also developing elements of
professional media education. Growing out of the ancient traditions of
rhetoric and debate, speech had a natural interest in film and broadcasting as extensions of the oral tradition—of the theater and of the podium.
As a result, such departments began to offer professional courses in the
production and performance aspects of radio and motion pictures; and,
in keeping with their roots in oral interpretation, to pay increasing attention to matters of content criticism, behavioral and social impact, and
media law, regulation, and economics.

Eventually, journalism at most major research universities came to define itself more broadly as mass communication. For the most part
"telecommunications" as an academic field tended to emerge as a formal
extension of "radio and television" or "broadcasting" programs within the
communication nexus, and there has been a trend during the post-World
War II period to incorporate most aspects of media and speech study, including telecommunications, under one joint interdisciplinary umbrella,
as in schools or colleges of communication.[1] Regardless of how the structure emerged, what is important to emphasize is that as the academy fostered the development of professional and scholarly study of journalism,
broadcasting, film, advertising, and cable television, it was also promoting a broader inquiry into the entire realm of the electronic media and the
newer communications technologies.

As the communication curricula developed, the interlinkages among
the various technologies and industries and the relevant social, cultural,
and political conditions became increasingly apparent. The transition

---

[1] The fact that most university departments or programs with the name telecommunications
derive from the broadcasting and communication heritage is ignored in some "official"
guides where the field is misrepresented as being associated exclusively with engineering
(see Peterson's, 1993).

from journalism history to communication history, for instance, began when it became clear that to understand the history of the newspaper it was necessary to pay more attention to the parallel history and shaping influences of the telegraph and the rise of an electronic communications network (Czitrom, 1982; Marvin, 1983, pp. 24–26). In studying the emergence of the film industry, it became essential to consider related models of industrialized research and development, monopolistic tendencies, and antitrust issues associated with the development of the telephone and the general U.S. industrial order (Jowett, 1976; Smythe, 1957). In trying to explain the emergence of the commercial radio industry, it was crucial to note closely the role of AT&T (American Telephone & Telegraph Co.) and the entire early consumer electrical appliance industry, as represented by General Electric and Westinghouse, in the formation of the RCA/NBC combination, which had such a lasting impact on the nature of American broadcasting (Barnouw, 1966).

Characteristics of Media and Communication Education

In the end, then, although there were two different tracks by which media and communication education came into being, there were several common characteristics that defined it and that bear directly on how one might think about telecommunications as an organized area of research and study.

First, rather quickly in their formative years, the principal strains of communication education emerged as a thoroughly *interdisciplinary* field. They grew out of the humanities initially, but they rapidly began to take increasing cognizance of relevant theory and applied practice in the social and behavioral sciences, business, law, and the media professions. Their interdisciplinarity was characterized by an integrative relationship among the contributing perspectives. This was much more than a multi-disciplinary approach, wherein various courses are collected from different departments or schools without, however, any dialogue among the relevant disciplines they represent. Journalism and speech provided a forum for active discourse among the relevant liberal arts and professional fields struggling to make sense of all aspects of the complex modern phenomena of communication.

Second, communication education developed a line of inquiry that opened up the possibilities for systematic and sustained *criticism* of the press and the mass media within the very curricula that were also preparing reporters, editors, broadcasters and filmmakers for work in their professions. Much more so than most other fields of professional education and at a much earlier stage in its intellectual development, mass communication studies formally introduced notions of social responsibility and ethical practice into its required course work. Questions about the impacts of the press and media in society and what might be their shortcomings and contradictions were, in time, just as much a part of a media practitioner's education as training in basic skills.

Third, the universe of media education had expanded considerably to incorporate *all forms of contemporary media technology*. It came to include not only the older models of mass media broadcasting and common carrier telecommunications, but also the entire realm of modern information technology. Media students were increasingly expected to be aware of and familiar with the differential capacities and implications of the diverse methods and industries of communication. Typically grounded in writing, speaking, and production skills associated with the various media, communication curricula tended to be increasingly sensitive to explorations of how changing forms of communication altered capacities for professional expression and public cultural experience.

Fourth, mass media studies were firmly grounded in a well-developed sense of how communication is *socially constructed*. Of all the professional fields in the academy, journalism and broadcasting programs were the closest to engineering in their concerns with changing technology, but because of their origins in the humanities and social sciences, such programs also developed a tradition of inquiry into the histories and guiding forces of the specific communications technologies and into their economic, social, and policy implications. Those considerations were closely associated with issues in the history, philosophy, and ethics of technology generally, all feeding back into the social criticism strains of communication.

Social Models of Communication Research

Initially the dominant social models were behavioral and functionalist, and most of what passed for advanced communication research seemed to be focusing almost exclusively on effects problems. However, the situation was never as hegemonic as it appeared. In various universities and research centers, other approaches were being developed. For instance, questions about press history led some to broader considerations of intellectual and social history and the role therein of all media (Jensen, 1957). What originally were prescriptive concerns about media law and regulation led some scholars to more critical inquiries about the contradictions in communication policy and the public purposes of the media (Rowland, 1986), combining with the parallel professional media education traditions of social responsibility and ethics. Interests in the social experience of modern communication led to closer study of their contents, to how communication works within culture and to the complicated problems of meaning and interpretation (Curran, Gurevitch, & Wollacott, 1982; Hall, Hobson, Lowe, & Willis, 1980). In all these instances and many others, it turned out that there were parallel sets of new or newly revived questions being raised in the parent disciplines and professional fields. Problems of critical theory were reemerging in literature, all the social sciences, law, and even economics, as were questions of cultural groundings in textual interpretation and social inquiry. Because of their focus on the mediated process of communication—symbolic construction—mass media studies began to discover and even shape aspects of these approaches themselves. Due to their interdisciplinary nature and their close contact with the ferment otherwise occurring in the cognate fields, this process of paradigmatic review in communication was particularly intense ("Ferment," 1983).

In any case, it had dramatic implications for the very notion of communication, moving it away from the dominance of linear, transmission, and atheoretical models of mid-century toward interpretive, ritualistic, and more historically and theoretically conscious approaches of both an older and newer era (Carey, 1989). If they centered on anything, these approaches tended to recognize that all aspects of communication—the technologies as well as their industrial structures, their messages and content, and even their public uses—are all generated by human social activity. They came to a well-articulated understanding of how media systems and communications technologies are the products of given societies in particular places and particular times, imbued with certain belief patterns, values, and institutional constraints; how they are neither natural nor neutral (Williams, 1974).

To look at it another way, the history of the various journalism, mass communication, and speech communication approaches has provided the time and experience to develop a well-established pattern of inquiry into all aspects of modern communication. Working at the interdisciplinary intersection among the humanities, social sciences, and relevant professional fields, communication research has been able to establish a perspective on the communication process that sees all its principal elements in relationship to one another.

This approach is concerned with the way communication as a whole is the study of the creation of meaning in people's lives. It suggests that the study of any aspect of communication, including telecommunications, focuses not just on the media and their technologies, but also on the way they are embedded in the entire array of practices and forms that constitute human communication and culture. It therefore tends to take cognizance of the entire process—a triangle whose points include the institutions and technologies, the products and contents of those systems, and the social conditions of their use and consequences—all cast against a background of interdisciplinary inquiry. This relationship is expressed graphically in Figure 1.

This diagram demonstrates that contemporary communication research pays careful attention to the set of broader paradigmatic perspectives that are drawn from the predominant undercurrents of debate in the social sciences and humanities. Communications technology and systems are thereby considered against several of the emerging intellectual constructs

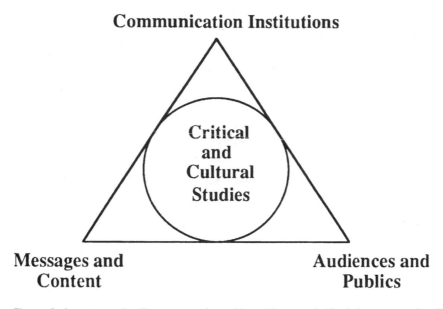

Figure 1. A communication research and teaching model for telecommunications.

in the modern academy—against larger questions about human life and social order; beliefs, values, ethics, and meaning; and ideology, philosophy, and culture.

The *institutional* aspects of this study include questions about media organizations and structures, ownership and control, resources and technologies, and professional standards and models. The media and means of cultural production are studied as national and transnational institutions bound by exterior and interior realities and engaged in the social construction of media content and services as foundries for the creation of meaning. That portion of the analysis tends to see the technologies and industries of telecommunications in much the same way it perceives the mass media generally (the press, broadcasting, cable, film, and the information industries) as the products of human economic and political endeavor in given times and places subject to specific social forces.

This approach attends to the *content and product* of the systems of human technology in a similar manner. It is oriented toward criticism and interpretation, exploring manifest and latent meanings in all forms of expression and information in the society. Although in its applied forms this portion of the triangle considers the skills of producing communication, it is also heavily engaged in the analysis of that material and in contributing to the scholarly capacity to discuss popular culture in a disciplined manner, examining linkages between the institutional sources of communication and the social conditions of their use.

The third part of the triangle considers communication as both private and public *social experience,* pointing to the place of the media and telecommunications in regular human life. In this respect, communication research is concerned with the way in which people use and are affected by the modern communication environment, the impact of the media and technologies on their lives. But this domain of inquiry goes well beyond the simple models of effects. It also examines the broader social, cultural, and ideological conditions in which people live and how in constructing those conditions people themselves shape and affect the communication systems of which they are a part.

The Required Syllabus for Telecommunications

What then does this history mean for the contemporary study of telecommunications in a university environment? To begin with, such inquiry has to be much more than instruction in the applied technical and management skills about such things as system load capacity, network design, and efficiency analysis. It cannot reside solely, nor even largely, in the "gee whiz" school of technology study that tends to celebrate the latest bells and whistles. It must have at its core—not on its margins—a rich domain of theoretical inquiry that raises fundamental questions about the changing nature of contemporary society, economics, and industrial process, and even cultural experience. To be part of a mature interdisciplinary field, telecommunications must bring to bear precisely the sorts of questions about technology and society, social responsibility and ethics, the political economics of telecommunications history, and the changing cultural forms of an information society that have long been pursued in communication research.

No student of any program in telecommunications should graduate without some exposure to the diverse work on the rise and characteristics of electronic culture by Harold Adams Innis (1950, 1951), Walter Ong (1967), or Joshua Meyrowitz (1985). They should be familiar with the debates in the philosophy of technology as reflected in the notion of *techné* as *poiesis,* in technology as a revealing, in Martin Heidegger (1977), in the analysis of *la technique* and the problem of technological determinism in Jacques Ellul (1980), and in critiques of the mythology of technology as value-free and politically neutral in Clifford Christians (1989) or Arnold Pacey (1983). They should be exposed to the analysis of the interrelationships among science, technology, and industrial policy by David Noble (1977) and to the struggle to redefine the rational society and reclaim the public sphere in Jürgen Habermas (1970). None should have failed to consider the critiques of conventional marketplace communication economics and the analysis of the global culture and information industries represented by Dallas Smythe (1957, 1982) or Nicholas Garnham (1990), or of media ownership patterns by Ben Bagdikian (1990). They should know Robert Horwitz's (1989) interpretation of the contradictions in telecommunications deregulation and Daniel Schiller's (1982) and Jeremy

Tunstall's (1986) analyses of the interactions among the government, broadcast, telephone, and computer industries, and the implications of such sources for readings of the AT&T divestiture and modified final judgment decisions. They should all have some clear understanding of the arguments wrapped up in the debates about dependency and the New World Information Order (International Commission, 1980; Roach, 1987) and how Anthony Smith (1980) approaches the problems of technological convergence and global information policies. They must be aware that there are two International Communication Associations (ICAs) and that there are important implications for research and teaching programs that are oriented more toward one than the other.[2]

At its core the academic field of telecommunications must transcend the problem of technology. Surely its research and teaching curricula must provide some basis of technical understanding. But, as a matter of intellectual and moral principle, and as was learned long ago in communication, telecommunications must never become the simple creature of the technologies it studies and the institutions that promote it. It is incumbent upon telecommunications to foster a critical social understanding of the changing information technologies, to understand not only their promises but also their limitations and vulnerabilities, and to apply that same interpretive discipline on the parallel problems in telecommunications policy. These are the sorts of approaches and questions already built into a generation or more of undergraduate and graduate education in communication.[3] Many telecommunications research and teaching programs are already imbued with such values—largely because of their roots in the journalism, speech, and mass communication research traditions—and as the terms of communication research have themselves changed, the association with that field has become even more valuable. All other programs carrying the name *telecommunications,* and their supervising university academic administrations, would profit by closely examining that heritage and drawing more deliberately from its experience.

References

Bagdikian, B. (1990). *The media monopoly* (3rd ed.). Boston: Beacon Press.

Barnouw, E. (1966). *A tower in Babel.* New York: Oxford University Press.

[2] One ICA is, of course, the academic association of communication scholars that sponsors the *Journal of Communication*; the other is principally a trade organization of major telecommunications industry interests that, among other things, sponsors some of the more technically oriented, engineering-based, academic telecommunications programs.

[3] A useful corrective might be to change the name of the field from telecommunication*s* to telecommunication. A generation ago a similar change in communication, and the related debate about the implications of dropping the "s," did much to clarify the social versus technical nature of that field.

Brubaker, J. S., & Rudy, W. (1976). *Higher education in transition: A history of American colleges and universities, 1636–1976* (3rd ed.). New York: Harper & Row.

Carey, J. W. (1989). *Communication as culture: Essays on media and society.* Boston: Unwin Hyman.

Christians, C. (1989). A theory of normative technology. In E. F. Byrne & J. C. Pitt (Eds.), *Technological transformation: Contextual and conceptual implications* (pp. 123–140). Dordrecht, the Netherlands: Kluwer Academic Publishers.

Commission on Freedom of the Press. (1947). *A free and responsible press: A general report on mass communication: Newspapers, radio, motion pictures, magazines, and books.* Chicago: University of Chicago Press.

Curran, J., Gurevitch, M., and Wollacott, J. (1982). The study of media: Theoretical approaches. In M. Gurevitch, T. Bennett, J. Curran, & J. Woollacott (Eds.), *Culture, society and the media* (pp. 11–29). London: Methuen.

Czitrom, D. J. (1982). *Media and the American mind: From Morse to McLuhan.* Chapel Hill: University of North Carolina Press.

Ellul, J. (1980). *The technological system.* New York: Continuum.

Ferment in the field: Communications scholars address critical issues and research tasks of the discipline [Special issue]. (1983). *Journal of Communication, 33*(3).

Garnham, N. (1990). *Capitalism and communication: Global culture and the economics of information.* London: Sage.

Habermas, J. (1970). *Toward a rational society: Student protest, science, and politics* (J. J. Shapiro, Trans.). Boston: Beacon Press.

Hall, S., Hobson, D., Lowe, A., & Willis, P. (Eds.). (1980). *Culture, media, language.* London: Hutchinson.

Heidegger, M. (1977). *The question concerning technology and other essays* (W. Lovitt, Trans.). New York: Harper & Row.

Horwitz, R. B. (1989). *The irony of regulatory reform: The deregulation of American telecommunications.* New York: Oxford University Press.

Innis, H. A. (1950). *Empire and communications.* Oxford: Clarendon Press.

Innis, H. A. (1951). *The bias of communication.* Toronto: University of Toronto Press.

International Commission for the Study of Communication Problems. (1980). *Many voices, one world: Communication and society, today and tomorrow: Towards a new, more just and more efficient world information and communication order.* London: K. Page.

Jensen, J. W. (1957). *Liberalism, democracy and the mass media.* Unpublished doctoral dissertation, University of Illinois College of Communications, Urbana.

Jowett, G. (1976). *Film: The democratic art.* Boston: Little, Brown.

Kohlstedt, S. G. (1988). The phenomenon of professions. In P. T. Marsh (Ed.), *Contesting the boundaries of liberal and professional education: The Syracuse experiment* (pp. 44–53). Syracuse, NY: Syracuse University Press.

Marvin, C. (1983). Space, time, and captive communication history. In M. Mander (Ed.), *Communication in transition* (pp. 20–38). New York: Praeger.

Meyrowitz, J. (1985). *No sense of place.* New York: Oxford University Press.

Noble, D. F. (1977). *America by design: Science, technology and the rise of corporate capitalism.* New York: Alfred A. Knopf.

Ong, W. J. (1967). *Presence of the word*. New Haven: Yale University Press.

Pacey, A. (1983). *The culture of technology*. Cambridge, MA: MIT Press.

Peterson's annual guides to graduate study. Book 5: Graduate programs in engineering and applied sciences. (1993). Princeton, NJ: Peterson's Guides.

Roach, C. (1987). The U.S. position on the New World Information and Communication Order. *Journal of Communication, 37*(4), 36–51.

Rowland, W. D., Jr. (1986). American telecommunications policy research: Its contradictory origins and influences. *Media, Culture and Society, 8*(2), 159–182.

Schiller, D. (1982). *Telematics and government*. Norwood, NJ: Ablex.

Siebert, F. S., Peterson, T., & Schramm, W. (1956). *Four theories of the press*. Urbana: University of Illinois Press.

Smith, A. (1980). *The geopolitics of information: How western culture dominates the world*. Boston: Faber & Faber.

Smythe, D. W. (1957). *The structure and policy of electronic communication*. Urbana: University of Illinois Press.

Smythe, D. W. (1982). Radio: Deregulation and the relation of the private and public sectors. *Journal of Communication, 32*(1), 192–200.

Tunstall, J. (1986). *Communication deregulation: The unleashing of America's communication*. Oxford: Basil Blackwell.

Williams, R. (1974). *Television: Technology and cultural form*. New York: Schocken Books.

Creating Imagined Communities: Development Communication and the Challenge of Feminism

by H. Leslie Steeves, University of Oregon

The failures of the first three development decades have been well documented. In efforts to reduce failures, communication scholars and practitioners have proposed strategies to increase recipient participation in projects. Their proposals have not been adequately implemented, and many groups are still bypassed or even harmed by development efforts. Women are usually the most neglected in these groups.

This essay draws on three areas of critical scholarship, each initiated a little over 20 years ago, to outline an expanded vision for development communication. Each, in different ways, addresses the failures of development programs under modernization. The first has been called *women in development* (WID), and emerged in part from the work of Danish economist Ester Boserup. Boserup (1970) demonstrated that women have crucial economic roles, yet are most excluded from the benefits of development and suffer the most from its negative side effects. The second area includes *grass-roots critiques of development,* of which Paolo Freire's 1970 book, *Pedagogy of the Oppressed,* is especially well-known. Freire, a Brazilian educator, argued that gaining freedom from oppression is a central human task and that emancipatory dialogue is necessary for liberation. The third area, *political economy of communication,* initiated global discussions about uses of power in communication industries and systems. Its early proponents, who began this work in the 1960s, included Canadian economist Dallas Smythe and U.S. political scientist Herbert Schiller.

Most of the scholars associated with these areas of critique share similar values about the equality of human beings and the need for a new, more just, more compassionate, and more inclusive world order. Taken together, the lines of scholarship they helped create also have produced much

H. Leslie Steeves is an associate professor in the School of Journalism and Communication at the University of Oregon, Eugene, Oregon. The author thanks Rashmi Luthra, Angharad Valdivia, Janet Wasko, and Nancy Worthington for their generous suggestions. She also is grateful to Rebecca Arbogast and Archie Smith, Jr. for many conversations and their influence on her work.

evidence for a number of key assumptions. They include the following: (a) to succeed, development communication must liberate; (b) development communication planning must proceed from the perspective of women; (c) women's perspective can be understood only in context; (d) context is understood in dialogue; and (e) dialogue must engage many levels and types of power. These assumptions suggest the necessity of re-thinking development and communication from the standpoint of women and in a way that is historically, contextually, and methodologically appropriate. While this is not necessarily inconsistent with contemporary revisions of early, one-way development communication models, I believe that the insights highlighted here must be prioritized, indeed centralized, in a revised model and in projects. Neglecting these insights will, I argue, continue to result in theories, policies, and projects that are ineffective, if not instruments of continued oppression.

Development Communication Must Liberate

This assertion necessitates a holistic, systemic understanding of development and communication as ongoing and inseparable processes defined in relationship with others and with the environment. As critical and feminist scholars argue, communication is not just the transmission of messages from A to B, but also the shared and ever-changing meaning that is created via relationships of many types and levels, within particular historical, economic, political, and cultural contexts. Similarly, development is not the simple progression of society from traditional to modern—or from totalitarian to democratic, depending on the biases of those directing development. Development is a multifaceted, relational process that entails many levels and layers of communication and always has a unique and complex context.

The fact that systemic understandings of communication and development have not been dominant in scholarship and practice has been addressed by those in the political economy tradition and in the grass-roots tradition. Sussman and Lent (1991, pp. 3–17) trace the notion of development communication as the transmission (or diffusion) of technological modernization from the ethnocentric writings of Daniel Lerner, Wilber Schramm, Ithiel de Sola Pool, Lucien Pye, and others to modern-day technology enthusiasts, including, for example, Tom Forester, Meheroo Jussawalla, and John Naisbitt. At the global level, this view is manifest in capitalist expansionism and policies of free flow, free trade, and structural adjustment that benefit the upper classes. More locally, this view is consistent with the use of marketing experts and advertising firms to research and plan persuasive, social marketing campaigns.

The *transmission–persuasion* tradition remains strong both because it appeals to the profit motive and because its measures of success are apparent. "Positive" outcomes are gauged by numerical indicators: change in GNP, technological growth, or numbers of people persuaded to adopt a new technology, such as contraception. All of these indicators translate into profit, accompanied by an "unrestrained eagerness to possess" and a "strictly materialistic concept of existence" (Freire, 1970, p. 44).

Freire and many in the traditions considered here believe that financial and technological indicators mean little if most people remain oppressed. Hence, to be positive, development must support the universally human quest for "freedom and justice," a struggle that is "thwarted by injustice, exploitation, oppression, and the violence of oppressors" (Freire, 1970, p. 28). This struggle must aim to liberate both oppressors and oppressed, the former because they "do not perceive their monopoly on having more as a privilege which dehumanizes others and themselves" (p. 45). Julius Nyerere (e.g., 1976), who also makes this argument, points out that liberation is fundamentally an individual transformative process leading to expanded consciousness and power (p. 9). It is a personal choice that cannot be imposed from the outside. At the same time, Nyerere and Freire assume that people do not make such choices in isolation but require active and reflexive engagement with others and the natural environment. They further assume that emancipatory power must come from the weakness of the oppressed, never the tyranny of oppressors (e.g., Freire, 1970, p. 28). Hence, development must begin from the perspective of the *most* oppressed in each context.

Though both Nyerere and Freire denounce gender oppression, this is a relatively minor concern in their writings. Very little scholarship related to communication in developing countries, including work emphasizing liberation themes and/or critiquing macro-level power imbalances, has addressed gender issues.[1] Although some development communication scholars recognize myriad nation- and class-related biases in projects,[2] most have failed to consider the reality of gender oppression. However, overwhelming evidence that gender adds a powerful layer of oppression to class, race, and nationality—that the liberation of the poor or other groups does not automatically result in the liberation of women—makes it important to revise Freire's and related perspectives in light of this evidence. Feminist scholars argue that the greatest neglect and harm in development are experienced by women, that justice demands greater attention to women, and that development epistemologies must be altered for consistency with women's ways of knowing. Hence, they advocate a systemic perspective that prioritizes women's needs.

[1] For examples of this scholarship, see citations and/or essays in Gallagher (1992), Kramarae and Treichler (1988), Rush and Allen (1989), Steeves and Arbogast (in press), and Traber and Lee (1991).

[2] See especially Melkote's (1991) detailed discussion of pro-literacy, pro-innovation, pro-persuasion, pro-source, and pro-mass media biases, among other biases.

Women's Perspective as a Basis for Development Communication

Boserup (1970) used data from Africa to demonstrate that women's historically productive roles in agriculture were harmed by colonial and later aid policies that focused on men, who were encouraged to work for salaries on nonfood cash crops, in mines or in cities. Boserup's argument for integrating women in development has been credited with precipitating a movement that led to the 1975–85 U.N. Decade for Women, as well as the creation of women-specific policies and programs in most aid agencies (e.g., Jaquette, 1990, p. 55). Many studies show, however, that these changes have been more cosmetic than substantive and that women continue to be neglected. This certainly has been the case in communication as well as other aspects of development.

The argument I make here follows, in part, from Freire's concern with those who are most oppressed. Women's perspective must ground development communication because women usually experience the most oppression and injustice and hence are in the best position to inform development strategies. "Oppression" refers to all the systematically related "forces and barriers" that function to restrict options, immobilize, mold, and reduce (Frye, 1983, pp. 4, 7). Activists confronting oppression usually argue ethically, politically, and/or legally that it is unjust in the sense that rights are undeservedly restricted or denied. Claims for justice are typically based on arguments about equality or earned merit, or sometimes, welfare based on need.

Boserup's argument is for justice based on equality and merit (Jaquette, 1990). Critics have challenged Boserup's emphasis on integrating women in capitalist economic development, which ignores women's reproductive work in measures of production (or merit) and exploits their labor (Mohanty, 1991b; Sen & Grown, 1987). Critics also claim that Boserup neglects urban women and cultural, nonmaterial obstacles to women in development (Valdivia, 1991). These criticisms suggest a more complex meaning of justice for women. This is the emphasis of an international group of activists, researchers, and policymakers that formed during the Decade for Women. The group, Development Alternatives with Women for a New Era (DAWN), was introduced by Sen and Grown (1987) and aims to develop "frameworks and methods to attain the goals of economic and social justice, peace, and development free of all forms of oppression by gender, class, race, and nation" (p. 9).

According to DAWN, development must begin with women because women provide the best vantage point for evaluating development for at least three reasons: women are vital to the reproduction and survival of human societies; women's work outside of the home is varied and widespread; and antipoverty programs should prioritize the poorest group, usually women (Sen & Grown, pp. 23–24). In addition, a focus on women may help draw attention to the global "chain of oppression," a chain that ultimately points to "economic, political, and cultural processes that reserve resources, power, and control for small groups of people" (p. 20).

DAWN's argument is for justice based on equality and merit, with merit defined in recognition of women's *complete* roles. The group seeks not only to liberate women, but to change unjust power structures in the process. This is consistent with Freire's belief that global justice requires a focus on the most oppressed. Feminist theorist Alison Jaggar (1989) has argued similarly that the most oppressed are in the best position to reveal societal injustice and provide "the possible beginnings of a society in which all could thrive" (p. 162). Assuming a central role for women in development, who are they and what are their circumstances?

Context Defines Women's Perspective

There has been an unfortunate tendency in much scholarship and practice to treat Western women as the norm and Third World women as monolithically oppressed by forces of which the former have long been free: poverty, dependency, ignorance, illiteracy, tradition, violence, reproductive constraints. This denies the complexity of Third World women's experience. It also undermines Western women's need to examine their own oppressive contexts and their participation in unjust global structures. Mohanty (1991b) uses the term *discursive colonization* to describe feminist writings that engage in "ethnocentric universalism" by neglecting the heterogeneous reality of Third World women's lives (pp. 53–55). However, Mohanty and others (e.g., the DAWN collective) do believe that Third World women have an urgent need to develop coalitions to seek voice. The basis for such coalitions must be a common *context* of struggle against oppression by gender, class, race, ethnicity, and nation, including colonization (Mohanty, 1991a, p. 7; Sen & Grown, 1987, p. 19). This context includes women in industrialized countries who also experience many forms of domination. Mohanty uses the idea of "imagined communities" to highlight the *political* (not biological, cultural, regional, or national) basis for forming coalitions:

> *Thus, it is not color or sex which constructs the ground for these struggles. Rather it is the way we think about race, class, and gender—the political links we choose to make among and between struggles. Thus, potentially, women of all colors (including white women) can align themselves with and participate in these struggles. However, clearly, our relation to and centrality with particular struggles depend on our different, often conflictual, locations and histories. This, then, is what provisionally holds the essays in this text on "third world women and the politics of feminism" together: imagined communities of women with divergent histories and social locations, woven together by the political*

threads of opposition to forms of domination that are not only pervasive but also systemic. (1991a, p. 4)[3]

Given this understanding of Third World women's context and activism, is the language of feminism appropriate? The term *feminism* is exceedingly problematic among many Third World women, who feel they have nothing in common with the Western feminisms that the term connotes (e.g., Kishwar, 1991). However, if feminism is understood to mean a political movement that includes many forms and levels of oppression, while accounting for differences in collective and individual experiences, it is clear that women everywhere have engaged in feminist struggles, as have some men. Hence, many Third World activists continue to use the term, while always positioning their views against, for example, a liberal emphasis on individual rights.

In keeping with a complex view of feminism, scholars addressing the situation of Third World women sometimes employ a socialist feminist perspective (e.g., see Mohanty, 1991a, p. 5; Steeves & Arbogast, 1993). Yet many feminist scholars who reject categories or prefer a different category are sensitive to connections among forms of oppression and the need for a critical, historical, and reflexive approach in dialogue with all groups involved. The label is not what is most important, and may at times be divisive (Gallagher, 1992, p. 2). What is important is a political consciousness of oppression, its varied forms and sources, and a willingness to transform one's understandings (and labels) in dialogue and collective action.

I believe the idea of dialogue is crucial to the problem posed by context. How can Third World women's context be understood by others—by upper-class women in the Third World, by minority women in industrialized countries, by white Western feminists, by development communication specialists? If these groups cannot communicate, there is little hope that genuine understanding can penetrate upward—to those who wield the greatest power.

Context Is Understood in Dialogue

Since the mid-1970s nearly everyone writing about development communication has assumed the demise of the dominant paradigm and its replacement by a two-way participatory model. The development support communication (DSC) perspective emphasizes ongoing participation by project recipients (e.g., Melkote, 1991). Many scholars have argued that

---

[3] My use of "imagined communities" was inspired by Mohanty, who borrowed and extended the concept first used by Anderson (1983) in his critical examination of the nation as an imagined political community.

participation requires broad media accessibility, suggesting the use of small-scale or "intermediate" communication technologies. Social marketing campaigns, which have become common, stress the importance of research at every stage (planning, process, and evaluation research) to ensure that projects remain relevant to recipients' needs (e.g., Manoff, 1985).

However, several overlapping problems have remained. First, development communication projects still frequently fail. Second, most projects still follow a one-way, transmission–persuasion model. Third, theories of participation and interactivity advanced thus far have been overly simplistic. One major inadequacy is the assumption that participants are equal, an assumption that ignores huge power differentials in Third World communication situations (Gonzalez, 1989, p. 309). Hence, what is needed is a model of communication that combines dialogic participation with a realistic appraisal of society, an appraisal committed to reducing power inequities that sustain oppression.

This essay cannot explicate the details of such a model or vision of communication, which would likely require drawing on theories of, for example, political economy (Karl Marx); hegemony (Antonio Gramsci, Raymond Williams); ideology and representation (Louis Althusser, Stuart Hall); language, culture, and power (Jürgen Habermas); and knowledge and power (Michel Foucault). Feminists making use of such theories must simultaneously challenge the theories' misogynous assumptions or draw on the works of others who have done so. Many feminists in cultural studies have utilized some combination of these and related traditions to ground their arguments. Of the feminist scholars concerned with Third World women and communication, Gayatri Spivak (e.g., 1988) is noteworthy in her sophisticated and critical use of Marx, Foucault, Althusser, Gramsci, and others to examine the question: "Can the subaltern speak?"

Those who assume the importance of dialogue while challenging unequal power relations include feminists and Third World scholars and practitioners concerned with empowerment and liberation as the goal of communication. Their arguments overlap considerably. Many have critiqued methods of social science because they are often practically inappropriate and also inconsistent with the ways of knowing that are meaningful to people living in non-Western cultures, to women, and to minority groups. Feminist and Third World epistemologists have stressed the necessity of grounding research and practice, reflection and action (i.e., praxis), in the lived experience of people. Haraway (1988) uses the idea of "situated knowledges" to describe this approach, in contrast to researcher-defined "objective" approaches that lead to "gross error and false knowledge of many kinds" (pp. 592–593).

In fact, there is evidence that development communication campaigns driven by objective data gathering (e.g., social marketing campaigns) have failed to grasp women's needs, even when women have been important publics.[4] However, understanding one another's lived experiences or situated knowledges requires a deep level of sharing—or dialogue—that will create emancipatory meaning and solutions, as the context of participants' lives and beliefs is increasingly revealed. Freire (1970, 1973) introduced the idea of "conscientização"—that is, "the *development* of the awakening of critical awareness" (1973, p. 19, emphasis in original)—to describe the radical outcome of dialogue. Lather (1986, pp. 266–267) uses the term "emancipatory theory," meaning "an expression and elaboration of politically popular feelings rather than an abstract framework imposed by intellectuals." Freire argues that dialogue, above all, is an act of love, requiring courage, humility, trust, hope, and critical thinking (1970, pp. 78–80). This is risky because it leads to expanded consciousness, hence loss and change, a point also made by feminist scholars (e.g., Jaggar, 1989, p. 161).

Given a commitment to search for emancipatory solutions dialogically, by what strategies is this achieved? First, it is important to recognize that dialogic communication, or interactivity, is a complex process with many multidimensional components, including feedback (e.g., Gonzalez, 1989). Second, no strategy is optimal across situations, and most contexts call for combined approaches to counteract biases. Additionally, dialogic understanding can never be perfect but is always a matter of degree. Specific techniques especially likely to facilitate sharing, increased awareness, and a reduced distance between dialogue participants include Freire's (1970, 1973) interpersonal and small group methodologies that help reveal, codify (illustrate), and critically address concepts that hold meaning for participants. Codifications may be verbal or utilize visuals. McLean (1992) argues that focus group methods may be appropriate in African contexts, where the group and not the individual is already the framework for attitudes and behaviors. Many scholars advocate using traditional forms of communication, such as theater, storytelling, and song, to increase understanding. In addition, the importance of working with indigenous groups, including women's groups, has been widely recognized.

While these techniques are significant, they are only techniques. Grassroots media and indigenous groups can be utilized in one-way persuasive campaigns, as can interpersonal consciousness-raising techniques. Much more significant is the commitment of participants to dialogue and liberation. But of what use is local commitment and dialogue if constrained or

---

[4] Luthra (e.g., 1991) examined the Family Planning Social Marketing Project of the U.S. Agency for International Development (U.S.A.I.D.) in Bangladesh and discovered commercial biases that resulted in a failure to grasp important health and communication needs of poor women. Worthington's (1992) study of the public health communication component of U.S.A.I.D.'s AIDS Technical Support Project in East Africa yielded similar findings.

sabotaged by larger power structures? Can emancipatory struggle cross levels of economic, political, and ideological power?

Challenging Power Relations

At the global level, the profit-driven interests of transnational corporations, donor agencies, and national governments are responsible for rapidly expanding media and information technologies, with little consideration of use and access by the majority of people, including women and the poor. Moreover, as governments acquire complex new technologies, likely consequences include a greater debt burden, a need for profitable services to repay the debt, and a reduction of public communication and information services that do benefit marginalized groups. Additional consequences for women, documented in detail elsewhere, are varied and include continued and increased exploitation as workers in transnational plants; increased exposure to oppressive, Western gender representations, including advertising and pornography; and decreased access to technological policy-making, as corporations merge and power is consolidated.

Given these realities and assuming all power is incomplete, what is the contested terrain where disenfranchised women and others in developing countries may find an opening for dialogue across wide gaps of interest— by class, gender, race, and nation? Within large aid agencies, the failures of top-down and trickle-down development in the 1960s and 1970s, followed by overlapping crises of food, population, and environment in the 1980s, have resulted in an enormous recent interest in sustainable development—that is, environmentally sensitive projects that improve people's quality of life without depleting nature. The growing recognition that women are the gatekeepers to the success of these efforts has resulted in increased funding for grass-roots projects involving women.[5] This recognition of women's central role in addressing massive global problems is long overdue. But questions remain about women's actual experiences in these projects: Are women being exploited to serve interests other than their own?[6] Do competing forces (global, national, and local) act to neutralize or sabotage any gains made by women?[7]

---

[5] For examples, see recent reports of multilateral agencies such as the World Bank, the U.N. Development Program, and the U.N. Fund for Population Activities.

[6] Feminist critics of fertility and environmental policies and projects have noted the tendency to blame women for these global problems and then inundate them with technologies (e.g., Jaquette & Staudt, 1985).

[7] For example, International Monetary Fund and World Bank programs of "structural adjustment" made stipulations on participating countries, including proportions of GNP to be spent on social welfare. These stipulations have negated many gains made by women at the grass roots (e.g., Vickers, 1991).

Communication experts must likewise ask whether women are heard beyond the grass roots. This relates to issues discussed previously—of commitment, courage, and risk, as well as incentive. If dialogic risks are accepted, the mechanics of dialogue pose a major challenge. Possibilities include the creative use of video technology, a strategy being used with some success by groups such as the Worldview International Foundation, based in Sri Lanka. But why should multilateral agency (e.g., World Bank) executives engage in dialogue with poor women? It seems unlikely without substantial persuasive efforts via coalition building among women and men with common political interests.

Here there is some basis for optimism because of greatly increased organizational activity among women in developing countries since the start of the Decade for Women (March & Taqqu, 1986). The global ecofeminism and feminist peace movements have further encouraged this activity. Women's media and information projects constitute a crucial area of activism. Besides creating alternative resources, women's groups around the world are trying to directly influence national and local laws and policies. Communication and information policy-making in developing countries has been encouraged—especially by critical scholars and by UNESCO—as a means of curbing cultural and technological expansionism and imbalances in information flows. However, as yet few developing countries have such comprehensive policies. Where policies are under discussion, there is an urgent need to include representatives of women, the poor, and minority ethnic groups. Once women gain a voice in policy discussions, important areas that might be addressed include women's ownership of and employment in media and information industries; women's access to useful information through communication; representations of women in media content, both imported and locally produced; and the gender implications of technologies selected, including their creative potential for coalition building and their accessibility to oppressed groups. Such gender-sensitive policy-making, alongside continued global networking and lobbying, will be crucial in monitoring and challenging the purportedly sustainable practices of dominant organizations.

Struggles for development and liberation must begin, I have argued, in dialogue with women. Dialogue is never easy, requiring self-reflexive listening and the courage to change. Yet only in this manner can Mohanty's imagined communities emerge and work to create a global community committed to mutual and environmental respect and care.

References

Anderson, B. R. O'G. (1983). *Imagined communities: Reflections on the origin and spread of nationalism*. London: Verso.

Boserup, E. (1970). *Women's role in economic development*. New York: St. Martin's Press.

Freire, P. (1970). *Pedagogy of the oppressed*. New York: Continuum.

Freire, P. (1973). *Education for critical consciousness*. New York: The Seabury Press.

Frye, M. (1983). *The politics of reality: Essays in feminist theory*. Trumansburg, NY: The Crossing Press.

Gallagher, M. (1992). Women and men in the media. *Communication Research Trends, 12*(1), 1–36.

Gonzalez, H. (1989). Interactivity and feedback in Third World development campaigns. *Critical Studies in Mass Communication, 6,* 295–314.

Haraway, D. (1988). Situated knowledges: The science question in feminism and the privilege of partial perspective. *Feminist Studies, 14*(3), 575–599.

Jaggar, A. M. (1989). Love and knowledge: Emotion in feminist epistemology. In A. M. Jaggar & S. R. Bordo (Eds.), *Gender/body/knowledge: Feminist reconstructions of being and knowing* (pp. 145–171). New Brunswick, NJ: Rutgers University Press.

Jaquette, J. (1990). Gender and justice in economic development. In I. Tinker (Ed.), *Persistent inequalities* (pp. 54–68). New York: Oxford University Press.

Jaquette, J., & Staudt, K. (1985). Women as "at risk" reproducers: Biology, science, and population in U.S. foreign policy. In V. Sapiro (Ed.), *Women, biology, and public policy* (pp. 235–268). Beverly Hills, CA: Sage.

Kishwar, M. (1991). Why I do not call myself a feminist. *Manushi, 61*(2), 2–8.

Kramarae, C., & Treichler, P. (Eds.). (1988). Women unaligned aligned. *Women & Language, 11*(2).

Lather, P. (1986). Research as praxis. *Harvard Educational Review, 56*(3), 257–276.

Luthra, R. (1991). Contraceptive social marketing in the third world: A case of multiple transfer. *Gazette, 47*(3), 159–176.

Manoff, R. K. (1985). *Social marketing: A new imperative for public health*. New York: Praeger.

March, K., & Taqqu, R. (1986). *Women's informal associations in developing countries: Catalysts for change?* Boulder, CO: Westview Press.

McLean, P. E. (1992). Methodological shortcomings of communication research in Southern Africa: A critique based on the Swaziland experience. In S. T. K. Boafo & N. A. George (Eds.), *Communication research in Africa: Issues and perspectives* (pp. 89–108). Nairobi, Kenya: African Council for Communication Education.

Melkote, S. R. (1991). *Communication for development in the Third World: Theory and practice*. Newbury Park, CA: Sage.

Mohanty, C. T. (1991a). Introduction: Cartographies of struggle: Third World women and the politics of feminism. In C. T. Mohanty, A. Russo, & L. Torres (Eds.), *Third World women and the politics of feminism* (pp. 1–47). Bloomington: Indiana University Press.

Mohanty, C. T. (1991b). Under western eyes: Feminist scholarship and colonial discourses. In C. T. Mohanty, A. Russo, & L. Torres (Eds.), *Third World women and the politics of feminism* (pp. 51–80). Bloomington: Indiana University Press.

Nyerere, J. K. (1976). Declaration of Dar es Salaam: "Liberated Man—the Purpose of Development." *Convergence, 9*(4), 9–17.

Rush, R. R. & Allen, D. (Eds.). (1989). *Communications at the crossroads: The gender gap connection*. Norwood, NJ: Ablex.

Sen, G., & Grown, C. (1987). *Development, crises, and alternative visions: Third World women's perspectives*. New York: Monthly Review Press.

Spivak, G. C. (1988). Can the subaltern speak? In L. Grossberg & C. Nelson (Eds.), *Marxism and the interpretation of culture* (pp. 271–316). Urbana: University of Illinois Press.

Steeves, H. L., & Arbogast, R. A. (1993). Feminism and communication in development: Ideology, law, ethics, practice. In B. Dervin & U. Hariharan (Eds.), *Progress in communication sciences, 11* (pp. 229–277). Norwood, NJ: Ablex.

Sussman, G., & Lent, J. A. (1991). Introduction: Critical perspectives on communication and Third World development. In G. Sussman & J. A. Lent (Eds.), *Transnational communications: Wiring the Third World* (pp. 1–26). Newbury Park, CA: Sage.

Traber, M., & Lee, P. (Eds.) (1991). Women's perspectives on communication. *Media Development, 38*(2).

Valdivia, A. N. (1991). Gender, press and revolution: A textual analysis of three newspapers in Nicaragua's Sandinista period, 1979–1988. (Doctoral dissertation, University of Illinois at Urbana-Champaign, 1991). *Dissertation Abstracts International, 52*(3), 726A.

Vickers, J. (1991). *Women and the world economic crisis.* London: Zed.

Worthington, N. (1992). *Gender and class in AIDS education: An analysis of the AIDSCOM project in Africa.* Unpublished master's thesis, University of Oregon (Journalism).

Scholarship as Silence

by David Docherty, British Broadcasting Corporation, David Morrison, University of Leeds, and Michael Tracey, University of Colorado at Boulder

In the past decade or two the major industrial societies have seen the assertion of a massive hegemony of the market, the reconstructing of the instruments of culture and information to match a notion of the sovereignty of the individual consumer. These events and processes are of considerable proportions and yet, in country after country, academic mass communication researchers have had little to say to those making policy. The field has in fact become largely a culture in exile from everything but itself, a cloistered world of ritual and enclosed performance within the isolation of the annual conference and the refereed journal. The question left dangling is what should, and can, be the proper relation of the scholar, intellectual, theorist, researcher to a changing world? Is our purpose to understand, influence, and shape, or is it only to understand, or is it indeed to avert our gaze from the visible movement of contemporary history and to peer at the less obvious, the more arcane, the obscure?

The issue is not new. In 1944, C. Wright Mills wrote: "Intellectuals are suffering the tremors of men who face overwhelming defeat. They are worried and distraught, some only half-aware of their condition, others so painfully aware of it that they must obscure it by busy work and self-deception" (quoted in Horowitz, 1967, p. 292).

Mills was, to use Jacoby's term (1987), a "public intellectual," someone who sought to address a general and educated audience. The field of mass communications—indeed, of the academy in general—is now ironically and overwhelmingly peopled by private intellectuals. The character of the privacy is variable, but it is interior, often self-serving, sometimes obsessively careerist, assured, technically proficient. Technique, of course, has long been important in the field of mass communication. There is the passionless, almost dehumanized technique of the empiricist for whom method and the accretion of data seem more important than the construction of a meaningful question, and who seems not to understand that facts should discipline the imagination, not replace it. And then

David Docherty is head of Television Planning and Strategy for the British Broadcasting Corporation. David Morrison is research director at the Institute of Communication Studies, University of Leeds. Michael Tracey is professor and director, Center for Mass Media Research, School of Journalism and Mass Communication, University of Colorado at Boulder.

there is the ideological technician, conversant in this or that theoretical scripture, disciple of this or that theoretician, for whom too often scholarship is a mask behind which sits an isolated, powerless, broken heart lost somewhere amidst the firefly moment of the 1960s or failed dreams of radical social change, all the while cocooned within a righteous certainty that avoids what Thompson once called "the collisions of evidence and the awkward confrontations of experience" (Thompson, 1980, p. 15). Both postures put too much distance between academic inquiry and what we will, in a deliberately contentious manner, call the real world that the policymaker inhabits. Both nurture the irrelevant culture of the cloister.

In solving a problem it is always useful to recognize that one exists in the first place. In 1992, Ellen Wartella, past-president of the International Communication Association (ICA), issued an important and interesting call to arms in an address to a symposium on communication research. She called for a fundamental "redirection of communication research and scholarship and a fundamental change of how we envision them, requiring us to put aside our paradigm debates and focus instead on the public nature of our enterprise" (Wartella, 1992). Wartella referred to three important public policy issues: the Children's Television Act of 1990, the development of Whittle Communications's "Channel One," and the 1992 congressional debate over funding appropriations for public broadcasting. She observed: "These are but a few of the current public issues which cry out for reasoned discussions and informed communication scholarship. Yet too often communication research is either not mentioned in the discussion, is irrelevant, or uninformative, or all three" (Wartella, 1992). Two questions immediately spring to mind: Do we have anything to say that is worth listening to and that might make even a bat's squeak of difference to the human condition, and if we do, do we care enough to say it?

The idea of taking our work into the public sphere was something we took for granted during our time at the Broadcasting Research Unit (BRU) in London. It was an article of faith that good social research was a means to an end, and that end was influence. With hindsight, however, we understand that despite unusually good access to policymakers in government and broadcasting, our ability to influence policy was, to say the least, tentative, nuanced, and, indeed, not infrequently ended in failure: Irrelevance can be an imposed condition as well as a choice.

That is the central dilemma that this paper addresses: Is a monkish condition of isolation and silence the inevitable fate of the mass communication scholar, or is there the possibility of influencing policy debates, if only we rip down the wall that separates the academy from society?

The issue we are raising is the familiar, hard choice between engagement and deliberate disengagement, between administrative and critical research. It is unclear whether Lazarsfeld or Adorno coined the phrase "administrative research." Certainly the relationship of critical theory, represented by Adorno, and administrative research, represented by Lazarsfeld, could never be easy. Lazarsfeld's attempt to unify the two traditions was doomed from the beginning since the implied stance within each tradition is inherently antithetical to the other. However, while Lazarsfeld never shirked the sobriquet "administrative scholar," he did prefer to see his work as benevolent administrative research. That is, it was critical of industry practices, but not critical of the overall social arrangement within which these practices occur. To a considerable extent, though not especially consciously, this was the tradition within which the BRU rested.

Adorno, on the other hand, who was studying radio from within critical theory, could in effect speak to no one, unless one considers other critical theorists an appropriate audience. That is the tradition and condition within which too much of the field of mass communication rests today. It is as if in following a certain line of theoretical argument, there has been a deliberate turning away from any desire to engage with policy and the institutions from which it emerges. For example, if one can detect an ideological fashion in much contemporary theory, it is to privilege individuals and deny their being influenced by forces external to the particularities of their lives. In other words, the issue of determination has all but disappeared. The most obvious evidence of this is in the increasing amount of research on the active viewer, the growth of ethnographic study in mass communication research, and the loud applause for the work of Carey and, particularly, Geertz.

In Carey's important and oft-quoted essay "Mass Communication and Cultural Studies" (1989), something of the problem becomes apparent. He acknowledges the significance of writers and social theorists such as Mills in pointing to "the structural conditions of life on this continent" (p. 39), in particular the concentration of power within a number of interlinked elites, which taken together constitute a power elite. Mills's work on this is highly empirical. Carey concludes, however, that the weakness of such theories is that they have "a relatively crude conception of culture" (p. 39), though he does not mention, for example, Mills's 1959 essay "The Cultural Apparatus," in which Mills argues that, given the structure of power in Western society, some *beings* are better placed than others not only to communicate but to create meaning (quoted in Horowitz, 1967, pp. 405–423).

For sophistication, Carey turns to Geertz and, with great erudition, slides inexorably and rapidly away from any engagement with the articulation of power through communication, and thus from any desire or possibility of engaging with its most obvious expression, policy. Consider, for example, this comment: "The entire imagery of culture as power—the opiate of the people, the hypodermic needle, the product of the environment—denies the functioning of autonomous minds and reduces subjects to trivial machines" (Carey, 1989, p. 52). Alongside this somewhat caricatured representation of the field, Carey offers a definition of cultural studies filtered through a Geertzian lens. Cultural studies, he says, "has far more *modest* objectives than other traditions. It does not seek to explain human behavior in terms of the laws that govern it; rather it seeks to understand it. . . . Cultural studies does not attempt to predict human behavior; rather it attempts to diagnose human meanings . . . to descend deeper into the empirical world" (Carey, 1989, p. 56). Here Carey invokes Geertz, who in turn quotes Weber, though the source is not referenced:

> *For many students of cultural studies the starting point, as with Geertz, is Max Weber. "Believing with Weber, that man is an animal suspended in webs of significance he himself has spun, I take culture to be those webs, and the analysis of it to be therefore not an experimental science in search of law but an interpretive one in search of meaning."* (Carey, 1989, p. 56)

The unreferenced allusion by Geertz is probably to Weber's essay "Objectivity in Social Science," in which Weber does *not* say what is claimed of him. He refers to *culture* as "a finite segment of the meaningless infinity of the world process, a segment on which human beings confer meaning and significance" (quoted in Hughes, 1974, pp. 308–309). He does not say that culture is a multiplicity of subjectivist constructions. Indeed, his work on, and deep concern with, rationality and bureaucratization suggest his firm belief that individuals could be subordinated to abstracted social constructions.

The relevance of this to our argument in this essay is to begin to suggest that maybe—we do not wish to be any stronger than that—an important part of the field of mass communication is in a state of denial. In proposing that cultural meanings are varied, multiple, and constructed from the ground up, not the top down, cultural studies ends up blind to the fact of the exercise of power in human affairs. That is why so much of the work done from this perspective is dogged by the "so what" question, and the question of whether we abandon the whole notion of ideology. Do we treat ideology as free-floating, there but not caused? Do we in fact see no equation between the institutionally based articulation of particular symbol systems and the formation of public consciousness? Or need we really go back to Mills's description of the management of symbols undertaken by what he referred to as "the observation posts, the interpretation centers, the presentation depots which in contemporary society are established by means of what I am going to call the cultural apparatus" that fashion "our standards of credibility, our definitions of reality, our modes of sensibility—as well as our immediate opinions and images" (Horowitz, 1967, pp. 406–407). As Mills might have said, if there are webs of significance, who's the real spider?

So here, in part at least, lies the answer to Wartella's challenge: The field has become increasingly confused as to where power lies in this society, or fearful of addressing centers of power, since in so doing one necessarily undermines the lionizing, the celebration, of ordinary folk. The reasons why would require a full-scale intellectual history, but the net result is that those who have pitched their tents on this particular part of the field seem to have little sense of or, outside of their departments, caring for the use of power within society. They have become curiously disengaged from the sources and uses of power, and thus can never properly confront its institutionalized expression, policy.

Our own experience working with the BRU in London suggests some of the ways in which one can, and cannot, influence policy and policy debates. It was quite clear that in order to engage with public policy debates we would have to have to play a numbers game. Clever thinking, elegant essays, treatises on history, disquisitions on philosophy, values and culture were important but not enough if we were to be taken seriously by those with power over policy.

An example of this was the government's establishment of a committee under the chairmanship of Alan Peacock to inquire into the future financing of the BBC—in particular, whether the BBC should be made to take advertising. In 1985, as the Peacock Committee sat, a number of press reports appeared which showed that about 70 percent of the public believed that the BBC should be made to take advertising, thus providing a kind of populist legitimacy to the demands of the Conservative Party's right wing and the major advertising agencies. We were, to say the least, skeptical.

With financial support from the then Greater London Council, the Independent Broadcasting Authority, and our own funds, we were able to support a new survey, commission National Opinion Polls (NOP) to undertake the field research, and produce a report, *Invisible Citizens* (Morrison, 1986), which systematically challenged the position of the advertisers. Before the book was published, however, a piece was prepared for *The Guardian's* media page, then edited by Peter Fiddick. This article presented, in very concise form, the report's main conclusion, which was that the public's attitude toward the license fee was far more complex than the polls had suggested. That article, and then the book—the publication of which we supported financially to ensure that, unlike many scholarly pieces, it appeared in the same century as the research on which it was based—were used by a number of those involved in debating the BBC's future funding. From one perspective, it was our most successful foray into the debates about the future of the BBC. We were, however, under no illusions that we had convinced the government of the virtues of public broadcasting. What the research did, though, along with arguments from other sources, was make it less easy to attack the license fee, a small but not unimportant victory.

The events around *Invisible Citizens* reflect some of the ways in which one can take social research and engage in important policy issues: pursue a highly defined question; have quick access to funds, a group of people who can share the labor, and modest ambitions; make sure that the findings are made available quickly; place the findings within a public forum such as a major newspaper.

In another piece of research, we were commissioned by the Home Office (the ministry of state with responsibility for broadcasting) to prepare a report based on extensive surveying of public attitudes about the future of radio. The result, *The Listener Speaks* (Barnett & Morrison, 1989), was an empirically based discussion about some difficult but important underlying issues, such as what people want from radio, as opposed to what they might be said to need. We did the research and held a large press conference with representatives of all the key press present; the official government publisher published copies and issued them to opinion formers in politics, the radio industry, Parliament, and the media.

It is possible to argue that as researchers we were as close as one can get in any structural sense to the policy process. Here was a major government policy initiative, a major funding initiative, and detailed negotiations with officials inside the Home Office who not only commissioned the research but took part in extensive discussion as to what might be the relationship between its findings and their drafting of policy. With that most exact of sciences, hindsight, can it be said that what we did made a difference? Probably not, at least not in any obviously definable manner. This was not because the relevant individuals did not pay close attention to the finding, nor even that they weren't genuinely interested. It certainly wasn't because the research was not up to scratch. It's just that the research could never be centrally relevant in a situation in which knowledge and understanding were not the primary bases for the formulation of policy. Broadcasting policy depended more on prescriptive judgments fashioned within an ideology overwhelmingly based on the market, rather than any readily apprehendable, analytically based conversation.

"Speaking to," or "conversation," presupposes a certain effectiveness in communication, and the existence of conditions in which one *can* be effective. The self-serving opacity, vagueness, vacuousness, and aridity of too much contemporary academic writing is widely recognized. Journals have become an industry defined more by the needs of career than thought, full of too many pieces in which the writer has little to say and no talent to say it. Too often obscurity of language or data presentation is worn as a badge which says that if you, dear reader, could only penetrate the inner sanctum of my mind you would indeed see the depth of my insight. Too often, when one does get there one finds an empty room.

Clearly, however, when it comes to the ability to communicate, there are exceptions. Much of the appeal of the likes of Carey and Geertz flows from the fact that they write—and speak—beautifully. Or read Hoggart, Williams, Thompson, or Hall at their best. All are fine, even brilliant, communicators—clear, precise, and full of insight. Indeed, one might argue that their skill with the pen is sometimes too effective, its bright intensity serving to blind rather than illuminate. One goes to other writers such as Habermas, knowing that although there may be no poetry along the way, the journey will be worth it. At least something does rest within *their* inner sanctum.

The viability of an idea or piece of research depends to an extent on the political and social condition within which it is expressed, and the ability to express is itself conditioned by circumstance. The clues are scattered throughout Jacoby's (1987) book, since on numerous occasions he points out that the public intellectuals, whom he so applauds, had in the 1940s and 1950s numerous outlets for their writings in magazines and public journals. As he notes—though he does not draw this conclusion—many, even most, of these outlets are now closed. The conveyor belt from the academic public intellectuals to the public mind ceased to function in the United States, until the conservative think tanks came along. At the same time, there had been an intellectual center of gravity in New York that decayed, particulates of which spread across the land. In reality, of course, there is no real center of gravity in the United States. There was no national press, only networks that were, in effect, based on a Lego-set construction of local stations; no national school curricula; no coherent national banking system; a federalist political system; and until the 1950s a highly limited infrastructure of national communications. Much of this has had to do with geography. But much of it also has had to do with desire—deconstruction was in the heart of the Founding Fathers. None of this is to deny the validity of Jacoby's observations that the academic was becoming professionalized and disciplined and thus, with the odd pragmatic exception, publicly castrated. It is, however, to suggest a serious practical problem that inhibits public discourse in the United States.

In Britain in the 1980s it was easier to play that game of the public intellectual. There was a national press and a national broadcasting system with which one could deal, and which were themselves interested in media research. Each major quality newspaper had media editors who were constantly looking for material. It became almost routine to pick up the phone and to propose, and to have accepted, a piece based on research we had under way or on some issue or other of relevance to the public debate about communications. In a similar vein, broadcasters returned time and time again to questions of communication policy—everything from questions of objectivity, bias, violence, and sex in the media to the most searching questions on the very nature of broadcasting, particularly in the light of challenges to the structural integrity of the public-service tradition in broadcasting. Indeed, of vital importance in nurturing mass communication research in Britain have been the research departments of the major broadcasting organizations. For example, Bob Towler and then Barrie Gunter at what was the Independent Broadcasting Authority, and Peter Menneer of the BBC's Broadcasting Research Department, have been important sponsors of academic research. In effect, there was—and remains—a highly developed public discourse about media issues in the United Kingdom in ways that have never existed in the United States. In that context it was relatively straightforward to work within the public sphere.

In this essay we have adopted what might appear a somewhat para-doxical position. On the one hand, we have suggested that the reason why important parts of the field have failed to engage, or even address, developments within communications is because of a rapid march up the blind alley of cultural studies. We have logically applauded the idea that mass communication researchers should become involved with the making of policy and with the public debates that go along with that process. At the same time we have pointed to the real difficulties of doing that. Those difficulties are a function of the fact that there is no obvious or easy relationship between ideas—the essential product of the re-searcher—and policy. Indeed, one might argue that the more important the policy, the more this is the case. The ability to influence, or even be engaged, is also clearly a function of a wider climate, which does or does not permit public discourse about any given issue. In short, we under-stand that it is not easy to do what we are suggesting should be done. That is, however, no excuse for not attempting to do it. At the very least there is a question of the social responsibility of academic inquiry: that those paid out of the public treasury should not keep their secrets too close to their chest.

However, a more significant point about social inquiry is, as Postman observed, that it exists not "to contribute to *our* field but to contribute to human understanding and decency . . . the so-called social sciences are sub-divisions of moral theology" (Postman, 1988, p. 17, emphasis added). If the central purpose of our work is only to further career, to preen, to seek the security of tenure or some prestigious chair, to find pleasure in seeing one's name in print, or to find comfort in disengaged radical cer-tainty as to the iniquities of the age; if this is all we are about, is it not an extraordinarily shabby form of narcissism, and a basic betrayal of a public trust and a public need to know?

References

Barnett, S., & Morrison, D. (1989). *The listener speaks: The radio audience and the future of radio*. London: Her Majesty's Stationery Office.

Carey, J. (1989) *Communication as culture: Essays on media and society*. Boston: University Hyman.

Horowitz, I. L. (1967). *Power, politics and people: The collected essays of C. Wright Mills*. New York: Oxford University Press.

Hughes, H. S. (1974). *Consciousness and society: The reorientation of European social thought 1890–1930*. St. Albans, United Kingdom: Paladin.

Jacoby, R., (1987). *The last intellectuals: American culture in the age of academe*. New York: Basic Books.

Morrison, D. E. (1986). *Invisible citizens: British public opinion and the future of broad-casting*. London: John Libbey.

Postman, N. (1988). *Conscientious objections: Stirring up trouble about language, technol-ogy and education*. New York: Random House.

Thompson, E. P. (1980). *Writing by candlelight*. London: The Merlin Press.

Wartella, E. (1992, September). Communication research on children and public policy. Symposium address for *Beyond agendas: New directions in communications research*, delivered at the Wichita Symposium, Wichita State University, Wichita, KS.

The Rise and Fall of Audience Research: An Old Story With a New Ending

by Sonia M. Livingstone, London School of Economics and Political Science

It may be that just as capitalist economies go from boom to slump every 10 years or so, the field of mass communication goes from polarization to convergence over a similar period. Perhaps more than any other field of social science research, mass communication research has been dominated by key theoretical and methodological oppositions that underlie the fierce debates and splits within the field. These oppositions include critical versus administrative research, the study of texts (which itself is conducted in very different ways) versus the study of audiences, and the use of qualitative versus quantitative methods. Topics of study that would be better integrated have been treated separately—for example, interpersonal and mass communications, film and television, and high and low culture. The last 10 years have seen a widespread and enthusiastic call for convergence in theoretical approaches and research traditions (e.g., Livingstone, 1990; Schroder, 1987). This call for convergence has focused on the mass communication audience and generated a body of research that draws upon diverse theoretical and methodological traditions which had hitherto been more concerned with texts or production. New approaches are being drawn into the field; for example, theories of literary reception from high culture are now being investigated empirically in relation to popular culture (Radway, 1984), and the theoretically significant notion of context-dependent meaning is being studied through ethnographic approaches to audiences (Seiter, Borchers, Kreutzner, & Warth, 1989; Silverstone & Hirsch, 1992).

Most radically, there has been some convergence between the major divisions within mass communication research: critical and administrative traditions. Fejes (1984) argued that while administrative research has tended to neglect the text, resulting in often crude assumptions about the sovereign viewer, critical research must face the converse problem, that of the disappearing audience. The legacy from critical theory has often

Sonia Livingstone is a lecturer in the Department of Social Psychology at The London School of Economics and Political Science.

meant that audience interpretation and activity have been assumed rather than examined, being supposedly predictable from theories of ideology and hegemony, and thus research has neglected concrete and local contexts. Integrating audiences and texts across these opposed traditions of research has, not surprisingly, generated considerable debate in the field.

This recent history is now being told in a variety of ways (e.g., see Blumler, McQuail, & Rosengren, 1990). For some, the failure of convergence could have been, and was, predicted from the start: Maybe audience research should become a less significant area of media research and maybe theory development should proceed without the problematic and time-consuming expectation of doing empirical research. For others, the convergence has been so successful that we can now all proceed with a harmony of pluralist aims, doing away with the impediments to research imposed by outdated divisions and prejudices. And some have been unaffected by the debates, continuing their research as before.

In his overview of the field of mass communications, Katz (1980) describes a history of oscillation between conceptions of powerful media and powerful viewers. Switching from one to the other roughly every decade, communication research first replaced the notion of a mass audience with that of a selective audience as part of a two-step flow of communications, replacing this later with the passive viewer of the behaviorist approach, which then gave way to the active viewer of uses and gratifications and social cognition. This history inspires a certain pessimism—maybe the active and interpretative viewer of the 1980s and 1990s is merely the next stage in a fashion cycle. Are the apparently new questions about open texts, subversive pleasures, audience reception, and so forth merely reworkings of earlier questions, soon to be replaced by their opposites? Curran (1990) suggests that many ideas recently hailed as new are indeed reworkings, a "new revisionism," of ideas current in the 1940s or earlier.

Katz's history is, of course, the history of the administrative approach, which, after sharing common origins with the critical approach in the Frankfurt School of the 1930s and 1940s, developed largely in isolation from it. However, the recent convergence, actual and potential, between administrative and critical schools of mass communication plays a key role in changing the pattern of oscillation—from a history of oscillating active and passive audiences to a dialectic process of theory development in the field. The convergence of the two schools over the issue of audience reception and related concepts of decoding, reading, and interpretation, as effected through numerous symposia and conferences (e.g., Seiter et al., 1989), has surely left a lasting legacy. These debates have not simply been about swinging from passive to active, text to audience, mass to public; nor have they been simply about generating new research questions—of pleasure, reception, interpretation, the domestic context of viewing. Rather, they represent attempts to transcend the old polarities altogether so they cannot be asserted again, questioning and rejecting a range of generalities that had previously structured the field for one or both schools.

Emerging Issues

The convergence of administrative and critical schools during the 1980s is, moreover, a historical phenomenon—many of their key issues, as discussed below, have become particularly problematic in recent years, prompting the need for radical rethinking. With the rapid and diverse developments in communication technologies, the ever-increasing interdependence of media and daily life, and the growing cross-cultural spread of mass media, there are now coming to the fore many issues that could not have been anticipated and whose analysis draws, necessarily, on contemporary developments in deconstruction and social theory. Let us now discuss some of these issues.

Despite the persistence of the labels, much mass communications research is now neither simply administrative nor critical. This is not to say that the convergence has been wholly successful; much research is proceeding as if no debates had taken place. Rather it is to argue that any research project should consider text, audience, and context; that the argument in favor of empirical investigation of any amenable theoretical development has largely been won; and that the ideological underpinnings of both the research process and its subject matter can be legitimately questioned by all. After all, much research from either tradition can be, and often is, used either to support or to critique the status quo.

Text and audience can no longer be seen as independent or studied separately. As audience reception and reader-response theories have made clear, text and reader are interdependent, mutually conceived, joint constructors of meaning. Rather than conceiving of powerful texts and passive viewers or of indeterminate texts and powerful viewers, what is required is a negotiated position that recognizes the complexity of the interaction between text and viewer, where encoding may differ radically from decoding. The attack on structuralism, where elite critics locate unique and determinate meanings "in" the text and where actual interpretations by readers are either neglected or regarded as misguided or incorrect, has changed the way we conceive of meaning.

Meaning emerges from the specific and located interaction between text and reader, where texts must be considered virtual until realized by actual—rather than ideal—readers. Texts attempt to position readers as particular kinds of subjects through particular modes of address, inviting readers to insert specific knowledge or perspectives into the interpretive flow. Readers may accept or neglect such textual invitations and constructions of subject positions, reading against the grain while avoiding aberrance, exploiting the inevitable degree of openness in the text, playing with textual conventions, and thereby jointly constructing different meanings on different occasions: "As the reader passes through the various perspectives offered by the text, and relates the different views and patterns to one another, he sets the work in motion, and so sets himself in motion too" (Iser, 1980, p. 106). To consider both text and audience is not simply a matter of including two discrete elements but of examining their interdependence.

The traditional separation of interpersonal and mass communication, assumed in both administrative and critical traditions, is untenable. Ethnographic research particularly has shown the significant ways in which family talk about, say, the royal family or racism inevitably takes place in a media-dominated environment (e.g., Billig, 1991; van Dijk, 1991) and, conversely, that the media must be located in the living room—a locus of domesticity and family interaction (Goodman, 1983; Liebes & Katz, 1990). The appropriation of communication technologies into domestic spaces raises issues of gender, culture, and power that frame the ways in which these technologies are experienced and used (Silverstone & Hirsch, 1992). The phenomenon of parasocial interaction, for example, means that we must now ask about rather than presume that we understand the overlapping processes that underlie both mass and interpersonal communication.

The diverse social contexts of viewing, the variable nature of viewers' involvement, and the proliferation of media technologies have transformed watching television into an activity that is essentially diverse and context-dependent. When Fogel argues that the talk show "has the hegemony over our contemporary dialogue-values" (1986, p. 153), thus undermining the separation of face-to-face and mediated communication, and when for Carbaugh, "Donahue" is a "cultural performance of individuality" (1988, p. xiii), undermining the separation of individual and mass, again it is not a matter of including both interpersonal and mass communications in research, but one of recognizing their mutuality, each acting to construct the other.

Two notions of "mass" have been questioned by recent research, that of the mass audience and of the mass media. When talking of the mass media, we must now specify the channel of interest, not because print and television are opposites but because different media and different channels are received in different ways, and these contexts must be explored. Similarly, the genre of a program is significant. As the soap opera debate (e.g., Livingstone, 1990) made clear, the category of television varies enormously with different genres; watching soap opera involves different audiences, patterns of involvement, domestic arrangements, and critical modes of interpretation compared with, say, watching the news.

While for some, television is still fundamentally a mass medium, Corner (1991) argues that recent audience research has itself evolved a new opposition, furthering either the *public knowledge* project (a focus on news and current affairs in relation to the politics of information and the viewer as citizen) or the *popular culture* project (a focus on fiction and drama in relation to the social problematics of taste and pleasure). The problem now is to reintegrate these projects and to undo the problems that have arisen from this, as well as other, polarizations in research.

Implicit in these two projects is a notion of diverse viewers and viewing styles, undermining the generic category of "viewer." Reception and ethnographic research have demonstrated that the mass audience is significantly heterogenous, not only in relation to gender, class, culture, and age, but also in relation to cognitions, involvement, and styles of viewing. In short, the "mass" of mass communications has been challenged, and theories and methods must adapt.

Working at the Boundaries

Any new body of research attracts criticism, and audience research is moving from a phase of enthusiasm to one of self-analysis. The notion of the active viewer can be taken too far, neglecting the constraining action of the text or treating trivial variations among readings as theoretically important. There is a problem in theorizing pleasure: Without adequate analysis of power relations among social groups, any divergence from a preferred reading of the text may be seen as a sign of political opposition to or subversion of the status quo. The boundaries of the text sometimes threaten to dissolve altogether, once we recognize problems of intertextuality, textual encrustations, and zipping and zapping across program flow. So does the notion of audience when the surprising diversity of viewing practices is revealed. Finally, we must integrate sociocognitively oriented work on the comprehension of narrative with interpretive work on the reception of connotative or ideological levels of meaning, without losing sight of the fundamentally social nature of reception processes.

All these issues concern the boundaries of theoretical arguments that, as with many theories, have been propounded in a simple and overextensive form and which, following empirical research, require some limitations and qualifications in their claims. There is, then, a boundary-placing exercise to be tackled over the next decade in order to make the most of developments in the last decade. Generally, the theoretical shift effected by the call to convergence within mass communications research represents a considerable achievement theoretically, and any sense of uncertainty about how to proceed which now exists should be regarded with excitement rather than gloom. The reconceptualization achieved may be exemplified through the issue of effects, a long-standing problem in mass communications.

Following a considerable critique of effects research through the 1970s and 1980s, this domain has become rather neglected of late. While a growing disappointment with both the theoretical resources and the empirical conclusions of effects research spawned a renewed interest in other aspects of mass communications, we have now argued ourselves into a position where questions of effect, as traditionally conceived, are not only too difficult to operationalize but don't even make sense. The required separation between cause and effect cannot be sustained once we allow text and audience or media and everyday life to become intertwined. If meanings are negotiated between text and reader rather than imposed by the former and submitted to or deflected by the latter, if everyday life is constituted within a media-dominated environment rather than affected by it, then we need new ways of asking about the social operation of power. Let us stop asking how audiences are affected by the mass media and start asking how particular audience groups engage in different ways with particular forms and genres of the mass media in different contexts. Whether or not sustainable generalities can or even should emerge from this study of the particular remains to be seen and debated over the next decade.

The potential of a mass communications that transcends old and unproductive polarities depends on the fruitfulness of the research questions posed in the field. Let me conclude this essay by identifying certain new and significant research questions that have emerged over the last 10 years and whose development may productively occupy the next 10 years.

The concept of audience reception requires elaboration, as it has tended to confuse interpretation of connotative meanings with comprehension of literal meanings (Livingstone, 1990), and so miscommunications and mistakes have not been adequately distinguished from divergent and creative meanings. The link between pleasure and reception requires further work, going beyond present uses and gratifications theory, to examine how pleasures may be gained from both familiarity and novelty, from closure and openness, from normativity and subversion. In relation to both reception and pleasure, however, the operation of textual constraints must be specified rather than neglected, for engagement with a text designed to generate subversive pleasures must, for example, surely be a different experience from one in which such pleasures are obtained only by reading against the grain. The constraints imposed by viewers are also significant, yet little is known of the role of prior social knowledge, genre expectations, and personal experiences of reception.

Text–reader relationships clearly differ according to the genre of programs, and yet so far rather few genres—notably, the news, current affairs, and the soap opera—have been studied in audience research. What of the situation comedy, the game show, the talk show, the sports programs, and many more? Different genres specify different "contracts" to be negotiated between the text and the reader (Livingstone & Lunt, 1994), which set up expectations on each side for the form of the communication (e.g., narrative, debate), its functions (uses and gratifications), its epistemology (e.g., the social realism of British soap operas, the "window on the world" of the news, the scientific factual approach of the documentary), and the communicative frame (e.g., the participants, the power of the viewer, the openness of the text, and the role of the reader).

If different genres result in different modes of text–reader interaction, these latter may result in different types of involvement (Liebes & Katz, 1990): critical or accepting, resisting or validating, casual or concentrated, apathetic or motivated. How shall we theorize this diversity of modes of interaction with the text? Further, how does viewer involvement depend upon family relations or other social dynamics surrounding viewing, or upon simultaneous engagement with multiple media, and conversely, how does it affect the face-to-face interaction and family conversation in which it is contextually embedded?

Broadening out the notion of context, we must ask further questions about cultural and societal contexts of viewing. Research has tended to pool studies from different countries toward the general goal of understanding the "audience"—for example, Liebes and Katz (1990) studied different ethnic groups in Israel and America; Seiter et al. (1989) studied German viewers; Livingstone (1990) studied British viewers; and many have studied American viewers. In the case of Liebes and Katz's research, the focus was specifically on cultural differences, but generally, while audience research has been conducted in very different cultures or countries, insufficient attention has been paid to the specific impact of a culture on the research findings obtained. Cultural contexts of viewing are often discussed in an ad hoc or post hoc manner, to explain specific results rather than to give a complex contextual understanding.

Finally, further methodological developments are sorely needed. Two "new" methodologies—which, of course, are both very old—have been used enthusiastically in recent audience research: the focus group and ethnography. Undoubtedly, these methods have demonstrated that viewers are active interpreters of texts and that viewing contexts vary widely in their impact on these interpretations. It is time for a sober assessment of these methods, given the problems that are raised through their use, and for their development and integration with other communication research methods. For example, how should the internal dynamics of focus groups be handled, how should the qualitative data that result be analyzed, how can "bottom-up" ethnography interface with theory, and what kind of contact with viewing contexts constitutes ethnography?

The field of mass communication, particularly in relation to the television audience, is not as fashion-led and mindlessly cyclical as it sometimes seems. In the last 10 years, we have identified two developments that have significant implications for the future of the field. First, the old polarities that had long structured the field have been finally transcended (or deconstructed) and cannot easily be returned to. Second, a new set of questions is emerging, both theoretical and methodological, which concern a range of particular issues and processes rather than generalities expressed in terms of the now untenable categories of viewer, media, effect.

References

Billig, M. (1991). *Talking of the royal family*. London: Routledge.

Blumler, J., McQuail, D., & Rosengren, K. E. (1990). (Eds.). Communications research in Europe: The state of the art. *European Journal of Communication* [special issue], *5*(2–3), 131–379.

Carbaugh, D. (1988). *Talking American: Cultural discourses on DONAHUE*. Norwood, NJ: Ablex.

Corner, J. (1991). Meaning, genre and context: The problematics of "public knowledge" in the new audience studies. In J. Curran & M. Gurevitch, (Eds.), *Mass media and society* (pp. 267–284). London: Methuen.

Curran, J. (1990). The new revisionism in mass communication research. *European Journal of Communication* [special issue], *5*(2–3), 135–164.

Fejes, F. (1984). Critical communications research and media effects: The problem of the disappearing audience. *Media, Culture and Society, 6*(3), 219–232.

Fogel, A. (1986). Talk shows: On reading television. In S. Donadio, S. Railton, & O. Seavey (Eds.), *Emerson and his legacy: Essays in honor of Quentin Anderson* (pp. 147–169). Carbondale: Southern Illinois University Press.

Goodman, I. R. (1983). Television's role in family interaction: A family systems perspective. *Journal of Family Issues, 4*(2), 405–424.

Iser, W. (1980). The reading process: A phenomenological approach. In J. P. Tompkins (Ed.), *Reader-response criticism: From formalism to post-structuralism* (pp. 103–129). Baltimore, MD: Johns Hopkins University Press.

Katz, E. (1980). On conceptualising media effects. *Studies in Communication, 1,* 119–141.

Liebes, T., & Katz, E. (1990). *The export of meaning*. Oxford: Oxford University Press.

Livingstone, S. M. (1990). *Making sense of television: The psychology of audience interpretation*. Oxford: Pergamon.

Livingstone, S. M., & Lunt, P. K. (1994). *Talk on television: Audience participation and public debate*. London: Routledge.

Radway, J. (1984). *Reading the romance: Women, patriarchy, and popular literature*. Chapel Hill: University of North Carolina Press.

Schroder, K. C. (1987). Convergence of antagonistic traditions? The case of audience research. *European Journal of Communication, 2,* 7–31.

Seiter, E., Borchers, H., Kreutzner, G., & Warth, E-M. (Eds.). (1989). *Remote control: Television audiences and cultural power*. London: Routledge.

Silverstone, R., & Hirsch, E. (Eds.). (1992). *Consuming technologies*. London: Routledge.

van Dijk, T. A. (1991). *Racism and the press*. London: Routledge.

Active Audience Theory: Pendulums and Pitfalls

by David Morley, University of London

It certainly seems that, over the last few years, the pendulum has swung again in the world of audience research. As we all know, in the bad old days, TV audiences were considered as passive consumers, to whom things happened as TV's miraculous powers affected them. Happily, so the story goes, it was then discovered that this was an inaccurate picture because, in fact, these people were out there, in front of the set, being active in all kinds of ways—making critical/oppositional readings of dominant cultural forms, perceiving ideological messages selectively/subversively, and so on. So it seems we needn't worry—the passively consuming audience is a thing of the past. As Evans (1990) notes, recent audience work in media studies can be characterized largely by two assumptions: (a) that the audience is always active (in a nontrivial sense), and (b) that media content is always polysemic, or open to interpretation. The question is what these assumptions are taken to mean exactly, and what their theoretical and empirical consequences are.

Hall's (1981) original formulation of the encoding/decoding model contained, as one of its central features, the concept of the *preferred reading* (toward which the text attempts to direct its reader) while acknowledging the possibility of alternative, negotiated or oppositional readings. This model has subsequently been quite transformed, to the point where it is often maintained that the majority of audience members routinely modify or deflect any dominant ideology reflected in media content (cf. Fiske, 1987), and the concept of a preferred reading, or of a structured polysemy, drops entirely from view. In this connection I have to confess a personal interest, as I have been puzzled to find some of my own earlier work (e.g., Morley, 1980) invoked as a theoretical legitimation of various forms of "active audience theory" (variously labeled as the "new revisionist" or "interpretivist" perspective by other critics). For any author to comment on the subsequent interpretation of their work is plainly an awkward enterprise, and when that work itself is substantively concerned with the ways in which audiences interpret texts, the irony is manifest. Nonetheless, I shall take this opportunity to comment on some recent debates in audience studies, and will argue that much active audience theo-

---

David Morley is a reader in communications at Goldsmiths' College, University of London.

ry is in fact premised on a heavily negotiated reading (if not a misreading) of some of the earlier work that is often invoked as its theoretical basis (see, Derrida, 1989; Norris, 1991; and Richards, 1960, for the relevant distinctions between variant readings and misreadings).

Rereading Audience Theory

For my own part, while I would argue that work such as the *Nationwide Audience* project, along with that of Ang (1985), Liebes and Katz (1991), and Radway (1984), offers counterevidence to a simple-minded *dominant ideology* thesis, and demonstrates that any hegemonic discourse is always insecure and incomplete, this should not lead us to abandon concern with the question of, as Martin-Barbero (1988) puts it, "how to understand the texture of hegemony/subalternity, the interlacing of resistance and submission, opposition and complicity" (p. 462). That was, and remains, precisely the point of studying audience consumption of media texts, a point that now, with the discrediting of some of the more "romantic" versions of active audience theory, is in great danger of being obscured—as demonstrated, for example, by Seamann's (1992) total failure to understand the point of what he describes as "pointless populism."

This is by no means to deny the existence of problems in contemporary audience theory. I would agree with Corner (1991) that much recent work in this field is marred by a facile insistence on the polysemy of media products and by an undocumented presumption that forms of interpretive resistance are more widespread than subordination, or the reproduction of dominant meanings (see Condit, 1989, on the unfortunate current tendency toward an overdrawn emphasis on the polysemous qualities of texts in media studies).

In a similar vein to Corner, Curran (1990) offers a highly critical account of what he describes as the "new revisionism" in mass communications research on media audiences. In brief, his charge is that while "this . . . 'revisionism' . . . presents itself as original and innovative, [it] . . . is none of these things" (p. 135), but rather amounts to "old pluralist dishes being reheated and presented as new cuisine" (p. 151). The history Curran offers is an informative one, alerting us to the achievements of scholars whose work has been unrecognized or neglected by many (myself included) thus far. However, my contention is that this is a particular history that could not have been written (by Curran or anyone else) 15 years ago, before the "new revisionism" (of which Curran is so critical) transformed our understanding of the field of audience research, and thus transformed our understanding of who and what was important in its history. I would argue that it is precisely this transformation that has allowed a historian such as Curran to go back and reread the history of communications research in such a way as to give prominence to those whose work can now, with hindsight, be seen to have prefigured the work of these new revisionists.

However, despite my differences with him about the general terms of his critique, I would agree with Curran that recent reception studies that document audience autonomy and offer optimistic/redemptive readings of mainstream media texts have often been taken to represent not simply a challenge to a simple-minded effects or dominant ideology model, but rather as, in themselves, documenting the total absence of media influence, in the "semiotic democracy" of postmodern pluralism.

As Curran observes, Fiske's (1986) celebration of a semiotic democracy, in which people drawn from a vast shifting range of subcultures and groups construct their own meanings within an autonomous cultural economy, is problematic in various respects, but not least because it is readily subsumable within a conservative ideology of sovereign consumer pluralism. The problem with the concept of semiotic democracy, as Murdock (1989) notes, is that this model of "perfect competition" is as "useless in understanding the workings of the cultural field as it is in economic analysis, since it is obvious that some discourses (like some firms in the market) are backed by greater material resources and promoted by spokespersons with preferential access to the major means of publicity and policymaking" (p. 438). Hence, as Hall (1989) argues, to speak of the cultural field is "to speak of a field of relations structured by power and difference" in which some positions are in dominance, and some are not, though these "positions are never permanently fixed" (p. 57).

Decoding to the Rescue

Budd, Entman, and Steinman (1990) argue that work of this kind now routinely assumes that "people habitually use the content of dominant media against itself, to empower themselves" (p. 170) so that, in their analysis, the crucial message of much contemporary American cultural studies media work is an optimistic one: "Whatever the message encoded, decoding comes to the rescue. Media domination is weak and ineffectual, since the people make their own meanings and pleasures." Or, to put it another way, "we don't need to worry about people watching several hours of TV a day, consuming its images, ads and values. People are already critical, active viewers and listeners, not cultural dopes manipulated by the media" (p. 170). While I would certainly not wish to return to any model of the audience as cultural dopes, the point Budd et al. make is a serious one, not least because, as they note, this "affirmative" model does tend to then justify the neglect of all questions concerning the economic, political, and ideological forces acting on the construction of texts (cf. Brunsdon, 1989), on the (unfounded) assumption that reception is, somehow, the only stage of the communications process that matters in the end. Apart from anything else, and at the risk of being whimsical, one might say that such an assumption does seem to be a curiously Christian one, in which the sins of the industry (or the message) are somehow seen to be redeemed in the afterlife of reception.

One crucial question concerns the significance that is subsequently given to often quite particular, ethnographic accounts of moments of cultural subversion in the process of media consumption or decoding. Thus, Budd et al. (1990) note that, in his account of the ways in which Aboriginal Australian children have been shown to reconstruct TV narratives involving blacks in such a way as to fit with and bolster their own self-conceptions, Fiske (1986) shows a worrying tendency to generalize radically from this very particular instance, so that in his account this type of alternative response, in quite particular circumstances, is decontextualized and then offered as a model for decoding in general, so that, as Budd et al. (1990) put it, "the part becomes the whole and the exception the rule" (1990, p. 179). (See also Schudson, 1987.)

The equivalence that Newcomb and Hirsch (1987) assert between the producer and consumer of messages, in so far as the television viewer matches the creator (or the program) in the making of meanings is, in effect, a facile one, which ignores de Certeau's (1984) distinction between the strategies of the powerful and the tactics of the weak—or, as Morley and Silverstone have argued elsewhere (1990), the difference between having power over a text, and power over the agenda within which that text is constructed and presented. The power of viewers to reinterpret meanings is hardly equivalent to the discursive power of centralized media institutions to construct the texts that the viewer then interprets, and to imagine otherwise is simply foolish. The problem, as Ang (1990) argues, is that while "audiences may be active, in myriad ways, in using and interpreting media . . . it would be utterly out of perspective to cheerfully equate 'active' with 'powerful'" (p. 247).

Between the Micro and the Macro

The boom in empirical media audience research in the 1980s was, in part, the result of the critique of overly structuralist approaches, which had taken patterns of media consumption to be the always-already-determined effect of some more fundamental structure—whether the economic structure of the cultural industries (Murdock & Golding, 1974), the political structure of the capitalist state (Althusser, 1971), or the psychic structure of the human subject (Lacan, 1977). However, a number of authors (e.g., Corner, 1991; Curran, 1990) have recently argued that the pendulum has now swung so far that we face the prospect of a field dominated by the production of micro (and often ethnographic) analyses of media consumption processes, which add up only to a set of micronarratives, outside of any effective macropolitical or cultural frame. It is then sometimes argued (Seamann, 1992) that the appropriate response is to abandon the blind alley of ethnography, and return to the eternal verities of political economy. While I have argued elsewhere (Morley, 1991) that ethnography must always itself be placed in a wider frame and, as indicated above, I have a number of reservations about much active audience theory, I nonetheless hold that the current backlash against microethnography is in danger of encouraging a return to macro political issues which is, in fact, premised on a malposed conception of the relation between the micro and the macro.

Thus, Corner (1991) argues that in recent research on the media audience, the question of media power has tended to be avoided, and that much of this new audience research amounts to "a form of sociological quietism . . . in which increasing emphasis on the microprocesses of viewing relations displaces . . . an engagement with the macrostructures of media and society" (p. 269). My own contention would be that this formulation is problematic, in so far as Corner implicitly equates the macro with the real and the micro with the realm of the epiphenomenal (if not the inconsequential). In the first place, Corner's analysis fails to recognize the gendered articulation of the divisions macro/micro, real/trivial, public/private, masculine/feminine—which is what much of the work that he criticizes has, in various ways, been concerned with (see, for instance, Gray, 1992; Morley, 1986; Radway, 1984). More centrally, Corner seems to invoke a notion of the macro that is conceptualized in terms of pregiven structures, rather than (to use Giddens's [1979] phrase) "structuration," and which fails to see that macro structures can be reproduced only through microprocesses.

It was precisely this realization that drove the initial shift in cultural studies work away from any notion of a mechanically imposed dominant ideology, and toward the more processual model of hegemony—as a better theoretical frame within which to analyze the reproduction of cultural power in its various forms (Hall, 1977). The whole point of that shift was to attempt to find better ways to articulate the micro and macro levels of analysis, not to abandon either pole in favor of the other. Nor, as Massey (1991) argues, should we fall into the trap of equating the micro (or local) with the merely concrete and empirical, and the macro (or global) with the abstract or theoretical. In all of this, we could do worse than heed Mills's (1959) strictures on the need to address the interplay of biography and history in the "sociological imagination."

All of this is particularly vital in the realm of media consumption, given the media's key role in articulating the public and the private, the global and the local, and in articulating global processes of cultural imperialism with local processes of situated consumption—where local meanings are so often made within and against the symbolic resources provided by global media networks.

To say that is not to offer any carte blanche defense of ethnography-as-where-it's-at. As Marcus and Fischer (1986) observe, the value of ethnography lies in reshaping our dominant macroframeworks for the understanding of some structural phenomenon such as the capitalist world system, so that we can "better represent the actual diversity and complexity of local situations" (p. 88) for which our theoretical frameworks try to account in general terms. Yet, in a reformulation of some of his earlier positions, Fiske (1990) cautions that any ethnography "runs the risk, which we must guard against at all costs, of allowing itself to be incorporated into the ideology of individualism" (p. 9). If ethnography is concerned with tracing the specifics of general, systemic processes—for instance, the particular tactics that various members of a given society have developed to make do with the cultural resources that society still offers them, then as Fiske notes, our concern must be with interpreting such activities in the broader context of that "larger system through which culture and politics intersect" (1990, p. 98).

References

Althusser, L. (1971). *Lenin and philosophy*. London: New Left Books.

Ang, I. (1985). *Watching "Dallas."* London: Methuen.

Ang, I. (1990). Culture & communication. *European Journal of Communication, 5*(2–3), 239–261.

Brunsdon, C. (1989). Text and audience. In E. Seiter, H. Borchers, G. Kreutzner, & E.-M. Warth (Eds.), *Remote control* (pp. 116–130). London: Routledge.

Budd, B., Entman, R., & Steinman, C. (1990). The affirmative character of American cultural studies. *Critical Studies in Mass Communication, 7*(2), 169–184.

Condit, C. (1989). The rhetorical limits of polysemy. *Critical Studies in Mass Communications, 6*(2), 103–122.

Corner, J. (1991). Meaning, genre and context: The problematics of "public knowledge" in the new audience studies. In J. Curran & M. Gurevitch (Eds.), *Mass media and society* (pp. 267–285). London: Edward Arnold.

Curran, J. (1990). The "new revisionism" in mass communications research. *European Journal of Communication, 5*(2–3), 135–165.

de Certeau, M. (1984). *The practice of everyday life*. Berkeley: University of California Press.

Derrida, J. (1989). *Limited Inc*. (2nd ed.). Evanston, IL: Northwestern University Press.

Evans, W. (1990). The interpretive turn in media research. *Critical Studies in Mass Communication, 7*(2), 145–168.

Fiske, J. (1986). Television: Polysemy and popularity. *Critical Studies in Mass Communication, 3*(2), 391–408.

Fiske, J. (1987). *Television culture*. London: Routledge.

Fiske, J. (1990). Ethnosemiotics. *Cultural Studies, 4*(1), 85–100.

Giddens, X. (1979). *Central problems in sociological theory*. London: Hutchinson.

Gray, A. (1992). *Video play-time: The gendering of a leisure technology*. London: Comedia/Routledge.

Hall, S. (1977). Culture, the media and the ideological effect. In J. Curran, M. Gurevitch, & J. Woollacot (Eds.), *Mass communication and society* (pp. 315–349). London: Edward Arnold.

Hall, S. (1981). Encoding and decoding in television discourse. In S. Hall, D. Hobson, A. Lowe, & P. Willis (Eds.), *Culture, media, language* (pp. 128–138). London: Hutchinson.

Hall, S. (1989). Ideology and communication theory. In B. Dervin, L. Grossberg, B. O'Keefe, & E. Wartella (Eds.), *Rethinking communication: Paradigm exemplars (Vol. 2)* (pp. 40–52). London: Sage.

Lacan, J. (1977). *Ecrits*. London: Tavistock.

Liebes, T., & Katz, E. (1991). *The export of meaning*. Oxford: Oxford University Press.

Marcus, G., & Fischer, M. (1986). *Anthropology as cultural critique*. Chicago: University of Chicago Press.

Martin-Barbero, J. (1988). Communication from culture. *Media, Culture, & Society, 10*(4), 447–465.

Massey, D. (1991). The political place of locality studies. *Environment & Planning (d): Society & Space, 9*(1), 267–281.

Mills, C. W. (1959). *The sociological imagination*. London: Oxford University Press.

Morley, D. (1980). *The "Nationwide" audience*. London: British Film Institute.

Morley, D. (1986). *Family television: Domestic leisure and cultural power*. London: Comedia.

Morley, D. (1991). Where the global meets the local. *Screen, 32*(1), 1–15.

Morley, D., & Silverstone, R. (1990). Domestic communications: Technologies and meanings. *Media, Culture, & Society, 12*(1), 31–55.

Murdock, G. (1989). Cultural studies: Missing links. *Critical Studies in Mass Communications, 6*(4), 436–440.

Murdock, G., & Golding, P. (1974). For a political economy of mass communications. In R. Miliband & J. Saville (Eds.), *The socialist register 1973* (pp. 205–234). London: Merlin Press.

Newcomb, H., & Hirsch, P. (1987). Television as a cultural forum. In H. Newcomb (Ed.), *Television: The critical view* (4th ed.) (pp. 455–471). Oxford: Oxford University Press.

Norris, C. (1991). *Deconstruction: Theory & practice* (rev. ed.). London: Routledge.

Radway, J. (1984). *Reading the romance: Women, patriarchy, and popular literature*. Chapel Hill: University of North Carolina Press.

Richards, I. (1960). Variant readings and misreadings. In T. Sebeok (Ed.), *Style in language* (pp. 241–253). Cambridge, MA: MIT Press.

Schudson, M. (1987). The new validation of popular culture. *Critical Studies in Mass Communication. 4*(1), 51–68.

Seamann, W. (1992). Active audience theory: Pointless populism. *Media, Culture, & Society, 14,* 301–311.

The Past in the Future: Problems and Potentials of Historical Reception Studies

by Klaus Bruhn Jensen, University of Copenhagen

One standard for assessing contemporary research on mass media audiences is the test of history: To what extent will our colleagues of the future be able to rely on the work of the present generation to gain a better understanding of the reception of mass media in their past, our present. This article suggests that historical reception studies offer a test case of what the explanatory value of communication theory is and could be.

The argument is developed within the framework of qualitative empirical reception studies, which during the past decade have contributed significantly to a convergence between social scientific and humanistic research traditions as well as to a rearticulation of the classic question of effects in terms of meaning, human subjectivity, and social action (Jensen & Rosengren, 1990). Reception analysis asks not only what media do to audiences, or even what audiences do with media, but how media and audiences interact as agents of, in semiotic terms, "the life of signs within society" (Saussure, 1959, p. 16).

I first offer a brief characterization of reception analysis and give examples of how qualitative methodologies have been employed to study media reception in the present. Next, I identify some forms of evidence that may be used to creatively fill the gaps in our knowledge about media reception in the past. Finally, I argue that it is incumbent on the field to develop data bases documenting media reception, which may broaden the scope of audience research in the future. Compared to other aspects of the communication process, which may be examined through traditional historical sources such as legislative documents, the records of media organizations, and some (admittedly imperfect) archives of media contents, *reception does not exist in the historical record*; it can only be reconstructed through the intervention of research. Whereas ratings and readership fig-

---

Klaus Bruhn Jensen is an associate professor and chairperson in the Department of Film and Media Studies, University of Copenhagen, Denmark. An earlier version of the article was presented as an invited lecture to a seminar of an ongoing project in media history, "Moving images in Norway," held in Oslo, Norway, in November 1991, and published in Danish in a series of working papers from the project, *Levende bilder* [Moving images], in January 1992.

ures presumably will survive, the social and cultural aspects of mass media reception are literally disappearing before our eyes and ears.

Media Reception in the Present

One means of studying mass media use in its historical and cultural contexts is the variety of qualitative methodologies, which have undergone a renaissance in recent communication research, not least within reception studies (Jensen, 1991b; Morley & Silverstone, 1991). Reception analysis may be defined as a qualitative form of audience-cum-content analysis, comparing the discourses of media contents and the discourses of the audience—as recovered through interviewing, observation, and textual analysis—in order to interpret and explain the process of their interaction in specific social contexts at a particular historical juncture. Media and audiences thus contribute to the production and circulation of meaning in society. History may be read as a discourse, which is articulated increasingly through the mass media, and rearticulated in part by their audiences.

Empirical reception studies draw on several theoretical traditions, some important sources being the lineage of Chicago School sociology, symbolic interactionism, ethnomethodology, and ethnography (Jankowski & Wester, 1991; see also Lindlof, 1987). Furthermore, the field of communication studies has been able to rely on well-established approaches in historical research, what are known as life history and oral history. The motivation underlying oral history is that crucial aspects of social life are not documented in the most readily available, written sources, but need to be complemented and corrected by oral evidence (for a survey, see Thompson, 1978). Similarly, life history assumes that individual biographies represent one relevant level of analysis, because the dynamic of macro and micro processes is played out in the lives of concrete social agents (Bertaux, 1981). In order to capture the perspective of everyday life, historians may have to become ethnographers of sorts, and oral history may further supply a perspective on history "from below"—from the unofficial, unrecorded viewpoint of the disempowered, as in labor history (Terkel, 1974). Finally, various so-called ethnographic approaches, deriving from the anthropological tradition, may produce what Geertz (1973) calls a "thick description" of media in everyday life, combining observation, interviewing, diaries, family photographs, and other evidence. Such approaches may be especially suited to exploring the nature of media use in different cultural contexts (for example, Lull, 1988).

One lesson of these research traditions is the relevance of combining several discourses, or forms of evidence, within a theoretical framework. This lesson may also be learned from some early work in the field of communication. In his classic study of the public response to Orson Welles's radio production of "War of the Worlds," Cantril (1940) showed the value of including unstructured conversations and other sources of information in order to arrive at an overall interpretation of the event. It is perhaps a weakness of contemporary audience research that a particular method may assume the status of a fetish. Moreover, as Curran (1990) notes, some recent research on audiences appears ignorant of comparable earlier studies. If media research forgets its own past, current research on media reception will inevitably suffer.

Previous studies illustrate the relevance of reception analysis for understanding the social implications of mass media at a particular historical juncture. For example, as part of the Mass Observation studies in Great Britain during the 1930s and 1940s, cinema audiences were examined. While the information on audiences cannot be related directly to the content of specific films, the source materials, comprising observation protocols and questionnaire responses, contain rich evidence for the historical study of media use during periods of economic crisis and war (Richards & Sheridan, 1987).

More recently, the British Film Institute in 1988 organized a large-scale documentation of one day in the life of British television, including program schedules, viewer responses, and commentary from media professionals. The materials were collected in a book edited by Sean Day-Lewis (1989) and reflect, among other things, on the changing role of television in Britain during a period witnessing deregulation and the growth of transnational satellite channels. One can only hope that this wealth of discourses about television reception will later become the basis of a case study in historical reception analysis.

Media Reception in the Past

While qualitative empirical reception studies have helped to broaden the agenda for studies of contemporary audiences, the quantity and types of sources that are available concerning the audiences of any given period in media history still tend to be limited. To some degree, the methodological imagination may compensate for the absence of traditional social scientific data from the past by, for example, examining other cultural discourses which traditionally have been the domain of humanistic scholarship. It is well known that historians often refer to literary works to support their interpretation of past societies and cultures. However, since literature and other high cultural forms often reflect the perspective of those at the top of the social hierarchy, communication researchers also must consider the discourses of popular culture.

In her study of the introduction of TV into American homes, Spigel (1990) examined the representation of television in magazines of the period. The premise of the study was that new media are introduced to the general public through old media, so that the analysis of magazine discourses would suggest the cultural agenda in which early American TV was placed. The findings show, among other things, that television was constructed as a new cultural resource that could either unite or divide the family at a time when a general restructuring of family and gender roles was in progress. It is interesting to note that, both in purely quantitative terms and in their mode of address, advertisements for TV sets were targeted especially to women. Whereas it would be an exaggeration to claim that television thus became a medium under female control, the advertising strategy may be interpreted as a way of accommodating television within the daily routine of women working in the home (and increasingly outside the home). A similar finding is offered in Moores's (1988) study of the introduction of radio into Great Britain, which methodologically relied on oral history. His analysis suggests that after initially being conceived as a technical toy for the father, radio gradually came to represent a routine activity to be subordinated to, or at least integrated with, the mother's other routines in the home.

Another study of British radio has emphasized how the temporal structures of the medium and of the everyday gradually were assimilated (Scannell, 1988). Examining the daily flow of programming, Scannell concluded that early radio contributed to a new experience of time. Not only did British radio of the 1930s establish a regular daily rhythm in its programming, but the specific content of programs at various times of day may have served as a new form of boundary rituals mediating between the private sphere of the home and the public sphere of work and school. The study goes on to suggest that the temporal structures of the media and the everyday should be conferred with superordinate temporal structures—for example, recurring national political and cultural events, as represented in the media. Historical time, as experienced by the audience-public, may be structured in part by the temporal arrangement of national events in the mass media. The methodological point is that program schedules, along with national statistics of time use, present sources of evidence regarding reception.

Seymour-Ure's (1989) study shows that by examining in depth the responses of a few audience members about whom much evidence is available, one may arrive at some fundamental conditions of reception that apply to the audience at large. Deploying an ingenious study design, Seymour-Ure examined how the interaction of prime ministers with television has changed over time. Focusing on prime ministers from Great Britain, Australia, and Canada, and conferring the age of each prime minister with the age of the medium in each national context, the analysis concludes that politicians are gradually socialized to use new media such as television, in part through their own media use. The analysis further concludes that the politician's degree of success depends, in large measure, on an ability to articulate his or her message within the characteristic form of TV at a given historical juncture. In this respect, it is correct to speak of certain prime ministers as "TV politicians," and it may be justified to speak of the audience as belonging to different "TV generations," to the extent that they are subject to a specific construction of politics and society through television. While both politicians and their voters thus are socialized by television, the difference remains that, for all practical purposes, only the politicians become actors *on* television, in addition to being recipients *of* television.

Media Reception in the Future

Perhaps the most important task for reception analysis lies in the future. Like other social scientific disciplines, most forms of communication research tend to focus on contemporary and modern issues, to the exclusion of historical developments or even the understanding of the present as a specific moment in social and cultural history. Communication research, I submit, should become both retrospective and prospective. It may become both of these things by asking what could be done to facilitate historical reception studies in the future.

This question implies a reformulation, to some extent, of the research agenda also for qualitative empirical reception analysis. Much of the effort in this domain over the last decade and a half has been focused on theoretical and methodological discussions, which might serve to consolidate, legitimate, and integrate reception analysis in the general enterprise of communication studies. One may conclude that this task has now been accomplished, insofar as interpretive modes of inquiry, such as qualitative reception studies, now are believed to constitute a "meaning paradigm" currently challenging traditional positions (Lowery & DeFleur, 1988, pp. 455–459)—even if no studies from the new paradigm are as yet considered fit for inclusion among the other "milestones" of the field. Having established the independent explanatory value of qualitative audience studies, current studies have begun to apply the tools of reception analysis to specific problematics, such as the audience response to politically and culturally controversial issues as disseminated by the mass media (for example, Corner, Richardson, & Fenton, 1990; Jensen, 1991a; Jhally & Lewis, 1992; Schlesinger, Dobash, Dobash, & Weaver, 1992).

One purpose of "applied" reception studies would be to produce source materials for future research. An example of such a project has been outlined by Neuman (1989) as "parallel content analysis." The basic idea is to conduct a parallel and continuous registration over time, first, of the content of a whole range of mass media and, second, of the content of the audience reception and uses of media. The premise, of course, is that both the thematic focus of media items and their formal articulation may, to some degree, be traced in the foci and rearticulations of the audience response. If conducted on a grand scale, the project might document some long lines of the interaction between mass media and the audience-public, such as processes of agenda setting and cultivation. The further assumption is that a whole range of content and audience studies may draw on the same data base, which may also facilitate collaborative and comparative studies across disciplinary and, perhaps, national and cultural boundaries.

Two caveats concerning the design must be addressed. First, data sets always carry a theoretical framework, explicitly or implicitly, which specifies the significance of the individual items and their interrelations. This might limit the value of a grand data base for a theoretically diverse field. Nevertheless, the development of the data base presumably could give rise to principled, interdisciplinary discussions about theory and methods, and it is likely that a common ground could be found, not for all, but for a substantial number of studies. Second, and more important in this context, it is decisive, as also suggested in Neuman's (1989) proposal, that qualitative data forms be included to enable future audience research to address precisely those contextual and discursive aspects of media use that have been recovered in recent reception analysis. When the proposal was critiqued and discussed in a research workshop on "time series measurement in communication effects research" at the 1991 meeting of the International Communication Association (ICA), all interventions in effect were articulated in the concepts and standards of quantitative survey methodology. Still, the construction of a common resource for qualitative and quantitative audience studies would provide a welcome opportunity to explore basic theoretical and methodological issues facing the field as a whole.

The second proposal, while growing out of the rather different perspective of critical theory and cultural studies, identifies a similar resource for research and for public debate. Green (1991) outlines the idea of general-purpose, multimedia data bases, which would include qualitative data on the audience reception, experience, and social uses of media, along with audience surveys and traditional data from media archives. There are, of course, many ethical and methodological problems associated with archiving, access, anonymity, and copyright in communication history (Schudson, 1991). Also, the problems multiply in the case of multimedia-cum-audience data bases, not least since qualitative studies produce especially sensitive data, such as interviews, diaries, and video observations. In principle, however, the accumulation of such data bases is of interest not just for historical research on the relative power of media and audiences, but also for contemporary public debate on what and whom the media are for—meta-communication on the social ends and means of mass communication. Reception should be documented, in part to empower audiences vis-à-vis media. Basing his vision on approaches that have been developed in oral history, community studies, and museums, Green suggests that one may

> *develop historical projects on media in relation to social change and the specific transformation of neighbourhoods and communities. Such a project, in addition to representing audience interests and concerns, could create a fund of knowledge that would be made*

> *available to the community as a whole through various print and*
> *electronic media forms. The project would constitute a cultural re-*
> *source facilitating meta-communication on media past and present,*
> *ideally feeding back into the media in the form of discussions, repeat*
> *screenings, and other formats.* (Green, 1991, pp. 230–231)

The economic cost of any of these resources will be high, the methodological difficulties considerable, and the political conflicts over inclusion of and access to data intense. But unless we as a field admit audiences to history, little will be heard from them in the future.

Rewriting History

To sum up, I have characterized qualitative empirical reception studies as an important complement and corrective to the still dominant traditions of survey and experimental audience research. Reception analysis offers insights into the interpretive processes and everyday contexts of media use, where audiences rearticulate and enact the meanings of mass communication. The life of signs within modern society is in large measure an accomplishment of the audience. Although reception may be less accessible for research than other aspects of the communication process, recent work has explored and documented the audience experience and social uses of media. Qualitative methodologies such as oral and life histories may serve to differentiate our understanding of media as sources of meaning and effect. Studying the past, we may turn to other media-related discourses—for example, the representation of new media in old media, the temporal structures of content, and the response of elite social actors to media. In the future, we may have a wider choice of sources that represent audience perspectives on the mass media.

I conclude with a perspective from the politics of research. If culture includes mass communication, it is indeed striking that only a very limited portion of mass communication, and an even smaller portion of the audience response, are preserved for the historical record through research and museums. The various forms of high culture and criticism still reign supreme in the conception of culture that is being passed on to present and future generations. Some redressing of this historical bias seems overdue in the name of both cultural diversity and scientific impartiality. Time may question current cultural standards and rewrite the history of contemporary popular culture. It is one of the responsibilities of communication research to produce the sources that would make a rewriting of history possible.

References

Bertaux, D. (Ed.). (1981). *Biography and society.* Beverly Hills, CA: Sage.

Cantril, H. (1940). *The invasion from Mars.* Princeton: Princeton University Press.

Corner, J., Richardson, K., & Fenton, N. (1990). *Nuclear reactions*. London: John Libbey.

Curran, J. (1990). The new revisionism in mass communication research: A reappraisal. *European Journal of Communication, 5*(2–3), 135–164.

Day-Lewis, S. (1989). *One day in the life of television*. London: Grafton.

Geertz, C. (1973). *The interpretation of cultures*. New York: Basic Books.

Green, M. (1991). Media, education, and communities. In K. B. Jensen & N. W. Jankowski (Eds.), *A handbook of qualitative methodologies for mass communication research* (pp. 216–231). London: Routledge.

Jankowski, N. W., & Wester, F. (1991). The qualitative tradition in social science inquiry: Contributions to mass communication research. In K. B. Jensen & N. W. Jankowski (Eds.), *A handbook of qualitative methodologies for mass communication research* (pp. 44–74). London: Routledge.

Jensen, K. B. (1991a). *News of the world: The reception and social uses of television news around the world*. Paris: UNESCO.

Jensen, K. B. (1991b). Reception analysis: Mass communication as the social production of meaning. In K. B. Jensen & N. W. Jankowski (Eds.), *A handbook of qualitative methodologies for mass communication research* (pp. 135–148). London: Routledge.

Jensen, K. B., & Rosengren, K. (1990). Five traditions in search of the audience. *European Journal of Communication, 5*(2–3), 207–238.

Jhally, S., & Lewis, J. (1992). *Enlightened racism: The Cosby show, audiences and the myth of the American dream*. Boulder: Westview Press.

Lindlof, T. (Ed.). (1987). *Natural audiences*. Norwood, NJ: Ablex.

Lowery, S., & DeFleur, M. (1988). *Milestones in mass communication research* (2nd ed.). New York: Longman.

Lull, J. (Ed.). (1988). *World families watch television*. Newbury Park, CA: Sage.

Moores, S. (1988). "The box on the dresser": Memories of early radio and everyday life. *Media, Culture, & Society, 10*(1), 23–40.

Morley, D., & Silverstone, R. (1991). Communication and context: Ethnographic perspectives on the media audience. In K. B. Jensen & N. W. Jankowski (Eds.), *A handbook of qualitative methodologies for mass communication research* (pp. 149–162). London: Routledge.

Neuman, W. R. (1989). Parallel content analysis: Old paradigms and new proposals. *Public Communication and Behavior, 2*, 205–289.

Richards, J., & Sheridan, D. (Eds.). (1987). *Mass observation at the movies*. London: Routledge Kegan Paul.

Saussure, F. de. (1959). *Course in general linguistics*. London: Peter Owen.

Scannell, P. (1988). Radio times: The temporal arrangements of broadcasting in the modern world. In P. Drummond & R. Paterson (Eds.), *Television and its audience* (pp. 15–31). London: British Film Institute.

Schlesinger, P., Dobash, R., Dobash, R., & Weaver, C. (1992). *Women viewing violence*. London: British Film Institute.

Schudson, M. (1991). Historical approaches to communication studies. In K. B. Jensen & N. W. Jankowski (Eds.), *A handbook of qualitative methodologies for mass communication research* (pp. 175–189). London: Routledge.

Seymour-Ure, C. (1989). Prime ministers' reactions to television: Britain, Australia and Canada. *Media, Culture, & Society, 11*(3), 307–326.

Spigel, L. (1990). Television in the family circle: The popular reception of a new medium. In P. Mellencamp (Ed.), *Logics of television* (pp. 73–97). Bloomington: Indiana University Press.

Terkel, S. (1974). *Working*. New York: Random House.

Thompson, P. (1978). *The voice of the past: Oral history*. London: Oxford University Press.

Reopening the Black Box: Toward a Limited Effects Theory

by Herbert J. Gans, Columbia University

The question of media effects is truly "the perennial black box of communications research," as the call for papers for this *Journal* issue puts it. Social life being multicausal, sorting out the causes of any event can only be approximate—and the effects of a large set of diverse institutions like the mass media even more so. As a result, the effects question also remains a virtually automatic source of intellectual vitality. Since no final answer can probably ever be achieved, the continuation of effects studies will assure the continued "fragmentation" of media research, thus preventing the development of a gangrenous consensus that kills off new ideas.

The effects question is also of major public importance, for people want to know whether the media on which they depend for information and entertainment have good or bad effects—on them, their children, and on America in general. Consequently, the social usefulness of media researchers is measured in part by the extent to which they try to answer such questions.

This is all to the good, for researchers who are not socially useful to the general public from time to time risk not being supported by governments, foundations, or commercial firms. If the researchers are academics, their books will not sell as well, their courses will not attract as many students, and their universities will then be more reluctant to allocate resources to them.

The interests of the general public create yet other intellectual reasons for reviving the study of media effects. In the absence of such study the public looks for other, generally less satisfactory, ways of answering such questions. For example, after the early years of effects research, when the so-called hypodermic or strong effects model was replaced by the limited effects model, and then by frustration when empirical studies did not produce significant effects of media exposure and usage, essayists serving the general public found answers elsewhere.

Many such essayists proposed what I think of as an *automatic effects*

Herbert J. Gans is the Robert S. Lynd Professor of Sociology in the Department of Sociology at Columbia University. Some ideas in this article come from unpublished papers written when he was a senior fellow at the Gannett (now the Freedom Forum) Media Studies Center, Columbia University.

theory, which argues that the media *must have* effects simply because they are all around us all of the time. Politically conservative writers resuscitated the strong effects model in order to attack what they perceived as overly liberal news and entertainment media that were out of sync with the allegedly conservative values of the general audience. Left critics attacked the media from a reverse but otherwise not very different perspective.

All of these analyses view the media as more influential than they really are. The believers in automatic effects imply that the media possess magic power. Critics from the Right and Left assume that the media so brainwash both the elite and the masses as to regularly bring about vast social, economic, or political changes. However, these approaches are not only wrong, but they raise false expectations about what the media can do—and meanwhile also blind people to the real holders of power. Conversely, blaming the media for consequences they have not caused turns them into scapegoats, which diverts attention from the real villains, if such exist, and "chills" people who work in the media. Consequently, increased research on the effects question would also help to produce a more thoughtful assessment of the influence of various media in American society.

Limiting Factors on Media Effects

The rest of this article discusses some of what ought to go into a more thoughtful assessment. It identifies and raises research questions about agents and structures that limit the potential effects of the mass media on the behavior and attitudes of people, and on the actions of institutions. I discuss eight limiting factors which seem to me the most important.

1. *When and how do media have what we call effects.* One still not fully resolved issue is the determination of when effects have actually taken place. Some researchers and lay writers equate effects with correlation, but correlations are not, and do not prove, causes. Simulated violence has long been popular on entertainment television (and in the older electronic mass media) but America's high rates of violence have other causes. Violent television could even itself be an effect of these other causes—even while concurrently acting in a cathartic fashion to help keep down actual violence.

Whether significant effects can even take place also needs further consideration. Much media content goes in one eye or ear and out the other, at least judging by how well people remember commercials or the names of high federal officials. There is no evidence of visible behavior change in the nation's courts on the day after "L.A. Law" is on television. Besides, courts and judges can and do enforce rules that suppress television role models. Even young children know where reality ends and entertainment begins, so that only the very gullible and pathological allow the former to affect them significantly. Otherwise, many among the millions of American and especially foreign youngsters raised on U.S. western and detective fare would have looked for weapons to kill local villains.

Perhaps there are covert long-term effects of the media, but if so, no researcher has yet glimpsed them. To be sure, life is different than before the mass media were invented, but so much else has changed that tracing what the media have caused would be difficult. Some observers believe they are amusing us to death, but in actual fact we do not even know who is amused by the television sitcoms.

Finally, most media effects are probably partial. For example, television is often said to have helped the civil rights movement win major political victories in the 1960s, but that help was also shaped by movement leaders' awareness of how the already existing popular support for the movement might be increased further by marches and demonstrations that television could show. Even so, the medium's effect would have been far less had Southern sheriffs not felt they could attack black marchers with dogs and cattle prods even while the TV cameras were rolling. Whether this reflected Southern white overconfidence, intransigence, or ignorance of television's political power is still not clear.

2. *"The media" is a buzzword, not a cause of effects.* If effects are to be studied seriously, they must be connected with, and traced to, overt and covert elements of *content, symbols, characters,* etc., in *formats, genres,* etc. transmitted by *specific* mass media as experienced by actual viewers with different perspectives and predispositions. Even the distinction between news and entertainment media is dubious, for what television producers construct as entertainment may be treated as informative by some parts of the audience. Tabloid TV's "infotainment" may in fact be neither. Perhaps much of the audience views it as morality plays about the violation of traditional familial and other social norms.

3. *Effects are limited by the intentions of the audience.* Intended effects are often the satisfaction of conscious wants, vaguely felt predispositions, or the workings of "selective perception," by which people tune out much, if not all, undesired content—and perhaps potential effects. Indeed, the common sense definition of media effects is largely limited to unintended ones.

Furthermore, such effects often go into operation only when they connect to social conditions external to the media. Television or movie "action" appears to evoke stronger effects among poor youngsters than more fortunate ones, just as cocaine seems to be far more addictive among very poor people than among middle-class users. Presumably, the emotional vulnerability that leads to addiction, to drugs, television, or violence is an effect of poverty itself.

Also, media research must leave more room for noneffects, when people "tune out" commercials, for goods and politicians, or if they treat routine sitcoms and their characters as surrogate or vicarious company. These, like regularly visiting friends or relatives, may have no visible effects at all.

4. *Intended effects often occur because media content is produced by organizations that need to attract and satisfy audiences.* Effects researchers sometimes forget this "fact of life," and its corollary: Surprises, or at least deliberate unintended effects on the audience, are eschewed most of the time, since they ultimately lead to the loss of goodwill for commercial as well as public media organizations. Nonetheless, even highly successful advertising campaigns attract or convert only a small proportion of people to a new brand or product.

News caters to the audience in more subtle ways than entertainment, but in virtually all cases it takes the form of a dramatic story, distanced to limit audience surprise. However, news effects are further limited by the fact that most of the audience, at least in the U.S., does not seem to have great need for national news, and only a minority pays attention to it regularly. In wartime, or in countries where the government affects people's lives directly and frequently, the audience's need for news is more immediate and intense.

Media researchers have spent a great deal of energy studying how political news and various kinds of political advertising affect voters and election outcomes, although now that election campaigners manage the news so completely, including by placing their candidates into situations where they can avoid journalists, the boundary between political news and advertising disappears. Furthermore, someday soon it may no longer be possible to study the effects of television on election campaigning, since nontelevision campaigning will have been eliminated.

Actually, the most important effect of political advertising may have little to do with the media. Which candidates the funders of campaigns decide to supply, and not supply, with money for political commercials and other support is often more important to election outcomes than the commercials themselves. Moreover, after the election, when the funders ask for favors from, or merely access to, the candidates they financed, they probably have a greater effect on national politics than the commercials they funded.

5. *News media effects are limited by the passivity of news organizations.* Journalists and the news media see their role as being mainly receptors: They usually receive news, which they then transform into stories. Even investigative reporters, who *are* making news, generally wait to receive "tips." As a result, news organizations remain primarily *messengers,* reports of events created by others.

If they report news about rising unemployment, for example, the effects they produce stem from the unemployment, not of their reporting it. Business people and politicians whose fortunes may be affected by rising joblessness sometimes appear to think that if such news could be suppressed, the effects of unemployment on worker or consumer confidence would be reduced. However, this, like other beliefs that society would be different if the news media could be made to disappear temporarily, is a faulty assumption to which strong-effects theorizing is particularly susceptible. (Equally faulty is the assumption that modern society would be very different without modern news media, as if the latter were not part and parcel of what makes society modern.)

In any case, most people judge the economy, including levels of unemployment, by how they, their relatives, and friends are faring. And when the media do play a role, we still need to discover how much effect belongs to the messages and how much to the way the messengers package the messages, and which they omit.

6. *Another limit on news effects is that news organizations are messengers for their major sources.* Journalists get most of their news from regular sources which, as study after study has shown, are usually speaking for political, economic, and other establishments. As a result, political news is not so much about politics as about what elected and appointed officials want to communicate about politics: the political performance they want to put on for their constituencies, and the political effects they would like to have on them. This is especially true when investigative reporting—or even time for normal legwork—is scarce, and conflicts within establishments are hidden or minimal enough to prevent journalists from finding regular and authoritative sources on several sides.

7. *Effects are limited further by the vast amount of news and the sparse amount of social change.* If news audiences had to respond to all the news to which they are exposed, they would not have time to live their own lives. In fact, people screen out many things, including the news, that could interfere with their own lives.

In addition, we know that most people do not make drastic changes in their lives unless they are exposed to unusual incentives or intense economic or social pressures that force them to change their ways involuntarily. Moreover, giant media firms, private or public, are almost always cowardly in the face of controversy—including support of change—for fear of alienating audiences. For example, sexual "liberation" came to network television characters and content long after it had arrived in the real world—and then in part because network television was trying desperately to survive against competition from already liberated movies, cable programming, and cassettes. Agenda-setting theory notwithstanding, sexually liberated viewers have indirectly affected network television, not the other way around.

8. *The mass media may have greater effects on institutions than on individuals.* Some of the limits on media effects may be relaxed in the cases of large and seemingly powerful institutions, especially when they run scared about the loyalty of their audiences, customers, and constituencies. Thus, the political parties have altered themselves comprehensively because of television, and precisely because they had lost most of their old constituencies. In fact, political parties now exist mainly to raise money for television commercials, and to hold annual conventions on which they can advertise themselves and their candidates on the small screen.

The Pentagon ran so scared after the news coverage of Vietnam that it virtually took over the news coverage of the Grenada, Panama, and Gulf wars—an impressive, if unappreciated, compliment to the effects it ascribed to uncensored television news. Institutions, especially large ones, seemingly move quickly to alter their ways if they feel that they must protect themselves against threats.

Effects of Researcher Ignorance

A very different factor influencing research on media effects is the continued ignorance of researchers about how people use, and live with, the mass media. Because media organizations are few and often accessible to researchers, we know a good deal about the production of news and entertainment content, and there are some quantitative and laboratory data about the audience. In recent years, social and cultural historians have also analyzed archival data to assess the long-range effects of major television genres, and important running news stories such as Watergate and Vietnam.

But researchers still know almost nothing about the processes by which people choose what to consume in the various media; how they consume it, with what levels of comprehension, attention, and intensity of affect; what, if anything, they talk about while using the media at home; whether and how their uses of various media connect to other aspects of their lives—and which; and what kinds of traces, if any, these media leave in their psyches and lives, and for how long.

Because media researchers make a living from the media, they play close attention to them, and probably closer and more intensive attention than anyone, including perhaps even many of the people involved in the creation of media content. As a result, media researchers may also be more affected by the media than anyone else, and it is possible that they project that effect on the "normal" consumers of media content. In some respects, this projection is all to the good, for it has stimulated researchers' imagination about the varieties of possible effects of the media on people and institutions. However, their projection could also have overstimulated their imagination, insofar as they may have overestimated the nature and extent of the roles that the media play in people's lives. Whether this is so can only be tested by studying how people use, and live, with various media.

This means getting close to the media audience—and nonaudience. One method is depth interviewers talking with people. Better still would be ethnographic community researchers who are able to be with people as they use—and ignore—the available media, and talk with them about these media, especially in relation to the other institutions that affect their work, family, and community lives.

This kind of research is slow and expensive, but while content analysis can report what analysts see in the content, and sample surveys, focus groups, and laboratory experiments can result in neat, bounded answers, these all maintain some distance from people and from the lived world of media use. Until researchers enter and understand that world sufficiently, and provide a bedrock of interview and ethnographic findings, media researchers cannot judge the validity and reliability of the more distanced methods. Nor can they begin to develop a proper assessment of the true effects of the media.

Realism and Romance: The Study of Media Effects

by Gaye Tuchman, University of Connecticut

In the late 1970s, when sociological researchers suggested that the meaning of news might be embedded in the context of its production, they invoked the twin concepts of hegemony and ideology. In essence, they argued, one might view news as a "means not to know"—an obfuscation that fragmented information, made obscure the connections among issues and events, and so strengthened the hegemony characteristic of American society (see, e.g., Fishman, 1980; Gitlin, 1980; Tuchman, 1978).

At the time, Elihu Katz (1980) noticed a potential irony in this research: For decades, researchers attached to the theorem of minimal effect had argued that the media reinforce existing dispositions. Now, while attacking the orientation implicit in much survey research, a new generation of researchers was arguing that the media impeded significant social change. That "media effect," Katz reminded, echoed the previous empirical findings.

For other researchers, the echo had a jarring pitch. First, the themes in the media themselves change over time (Gamson & Modigliani, 1989; Long, 1985). Second, if the media are hegemonic, if they so resonate with the interests of the powerful as to have a conservative impact (as also argued by cultivation theory), then what of the notion of human agency? Cannot and do not people resist? May they not create a "counterhegemony?" Do they not assess media messages against their own experiences to create their own "tool-kits" (Swidler, 1986), interpretations (Simonds, 1992), or oppositional cultures (Willis, 1990)?

These questions broach the significant issues raised by students of mass media within the past decade. But the contexts in which these issues have been discussed themselves vary—from the seeming celebration of the oppositional culture of white working-class young men (Willis, 1977; cf. McRobbie, 1980) to the careful dissection of group conversations about four public issues that have been featured in the American media over the past few decades (Gamson, 1992).

Two of these studies represent diametrical approaches toward the issue of agency: romanticism versus realism. Willis's *Common Culture* (1990)

Gaye Tuchman is a professor of sociology at the University of Connecticut, Storrs.

expresses such a romantic attachment to the agency of working-class youth that he fails to see how their deployment of technology may be read as an acceptance of dominant culture. Gamson's *Talking Politics* (1992) affirms the ability of working people to reason about complex political issues and their potential for collective action.

Why is Willis a romantic? Willis (with others, 1990) attacks "hegemony theory" for ignoring how working-class British youths use the media to express creatively their own "grounded aesthetic" and their own individuality. For instance, he emphasizes that youths use tape recorders to create their own programs or shop in second-hand clothing stores to create their own look (or identity). Willis stresses that the product of record companies is not a vinyl disk, but sounds that a youth may transform to display to him/herself and to others his/her own musical identity. In the process of personal production, the youth may thus engage in an individually and socially meaningful creative act. The act is socially meaningful in as much as a youth is using a musical vocabulary or "conventions" (Becker, 1982) shared with peers. For Willis, this creative production of new meanings embodies resistance to hegemonic meanings: It represents counterhegemony.

But does it? Is the identification of new forms of media use as creative acts an explication of resistance or of the persistence of a modern ideology—the right to an individual identity? By *the right to an individual identity,* I mean the notion that each of us is intrinsically different; each of us expresses his or her own "unique" personality; each of us assembles from the tool-kit of culture (Swidler, 1986) a presentation of self that makes us inextricably who we are.

Consider some examples of how Americans use technology to express this ideology. Some, but not all, owners of telephone answering machines may express their identity through the recorded message. A colleague who adores classical music has his machine answer with some phrases lifted from Bach. The recording of a friend who is a literary critic instructs callers to "leave a message after that postmodern tone, the beep." A college student has programmed her machine to play heavy metal rock; at the end of a few phrases and some drums, her recorded voice announces "DO IT!" and so supposes that all callers know what to do. Each of these people is working to create a meaningful message that humorously proclaims an essence of his or her being.

Grasp of that essence is socially grounded. Without shared understandings, the caller may not understand the referent in the owner's message. In all likelihood, the college student would not respond to the postmodern beep with a quiet chuckle. Postmodernism is not necessarily part of her shared stock of knowledge. Shared knowledge is important, as callers themselves indicate: When a machine spews out an unusual message, a caller may include in his/her own message a comment on the owner's recording. Willis might say each owner is formulating a grounded aesthetic appreciated by callers.

If one discounts the significance of the modern ideology of individualism, other (structural) factors assume importance. Concentrating on broad characteristics of the modern era, one may analyze each idiosyncratic message (or even the very use of telephone answering machines) as an instance of either mechanical reproduction or an instance of the depersonalization of the machine age represented by the substitution of a recorded message for the person who would otherwise answer the phone, just as the telephone is itself a substitute for personal contact.[1]

Dissatisfaction with depersonalization is not unique to the working class. It even appears in that postmodern bricolage meaningful to most members of western culture: the television commercial, the archetypical art form of the modern era (Schudson, 1984). Ads are a ubiquitous art form. Not only do youth sing commercials (Willis, 1990), as do American elementary school children (personal observation), but texts used in the third grade have pupils write advertisements as spelling exercises (personal observation).

Currently one advertisement (for an American airline) identifies the telephone and the fax with depersonalization. In the television ad, the owner of a firm reports that an old customer has canceled his account because he doesn't feel he knows the firm any more. The owner explains that the sales staff has been doing business by phone and fax and so has lost touch (personal contact) with its customers. He hands his sales staff airline tickets and tells each person whom he is going to visit to reestablish personal contact. As for the owner, he is flying to visit the old customer who had canceled his account.

The advertising example is theoretically meaningful because it plays on use of one modern device (air travel by the most common frequent flyer,

---

[1] Michael Gurevitch reminds me of another possible interpretation. One could argue that telephones facilitate contact. It is difficult to reach people who do not have a phone. Similarly, if no one answers the phone, it is difficult to leave a message for someone who does not have an answering machine.

But, I would counter: Machines transform the interaction. The kind of information conveyed via either telephone or answering machine differs in kind from face-to-face communication. When Michael and I communicate over e-mail, we are sending messages to each other's 10-years-ago self, for we have not seen one another in that time. We do not know whether the other with whom we are communicating has significantly changed (physically or psychologically) since we last met. Such information might be immediately apparent in a face-to-face interaction and might prompt either one of us to modify our message.

Machines may give their owners more or less control over machine-mediated interactions. When they do not wish to be disturbed, some people take their phone off the hook so that they will *not* know that someone is trying to reach them (and so be tempted to answer). Others put on their answering machine and monitor the calls. On the one hand, the machines enable a person to reach someone else. On the other, they introduce an element of rude uncertainty. If I phone someone who picks up after hearing me talk to the machine, I may feel complimented that he chose to spoke to me. If I know that someone uses her machine to monitor calls and she does not pick up the phone, I may wonder if she chose not to speak with me.

the businessperson) to criticize other contemporary mechanical devices: airplanes as critique of telephones and faxes—mechanical device as the solution to the problems created by mechanical devices (just as faxes are an economic solution to the expense of air travel). Meanings collapse within meanings. In the word once used by McLuhan, such advertisements represent an "implosion." They also affirm a well-known dictum: Mechanical devices are inextricably woven into the social (and structural) relations of contemporary industrial societies. By ignoring this common sociological knowledge and stressing how working-class youth use available materials to assemble an identity, Willis has romanticized working class youth as much as Rousseau and Chateaubriand idealized the "noble savage."

Gamson's (1992) tack is quite different. To understand how 188 participants in 37 focus groups interpreted four prominent political issues, he and colleagues first did massive content analyses of how these issues were portrayed in diverse media. Like the work of Graber (1988), his approach is realistic inasmuch as it insists that one cannot understand how people interpret the media without understanding the basic frames the media employ and how those frames change over time.

Analyzing working-people's conversations, Gamson emphasizes: "(a) People are not so passive, (b) People are not so dumb, and (c) People negotiate with media messages in complicated ways that vary from issue to issue" (1992, p. 4). He is aware that he is viewing these conversations in a sympathetic light:

> *The phrases* not so passive *and* not so dumb *refer to the way mass publics frequently appear in social science portraits. Of course, this is another case of whether the glass is half empty or half full. . . . Yet people read media in complicated and sometimes unpredictable ways, and draw heavily on other resources as well in constructing meaning.* (Gamson, 1992, pp. 4, 6)

Among these resources are experiential knowledge and popular wisdom. To be sure, the personal and the cultural are intertwined; together they exert an impact on interpretation. Those interpretations may construct "meanings quite different from the preferred reading" (Gamson, 1992, p. 124). The characteristics of the interpretations depend on the group, the resources invoked, and the issue, including how proximate an issue is to the lives of group members and how engaged they are with it. Understanding how working people interpret media discourse is, then, a rather complex task.

Ultimately, Gamson concludes that the revolution is not in sight: "I do not claim that these working people have the kind of political consciousness that makes one ready to take up arms against the world's injustices. . . . The raw materials are there, but this varies dramatically from issue to issue" (Gamson, 1992, pp. 175, 176). Seeing injustice "promotes personal identification with whatever group is being wronged spurs the search for agents who are responsible for the undeserved hardship that members of the recipient group suffer," but even the presence of an injustice frame does not suffice (Gamson, 1992, pp. 175, 176). In groups, personal, cultural, and integrated strategies all interact with media frames, and the causal arrows among them are bidirectional.

In essence, Gamson and Willis are making the same announcement: People make their own uses of media.[2] They may resist hegemony. But, in his more complicated and realistic tale, Gamson sees how the dominant culture exerts an influence on people's use and interpretation of media: Some groups accept preferred readings and, potentially, may even accede to reactionary social movements.

Ultimately, the task facing media researchers is continued exploration of the path that Gamson (1992) and others (Long, 1985; Radway, 1984; and Simonds, 1992) have identified. Yes, the media seem hegemonic. They offer preferred readings. But even relatively uneducated individuals and groups of working people have their own thematic understandings of the social and political world. Researchers must realistically take these into account.

References

Becker, H. (1982). *Art worlds*. Berkeley: University of California Press.

Fishman, M. (1980). *Manufacturing the news*. Austin: University of Texas Press.

Gamson, W. A. (1992). *Talking politics*. New York: Cambridge University Press.

Gamson, W. A., & Modigliani, A. (1989). Media discourse and public opinion on nuclear power. *American Journal of Sociology, 95,* 1–37.

Gitlin, T. (1980). *The whole world is watching*. Berkeley: University of California Press.

Graber, D. (1988). *Processing the news*. New York: Longmann.

Katz, E. (1980). On conceptualizing media effects. In T. McCormack (Ed.), *Studies in communications, 1* (pp. 119–142). Greenwich, CT: JAI Press.

Long, E. (1985). *The American dream and the popular novel*. Boston: Routledge & Kegan Paul.

McRobbie, A. (1980). Settling accounts with subcultures. *Screen Education, 34,* 37–49.

Radway, J. (1984). *Reading the romance*. Chapel Hill: University of North Carolina Press.

Schudson, M. (1984). *Advertising, the uneasy persuasion*. New York: Basic Books.

Simonds, W. (1992). *Women and self-help culture: Reading between the lines*. New Brunswick, NJ: Rutgers University Press.

Swidler, A. (1986). Culture in action: Symbols and strategies. *American Sociological Review, 51,* 273–286.

---

[2] This notion itself echoes uses and gratifications theory, but both the theoretical and methodological thrust are new.

Tuchman, G. (1978). *Making news*. New York: Free Press.

Willis, P. (1977). *Learning to labor*. New York: Columbia University Press.

Willis, P. (with Jones, S., Canaan, J., & Hurd, G.). (1990). *Common culture*. San Francisco: Westview Press.

Revealing the Black Box: Information Processing and Media Effects

by Seth Geiger, University of California, Santa Barbara, and John Newhagen, University of Maryland, College Park

Understanding how individuals process messages is central to any comprehensive theory of communication. As self-evident as this statement may seem, the conceptualization and measurement of mass media effects have generally ignored message processing issues. This article addresses some of the fundamental assumptions of an information processing approach to mass media effects, and the kinds of contributions we believe it brings to mass communication. In particular, we trace the conceptual and methodological innovations of an information processing perspective as they have been applied to the study of television in the decade since the publication of "Ferment in the Field."

This information processing approach contributes to inquiry into mass media effects in a number of ways. First, it presents a unique characterization of individuals and of media messages. Conceptually, it emphasizes the way people make sense of, attend to, and remember television messages. Television messages in turn are defined in terms of psychological dimensions and attributes that are likely to influence these processes. These message attributes reflect how a message is constructed and presented as well as the more typical concerns associated with content and genre. Second, it utilizes a particular methodology, relying on controlled experiments and implicit measures taken during viewing for gathering information about these mental processes.

Inquiry in individual-level mass media effects has been limited by conceptualizing the human processor as an impenetrable "black box" with unknowable processes taking place between message reception and the traditional outcomes of learning, attitudes, or behaviors. Instead, we see these component processes as both important outcomes and predictors in their own right. In short, the perennial black box of mass media effects can be better illuminated by examining the black box of human information processing that goes on within it.

Seth Geiger is an assistant professor in the Department of Communication at the University of California at Santa Barbara. John Newhagen is an assistant professor in the College of Journalism at the University of Maryland at College Park. Both authors contributed equally to this piece.

Theories About Messages

A definition of mass media messages from an information processing perspective focuses on the kinds of attributes that are likely to have an impact on the way individuals process, attend to, remember, and learn from the media. This is a departure from the way communication research has traditionally focused on messages in terms of the content of the message (e.g., violent/nonviolent; entertainment/news) and the program genres dictated by the media industry (e.g., situation comedy/action-adventure). In assessing the cognitive processing of messages, these categories in and of themselves may have very little to do with the way people actually process media messages. For example, the kinds of attributes that influence attention to a television program include the complexity of its language and images, the visual pacing of segments, the composition of shots, the use of special effects, and the degree of relatedness between shot segments within a narrative sequence. However, all of these attributes are independent of traditional content- and genre-based definitions.

From an information processing perspective, television is a complex and cognitively demanding stimulus. This statement, based on cognitive principles, is at first glance startling. Television is typically regarded as a mindless activity or as a powerful stimulus that overcomes the viewer (Singer, 1980). While television viewing may not always be characterized as a mindful, or conscious activity, it does involve a number of demanding mental operations. Television presents information on two sensory channels, audio and visual, in a continuous stream. This means that viewers must attend to both channels and relate them in a meaningful fashion in order to make sense of the message. Because television unfolds in real time, viewers must also construct a mental image incorporating the messages' characters, plots, and themes, and store and access this information from memory. This complexity is further reflected in the extensive array of production techniques, special effects, and modes of presentation used in message construction.

An example of an information processing approach to the conceptualization of messages can be seen in the study of how two conceptually distinct dimensions of television, content and structure, interact to influence the evaluation and memory for candidates in political advertisements (Geiger & Reeves, 1991). The advertisements' content focused either on the candidate's image or on some issue position or past action, while their structure was either fast-paced—containing a number of cuts, different locations, narrative voice and music—or relatively slow-paced. Candidates in dynamic-issue advertisements received the most positive evaluations, the best audio recognition, and the poorest visual recognition. Candidates in image advertisements were poorly evaluated, especially for the static-image spots, but received the best visual recognition. These results indicate that content and structure interact in important ways to influence the meaning and ultimately the way people remember television messages.

Theories About Individuals

Since mass media messages are conceptualized in terms of attributes that have some influence on how television is viewed and remembered, it is essential that this conceptualization reflect some understanding of the human processor. This cognitive perspective focuses on the component processes involved in how individuals make sense of the world around them. These processes include the perception of information, encoding, representation, memory, and problem solving (Anderson, 1985). Each of these processes imply discrete stages of mental operations that occur while an individual performs a given task, such as watching television. While there are a number of ways of conceptualizing information processing, all models essentially make a number of common claims. First, the total amount of information active at any one time is finite. Second, the likelihood for retrieval of information is determined by how well learned, or activated, an item is. These two characteristics reflect the processes of attention and memory, and are essential for the processing of complex information, such as watching television or reading a newspaper.

Processes of Attention
Attention has an intuitive appeal for communication researchers because it is obviously an important stage for any kind of intentional mass media effect, such as learning from television news or persuasion in a political campaign. Attention within mass communication research has typically been defined either in terms of interest in particular media content or the choice of a program (Chaffee & Schleuder, 1986). In cognitive terms, attention is the process of selectively allocating limited resources (Hirst & Kalmar, 1987). There are in fact two types of attention: controlled and automatic. Controlled attention is synonymous with mental effort and is dictated by the goals of the individual processor. Automatic attention does not require the use of limited resources, and is determined by attributes of the information. One of the basic components of a skill, such as comprehending a complex mass media message, is to make controlled tasks more automatic, and therefore more effortless, with practice.

To illustrate how these concepts can be applied to media effects research, television viewing can be treated as a complex activity that requires both controlled and automatic attention. Research has found strong support for an automatic orientation response occurring immediately following a cut (Lang, Geiger, Strickwerda, & Sumner, 1993). The amount of attention and effort following a cut depends on the relatedness of the new information to the previous information. If the new information matches expectations based on previously presented information, the amount of attention required will be less than when unrelated information is presented (Geiger & Reeves, 1993). Automatic attention also permits television viewers to make sense of the steady stream of sound and images, while simultaneously attending to the plot, the characters, and the meaning of the message.

Memory for Television

Like attention, memory for information is an implicit component of media effects research; in order for an effect to have occurred, some aspect of the message must have been retained. Current models of memory are based on two features that are important for how television information is organized and retrieved (Anderson, 1983). First, concepts are organized by their degree of association in a viewer's experience. Secondly, the time required for the activation and retrieval of an item depends on its relation to the items that are currently active. These two features have important implications for how viewers make sense of sequences of television. As a television program unfolds, it presents segments that build on previous information. Viewers make use of this previous information to anticipate subsequent information. This phenomenon occurs through the increase in activation strength for the concepts and themes presented in the program. As activation spreads between associated items, a network is established that facilitates the processing of new information as a result of the already activated associated concepts.

Cognitive models of memory have been applied most extensively to research in television news (Brosius, 1993; Davis & Robinson, 1986; Drew & Grimes, 1987; Newhagen & Reeves, 1992). This is, in part, a response to results indicating memory for television news to be very weak (Neuman, 1976). A cognitive perspective may make an important contribution to the understanding of this phenomenon. One experiment has assessed the influence of the order of story presentation on memory (Gunter, Clifford, & Berry, 1980). Memory decreased significantly from the first to the fourth story in a sequence when all four stories focused on the same category of news (e.g., domestic, foreign, politics, sports). When the category shifted between the third and the fourth stories, a dramatic increase in memory for the fourth story occurred. This pattern of results follows a "release from proactive interference," where one message interferes with the recall of subsequent messages that share the same category of information. Similar patterns of interference also have been found with images of death and suffering retroactively inhibiting recall for information (Newhagen & Reeves, 1992).

Ways of Knowing

The scope of claims about media effects is constrained by how concepts are operationalized and measured. Many interesting media effects have been beyond the operational reach of empirical scrutiny simply because of the individual's inability to talk about what goes on inside the black box of the mind. This is true to the degree that media effects research for the last half century has relied on self-report measures. For instance, asking respondents to indicate their own levels of attention has proved to be fruitful in surveys, where attention is conceptualized as interest in a particular type of program (Chaffee & Schleuder, 1986). However, attention as a measure of mental capacity and effort simply takes place below the threshold of awareness, or occurs too rapidly to be available for conscious report (Bower & Clapper, 1989). Further, introspection is heavily influenced by the observers' theoretical preconceptions.

Self-Report and Implicit Measures

Measures that are taken on-line, while an individual views television without any interruption, serve as one alternative to self-report measures. These covert measures are preferred because they maintain the integrity of the television viewing experience. Measuring attention through self-report requires viewers to consciously abandon the viewing process and respond to questions about their mental activity, such as "How much attention were you just paying to the message?" While this tactic might prove useful for larger message units, such as programs or commercial spots, it intrudes on the natural viewing processes for smaller units, some lasting only seconds or even milliseconds. Overt measures also assume that individuals are capable of reflecting on and articulating cognitive processes as they occur, an ability that is not evident in our research with individuals. For example, Newhagen (in press) tested memory for television news presentations of the Persian Gulf War containing 5-second censorship disclaimers on the screen. Despite the fact that many subjects could not verbally report having seen the disclaimers, they affected memory for prior and subsequent material.

On-line measures. One operational alternative to self-reports of attention is the response latency to a secondary task. Because processing resources are limited, the time it takes to respond to a concurrent task will increase as the primary task becomes more demanding. In the case of television, when more attention is required to watch a message, more resources are allocated to viewing and less are available for responding to the secondary task. Longer response times to performance of the secondary task indicate greater attention to the message.

Attention to a television presentation has been measured by the latency to respond to a secondary task such as a tone (Reeves, Newhagen, Maibach, Basil, & Kurz, 1991). In this experiment, subjects' latency to respond was longer (indicating more attention) to the spots containing a positive emotional valence, then to negative spots. Other psychophysiological measures, such as heart rate and galvanic skin response, also provide implicit measures of attention and arousal (see e.g., Lang, 1990).

Similarly, memory for information in a media presentation can be measured using the latency to a choice recognition task. The foundation for this measure is based on the assumption that items with the greatest memory strength will be recognized faster than items with weaker strengths and associations. Newhagen and Reeves (1992), for instance, tested television viewers' memory for news stories, half of which contained emotionally compelling images of human death and suffering. After viewing the news stories, subjects had their recognition memory tested using short visual and audio clips. Recognition was faster and more accurate for stories containing compelling images, indicating superior memory.

Repeated-measures designs and message variance. One final methodological departure from traditional mass media effects research involves the use of a number of different messages that share a particular feature of interest. By using multiple messages to represent the levels of each factor, this repeated-measures design strategy allows for the control and randomization of extraneous features. Since mass media messages are highly complex, there are inevitably a number of factors confounded in any message that are not of explicit interest to a particular experiment. These are controlled by randomizing these extraneous factors across a range of stimulus messages.

Two things should be emphasized about this strategy: First, it is statistically conservative, in the sense that the complexity of each new message included in a design will be reflected as error rather than as systematic variance. Simply put, the strategy guards against claims about systematic variance in unmeasured message attributes. For instance, in a study of negative images based solely on a television news story of a riot, subject attention might increase due to both the presence of the images *and* fast pacing. However, by including a range of stories involving different levels of pacing, the sum of the measured variance in attention is more likely to reflect only the concept of interest—in this case, negative images. Second, the fact that information processing studies frequently employ a repeated-measures design is largely a matter of historical coincidence. That is to say, nearly any theoretical perspective in communication research might profit in its generalizability to *other messages* if the repeated-measures design is used.

Understanding Media Effects

Over 20 years ago, Wilbur Schramm commented on how the understanding of the communication process was incomplete without an understanding of "the 'black box' of the central nervous system" (Schramm, 1971, pp. 24–25). From the time of Schramm's lament that the black box was out of the reach of communication theory until now, a number of important changes have taken place in social science. First, there has been a cognitive revolution in the areas of psychology and neuroscience that has impacted the social sciences in varying degrees (Gardner, 1985). Over the same period, the scientific method of empirical inquiry has been questioned in terms of its validity and applicability to human behavior (Giddens, 1976; Jennings, 1983). These questions have in turn been extended to the area of mass media effects research, whose critics claim the empirical paradigm is inappropriate for explaining a process as complex as the impact of the mass media on individuals and societies (Carey, 1989).

Two explanations have been posited for the discrepancy that exists between the strong effects theories of mass communication and a pattern of empirical results that support only limited effects. One is that there are indeed no strong effects. The alternative is that because of the complexity of the system in which they exist, they simply cannot be directly measured in a meaningful way. But both professional media practitioners and academic researchers continue to find either of these conclusions intuitively unlikely. Contrary to the two previous explanations, we contend that theoretical and operational shortcomings relying on self-report have simply failed to adequately explain how people process mass media messages, and that information processing approaches and implicit tasks provide one alternative.

It can be argued that the ability to ask interesting questions about the effects of mass media has always preceded the field's capacity to answer them. Perhaps the prototypical example can be seen in the work of Claude Innis (1964) and his student Marshall McLuhan (1964), whose provocative essays on the implications of electronic mass media are just now coming within the reach of empirical scrutiny. Indeed, Converse (1964) warned that "no intellectual position is likely to become obsolete quite so rapidly as one that takes current empirical capability as the limit of the possible in a more absolute sense" (p. 207).

We believe a truly comprehensive theory of communication should include comments from all levels of analysis. This includes a critical macrosocietal discussion of the implications of mass media in the context of culture and in the interaction of its social institutions. But to be complete, a comprehensive theory must necessarily include an explanation of the effects of messages on individuals as well. It is only across those levels of analysis that a truly global understanding of communication can be achieved. The fact that conceptual and methodological tools to look inside the black box of the human mind are only just becoming widely available to communication researchers should not be a cause for despair or abandonment. Quite the opposite, we believe that a cognitive approach to mass media effects, by revealing the intervening processes involved in making sense of media messages, can only make an increasingly important contribution to our understanding of communication.

References

Anderson, J. R. (1983). A spreading activation theory of memory. *Journal of Verbal Learning and Verbal Behavior, 22,* 261–295.

Anderson, J. R. (1985). *Cognitive psychology.* New York: W. H. Freeman.

Bower, G., & Clapper, J. (1989). Experimental methods in cognitive science. In M. Posner (Ed.), *Foundations of cognitive science* (pp. 245–300). Cambridge, MA: MIT Press.

Brosius, H. B. (1993). The effects of emotional pictures in television news. *Communication Research, 20,* 105–124.

Carey, J. (1989). Mass communication and cultural studies. In J. Carey, *Communication as culture* (pp. 37–68). Winchester, MA: Unwin-Hyman.

Chaffee, S. H., & Schleuder, J. (1986). Measurement and effects of attention to news media. *Human Communication Research, 13,* 76–107.

Converse, P. (1964). The nature of belief systems in mass publics. In D. Apter (Ed.), *Ideology and discontent* (pp. 206–261). New York: Free Press of Glencoe.

Davis, D. K., & Robinson, J. P. (1986). News story attributes and comprehension. In J. P. Robinson & M. R. Levy (Eds.), *The main source* (pp. 179–210). Beverly Hills, CA: Sage.

Drew, D. G., & Grimes, T. (1987). Audio-visual redundancy and TV news recall. *Communication Research, 14*(4), 452–461.

Gardner, H. (1985). *The mind's new science.* New York: Basic Books.

Geiger, S., & Reeves, B. (1991). The effects of visual structure and content emphasis on the evaluation and memory for political candidates. In F. Biocca (Ed.), *Television and political advertising* (pp. 125–144). Hillsdale, NJ: Lawrence Erlbaum Associates.

Geiger, S., & Reeves, B. (1993). The effects of cuts and semantic relatedness on attention to television. *Communication Research, 20,* 155–175.

Giddens, A. (1976). *New rules of sociological method.* New York: Basic Books.

Gunter, B., Clifford, B. R., & Berry, C. (1980). Release from proactive interference with television news items: Evidence for encoding dimensions within television news. *Journal of Experimental Psychology: Human Learning and Memory, 6,* 216–223.

Hirst, W., & Kalmar, D. (1987). Characterizing attentional resources. *Journal of Experimental Psychology: General, 116,* 68–81.

Innis, H. (1964). *The bias of communication.* Toronto: University of Toronto Press.

Jennings, B. (1983). Interpretive social science and policy analysis. In D. Callahan & B. Jennings (Eds.), *Ethics, social sciences, and policy analysis* (pp. 3–36). New York: Plenum.

Lang, A. (1990). Involuntary attention and physiological arousal evoked by structural features and emotional content in TV commercials. *Communication Research, 17,* 275–299.

Lang, A., Geiger, S., Strickwerda, M., & Sumner, J. (1993). The effects of related and unrelated cuts on television viewers' attention, processing capacity, and memory. *Communication Research, 20,* 4–29.

McLuhan, M. (1964). *Understanding media: The extensions of man.* New York: McGraw-Hill.

Neuman, W. R. (1976). Patterns of recall among television news viewers. *Public Opinion Quarterly, 40,* 115–123.

Newhagen, J. (in press). The relationship between censorship and the emotional and critical tone of television news coverage of the Persian Gulf War. *Journalism Quarterly*.

Newhagen, J., & Reeves, B. (1992). The evening's bad news: Effects of compelling negative television news images on memory. *Journal of Communication, 42*(2), 25–41.

Reeves, B., Newhagen, J., Maibach, E., Basil, M., & Kurz, K. (1991). Negative and positive television messages: Effects of message type and message context on attention and memory. *American Behavioral Scientist, 34,* 679–694.

Schramm, W. (1971). The nature of communication between humans. In W. Schramm & D. F. Roberts (Eds.), *Process & effects of mass communication* (pp. 3–53). Urbana: University of Illinois.

Singer, J. L. (1980). The power and limitations of television. In P. H. Tannenbaum (Ed.), *The entertainment function of television* (pp. 31–65). Hillsdale, NJ: Academic.

Framing: Toward Clarification of a Fractured Paradigm

by Robert M. Entman, Northwestern University

In response to the proposition that communication lacks disciplinary status because of deficient core knowledge, I propose that we turn an ostensible weakness into a strength. We should identify our mission as bringing together insights and theories that would otherwise remain scattered in other disciplines. Because of the lack of interchange among the disciplines, hypotheses thoroughly discredited in one field may receive wide acceptance in another. Potential research paradigms remain fractured, with pieces here and there but no comprehensive statement to guide research. By bringing ideas together in one location, communication can aspire to become a master discipline that synthesizes related theories and concepts and exposes them to the most rigorous, comprehensive statement and exploration. Reaching this goal would require a more self-conscious determination by communication scholars to plumb other fields and feed back their studies to outside researchers. At the same time, such an enterprise would enhance the theoretical rigor of communication scholarship proper.

The idea of "framing" offers a case study of just the kind of scattered conceptualization I have identified. Despite its omnipresence across the social sciences and humanities, nowhere is there a general statement of framing theory that shows exactly how frames become embedded within and make themselves manifest in a text, or how framing influences thinking. Analysis of this concept suggests how the discipline of communication might contribute something unique: synthesizing a key concept's disparate uses, showing how they invariably involve communication, and constructing a coherent theory from them.

Whatever its specific use, the concept of framing consistently offers a way to describe the power of a communicating text. Analysis of frames illuminates the precise way in which influence over a human consciousness is exerted by the transfer (or communication) of information from

Robert M. Entman is an associate professor of communication studies, journalism, and political science and chair of the program in Communications, Media, and Public Policy at the Center for Urban Affairs and Policy Research at Northwestern University, Evanston, IL. He gratefully acknowledges the comments of students in his "Mass Communication and Democratic Theory" seminar, especially Andrew Rojecki.

one location—such as a speech, utterance, news report, or novel—to that consciousness. (A representative list of classic and recent citations would include: Edelman, 1993; Entman & Rojecki, 1993; Fiske & Taylor, 1991; Gamson, 1992; Goffman, 1974; Graber, 1988; Iyengar, 1991; Kahneman & Tversky, 1984; Pan & Kosicki, 1993; Riker, 1986; Snow & Benford, 1988; Tuchman, 1978; White, 1987; Zaller, 1992.) A literature review suggests that framing is often defined casually, with much left to an assumed tacit understanding of reader and researcher. After all, the words *frame, framing,* and *framework* are common outside of formal scholarly discourse, and their connotation there is roughly the same. The goal here is to identify and make explicit common tendencies among the various uses of the terms and to suggest a more precise and universal understanding of them.

Of Frames and Framing

Framing essentially involves *selection* and *salience.* To frame is to *select some aspects of a perceived reality and make them more salient in a communicating text, in such a way as to promote a particular problem definition, causal interpretation, moral evaluation, and/or treatment recommendation* for the item described. Typically frames diagnose, evaluate, and prescribe, a point explored most thoroughly by Gamson (1992). An example is the "cold war" frame that dominated U.S. news of foreign affairs until recently. The cold war frame highlighted certain foreign events—say, civil wars—as problems, identified their source (communist rebels), offered moral judgments (atheistic aggression), and commended particular solutions (U.S. support for the other side).

Frames, then, *define problems*—determine what a causal agent is doing with what costs and benefits, usually measured in terms of common cultural values; *diagnose causes*—identify the forces creating the problem; *make moral judgments*—evaluate causal agents and their effects; and *suggest remedies*—offer and justify treatments for the problems and predict their likely effects. A single sentence may perform more than one of these four framing functions, although many sentences in a text may perform none of them. And a frame in any particular text may not necessarily include all four functions.

The cold war example also suggests that frames have at least four locations in the communication process: the communicator, the text, the receiver, and the culture. *Communicators* make conscious or unconscious framing judgments in deciding what to say, guided by frames (often called schemata) that organize their belief systems. The *text* contains frames, which are manifested by the presence or absence of certain keywords, stock phrases, stereotyped images, sources of information, and sentences that provide thematically reinforcing clusters of facts or judgments. The frames that guide the *receiver's* thinking and conclusion may or may not reflect the frames in the text and the framing intention of the communicator. The *culture* is the stock of commonly invoked frames; in fact, culture might be defined as the empirically demonstrable set of common frames exhibited in the discourse and thinking of most people in a social grouping. Framing in all four locations includes similar functions: selection and highlighting, and use of the highlighted elements to construct an argument about problems and their causation, evaluation, and/or solution.

How Frames Work

Frames highlight some bits of information about an item that is the subject of a communication, thereby elevating them in salience. The word *salience* itself needs to be defined: It means making a piece of information more noticeable, meaningful, or memorable to audiences. An increase in salience enhances the probability that receivers will perceive the information, discern meaning and thus process it, and store it in memory (see Fiske & Taylor, 1991).

Texts can make bits of information more salient by placement or repetition, or by associating them with culturally familiar symbols. However, even a single unillustrated appearance of a notion in an obscure part of the text can be highly salient, if it comports with the existing schemata in a receiver's belief systems. By the same token, an idea emphasized in a text can be difficult for receivers to notice, interpret, or remember because of their existing schemata. For our purposes, schemata and closely related concepts such as categories, scripts, or stereotypes connote mentally stored clusters of ideas that guide individuals' processing of information (see, e.g., Graber, 1988). Because salience is a product of the interaction of texts and receivers, the presence of frames in the text, as detected by researchers, does not guarantee their influence in audience thinking (Entman, 1989; Graber, 1988).

Kahneman and Tversky (1984) offer perhaps the most widely cited recent example of the power of framing and the way it operates by selecting and highlighting some features of reality while omitting others. The authors asked experimental subjects the following:

> *Imagine that the U.S. is preparing for the outbreak of an unusual Asian disease, which is expected to kill 600 people. Two alternative programs to combat the disease have been proposed. Assume that the exact scientific estimates of the consequences of the programs are as follows: If Program A is adopted, 200 people will be saved. If Program B is adopted, there is a one-third probability that 600 people will be saved and a two-thirds probability that no people will be saved. Which of the two programs would you favor? (1984, p. 343)*

In this experiment, 72 percent of subjects chose Program A; 28 percent chose Program B. In the next experiment, *identical options* to treating the same described situation were offered, but framed in terms of likely deaths rather than likely lives saved: "If Program C is adopted, 400 people will die. If Program D is adopted, there is a one-third probability that nobody will die and a two-thirds probability that 600 people will die" (Kahneman & Tversky, 1984, p. 343). The percentages choosing the options were reversed by the framing. Program C was chosen by 22 percent, though its twin Program A was selected by 72 percent; and Program D garnered 78 percent, while the identical Program B received only 28 percent.

As this example vividly illustrates, the frame determines whether most people notice and how they understand and remember a problem, as well as how they evaluate and choose to act upon it. The notion of framing thus implies that the frame has a common effect on large portions of the receiving audience, though it is not likely to have a universal effect on all.

Kahneman and Tversky's experiments demonstrate that frames select and call attention to particular aspects of the reality described, which logically means that frames simultaneously direct attention away from other aspects. Most frames are defined by what they omit as well as include, and the omissions of potential problem definitions, explanations, evaluations, and recommendations may be as critical as the inclusions in guiding the audience.

Edelman highlights the way frames exert their power through the selective description and omission of the features of a situation:

> *The character, causes, and consequences of any phenomenon become radically different as changes are made in what is prominently displayed, what is repressed and especially in how observations are classified. . . . [T]he social world is . . . a kaleidoscope of potential realities, any of which can be readily evoked by altering the ways in which observations are framed and categorized.* (1993, p. 232)

Receivers' responses are clearly affected if they perceive and process information about one interpretation and possess little or incommensurable data about alternatives. This is why exclusion of interpretations by frames is as significant to outcomes as inclusion.

Sniderman, Brody, and Tetlock (1991) provide a clear instance of the power of presence and absence in framing:

> *The effect of framing is to prime values differentially, establishing the salience of the one or the other. [Thus] . . . a majority of the public supports the rights of persons with AIDS when the issue is framed [in a survey question] to accentuate civil liberties considerations—and supports . . . mandatory testing when the issue is framed to accentuate public health considerations.* (p. 52)

The text of the survey question supplies most people with the considerations they use when they respond to the issue of AIDS testing (Zaller, 1992). Often a potential counterframing of the subject is mostly or wholly absent from a text, although, to use this instance, an audience member with a strong civil liberties philosophy might reject mandatory testing even if the poll framed AIDS strictly in public health terms.

Frames in Political News

This portrait of framing has important implications for political communication. Frames call attention to some aspects of reality while obscuring other elements, which might lead audiences to have different reactions. Politicians seeking support are thus compelled to compete with each other and with journalists over news frames (Entman, 1989; Riker, 1986). Framing in this light plays a major role in the exertion of political power, and the frame in a news text is really the imprint of power—it registers the identity of actors or interests that competed to dominate the text.

Reflecting the play of power and boundaries of discourse over an issue, many news texts exhibit homogeneous framing at one level of analysis, yet competing frames at another. Thus, in the pre-war debate over U.S. policy toward Iraq, there was a tacit consensus among U.S. elites not to argue for such options as negotiation between Iraq and Kuwait. The news frame included only two remedies, war now or sanctions now with war (likely) later, while problem definitions, causal analyses, and moral evaluations were homogeneous. Between the selected remedies, however, framing was contested by elites, and news coverage offered different sets of facts and evaluations. The Iraq example reveals that the power of news frames can be self-reinforcing. During the pre-war debate, any critique transcending the remedies inside the frame (war soon versus more time for sanctions) breached the bounds of acceptable discourse, hence was unlikely to influence policy. By conventional journalistic standards, such views were not newsworthy (Entman & Page, in press). Unpublicized, the views could gain few adherents and generate little perceived or actual effect on public opinion, which meant elites felt no pressure to expand the frame so it included other treatments for Iraqi aggression, such as negotiation. Relatedly, Gamson (1992) observes that a frame can exert great social power when encoded in a term like *affirmative action*. Once a term is widely accepted, to use another is to risk that target audiences will perceive the communicator as lacking credibility—or will even fail to understand what the communicator is talking about. Thus the power of a frame can be as great as that of language itself.

Benefits of a Consistent Concept of Framing

An understanding of frames helps illuminate many empirical and normative controversies, most importantly because the concept of framing directs our attention to the details of just how a communicated text exerts its power. The example of mass communication explored here suggests how a common understanding might help constitute framing as a *research paradigm*. A research paradigm is defined here as a general theory that informs most scholarship on the operation and outcomes of any particular system of thought and action. The framing paradigm could be applied with similar benefits to the study of public opinion and voting behavior in political science; to cognitive studies in social psychology; or to class, gender, and race research in cultural studies and sociology, to name a few. Here are some illustrations of theoretical debates in the study of mass communication that would benefit from an explicit and common understanding of the concept of frames.

1. *Audience autonomy.* The concept of framing provides an operational definition for the notion of *dominant meaning* that is so central to debates about polysemy and audience independence in decoding media texts (Fiske, 1987). From a framing perspective, dominant meaning consists of the problem, causal, evaluative, and treatment interpretations with the highest probability of being noticed, processed, and accepted by the most people. To identify a meaning as dominant or preferred is to suggest a particular framing of the situation that is most heavily supported by the text and is congruent with the most common audience schemata.

A framing paradigm cautions researchers not to take fugitive components of the message and show how they *might* be interpreted in ways that oppose the dominant meaning. If the text frame emphasizes in a variety of mutually reinforcing ways that the glass is half full, the evidence of social science suggests that relatively few in the audience will conclude it is half empty. To argue that the polysemic properties of the message conduce to such counterframing, researchers must show that real-world audiences reframe the message, and that this reframing is not a by-product of the research conditions—for example, a focus group discussion in which one participant can lead the rest, or a highly suggestive interview protocol (Budd, Entman, & Steinman, 1990).

Certainly people can recall their own facts, forge linkages not made explicitly in the text, or retrieve from memory a causal explanation or cure that is completely absent from the text. In essence, this is just what professors encourage their students to do habitually. But Zaller (1992), Kahneman and Tversky (1984), and Iyengar (1991), among others, suggest that on most matters of social or political interest, people are not generally so well-informed and cognitively active, and that framing therefore heavily influences their responses to communications, although Gamson (1992) describes conditions that can mitigate this influence.

2. *Journalistic objectivity.* Journalists may follow the rules for "objective" reporting and yet convey a dominant framing of the news text that prevents most audience members from making a balanced assessment of a situation. Now, because they lack a common understanding of framing, journalists frequently allow the most skillful media manipulators to im-

pose their dominant frames on the news (Entman, 1989; Entman & Page, in press; Entman & Rojecki, 1993). If educated to understand the difference between including scattered oppositional facts and challenging a dominant frame, journalists might be better equipped to construct news that makes equally salient—equally accessible to the average, inattentive, and marginally informed audience—two or more interpretations of problems. This task would require a far more active and sophisticated role for reporters than they now take, resulting in more balanced reporting than what the formulaic norm of objectivity produces (Tuchman, 1978).

3. *Content analysis.* The major task of determining textual meaning should be to identify and describe frames; content analysis informed by a theory of framing would avoid treating all negative or positive terms or utterances as equally salient and influential. Often, coders simply tote up all messages they judge as positive and negative and draw conclusions about the dominant meanings. They neglect to measure the salience of elements in the text, and fail to gauge the relationships of the most salient clusters of messages—the frames—to the audience's schemata. Unguided by a framing paradigm, content analysis may often yield data that misrepresent the media messages that most audience members are actually picking up.

4. *Public opinion and normative democratic theory.* In Zaller's (1992) account, framing appears to be a central power in the democratic process, for political elites control the framing of issues. These frames can determine just what "public opinion" is—a different frame, according to Zaller, and survey evidence and even voting can indicate a different public opinion. His theory, along with that of Kahneman and Tversky, seems to raise radical doubts about democracy itself. If by shaping frames elites can determine the major manifestations of "true" public opinion that are available to government (via polls or voting), what can true public opinion be? How can even sincere democratic representatives respond correctly to public opinion when empirical evidence of it appears to be so malleable, so vulnerable to framing effects?

Say there are three ways to frame an issue and one generates 40 percent approval, the others 50 percent and 60 percent, respectively. Approving the option with 60 percent support is not axiomatically the most democratic response because of the cyclical majority problem (Riker, 1986), which makes majority rule among several complex options mathematically impossible. Just as important, attempting to determine which of the differently framed opinions is the closest to the public's "real" sentiments appears futile, because it would require agreement among contending elites and citizens on which frame was most accurate, fair, complete, and so forth. A framing paradigm can illuminate, if not solve, such central puzzles in normative democratic theory.

Indeed, the concept of framing is important enough in the many fields of inquiry that use it to merit a book-length essay. The present effort, constrained by space limitations, offers not the definitive word on frames but a preliminary contribution. Equally important, this article exemplifies how the field of communication might develop from its wide ambit and eclectic approaches a core of knowledge that could translate into research paradigms contributing to social theory in the largest sense.

References

Budd, M., Entman, R. M., & Steinman, C. (1990). The affirmative character of U.S. cultural studies. *Critical Studies in Mass Communication, 7,* 169–184.

Edelman, M. J. (1993). Contestable categories and public opinion. *Political Communication, 10*(3), 231–242.

Entman, R. M. (1989). *Democracy without citizens: Media and the decay of American politics.* New York: Oxford University Press.

Entman, R. M., & Page, B. I. (in press). The news before the storm: The Iraq war debate and the limits to media independence. In W. L. Bennett & D. L. Paletz (Eds.), *Just deserts: The news media, U.S. foreign policy, and the Gulf War.* Chicago: University of Chicago Press.

Entman, R. M., & Rojecki, A. (1993). Freezing out the public: Elite and media framing of the U.S. anti-nuclear movement. *Political Communication, 10*(2), 151–167.

Fiske, J. (1987). *Television culture.* New York: Routledge.

Fiske, S. T., & Taylor, S. E. (1991). *Social cognition.* New York: McGraw-Hill.

Gamson, W. (1992). *Talking politics.* New York: Cambridge University Press.

Goffman, E. (1974). *Frame analysis.* New York: Free Press.

Graber, D.A. (1988). *Processing the news: How people tame the information tide* (2nd ed.). New York: Longman.

Iyengar, S. (1991). *Is anyone responsible?* Chicago: University of Chicago Press.

Kahneman, D., & Tversky, A. (1984). Choice, values, and frames. *American Psychologist, 39,* 341–350.

Pan, Z., & Kosicki, G. M. (1993). Framing analysis: An approach to news discourse. *Political Communication, 10*(1), 55–76.

Riker, W. H. (1986). *The art of political manipulation.* New Haven: Yale University Press.

Sniderman, P. M., Brody, R. A., & Tetlock, P. E. (1991). *Reasoning and choice: Explorations in political psychology.* Cambridge, England: Cambridge University Press.

Snow, D. A., & Benford, R. D. (1988). Ideology, frame resonance, and participation mobilization. *International Social Movement Research, 1,* 197–217.

Tuchman, G. (1978). *Making news.* New York: Free Press.

White, H. (1987). *The content of the form.* Baltimore, MD: Johns Hopkins University Press.

Zaller, J. R. (1992). *The nature and origins of mass opinion.* New York: Cambridge University Press.

Communication Research in the Design of Communication Interfaces and Systems

by Frank Biocca, University of North Carolina

In order to reconquer the machine and subdue it to human purposes, one must first understand it and assimilate it. So far we have embraced the machine without fully understanding it. (Mumford, 1934, p. 334)

In order to understand the phenomena surrounding a new technology, we must open the question of design—the interaction between understanding and creation. (Winograd & Flores, 1987, p. 4)

Communication research seems forever ordained to consider and reconsider the means of communication, the machinery with which we create meaning. In many ways, communication research is about how humans create techniques and technologies to turn each others' thoughts into each others' experiences.

Lately, our attention has been drawn to the circuits, interfaces, and fiber-optic cables that increasingly bind our minds together into giant networks—the giant sea of information and codified experiences recently dubbed "cyberspace" (Benedikt, 1992). The introduction of virtual reality technology spurs us to think once again about the relations between thought, the senses, and the machinery that facilitate communication expression and distribution (e.g., Biocca, 1992a; Biocca & Levy, in press). Before each new communication technology becomes an invisible part of our second nature, its novelty increases our awareness of how much communication is socially constructed (e.g., Biocca, 1987, 1988). As in *The Wizard of Oz,* the flutter of the curtain makes the human operator of the giant machinery of communication suddenly visible: He is us.

Frank Biocca is an associate professor and the director of the Center for Research in Journalism and Mass Communication at the University of North Carolina at Chapel Hill. This article began while the author was on leave at Stanford and the University of California at Berkeley. He would like to thank Byron Reeves, Don Roberts, Steve Chaffee, Richard Cole, and Todd Gitlin for making his stay in California possible, pleasant, and productive.

In this article[1] I will not propose another paradigm, method, or school. Rather, I seek something more basic and immediate. My words are motivated by a sense of urgency, a feeling that over the next 20 years the evolution of communication technologies may offer a critical—can I bring myself to write *historic?*—opportunity for researchers to positively shape the channels of mediated human communication. I wonder if communication research is well equipped to meet the challenge. I hope that it is. I fear that many research paradigms inadvertently condemn communication research to the role of spectator in the long march of communication technology. Communication research cheers or boos, but the march of communication technology goes on regardless. If we really believe that communication environments are socially constructed, then can communication research aggressively engage in their construction?

In this article I want to focus on communication research's role in the conception, infancy, and adolescence of a young medium. There must be a time in the development of a new communication medium, interface, or environment when its form, like that of a child, is still pliable, when the forces of its social construction are most fluid. At the birth of each new communication medium and system we should not ask whether the technology is a liberator or a criminal—the answer is always that it is both and neither. I'm reminded of the journalist from "ABC's Nightline" who asked me whether virtual reality was a "good thing" or a "bad thing." Rather, like Lewis Mumford, he might more profitably have asked how we can "subdue it to human purposes" (1934, p. 334).

First, I will sketch the social setting—an historic juncture in the communication environment. Then I will try to sketch out a possible research response, a "human factors approach" to communication research.

The Historic Juncture

It is evident that something new is happening in human communication. . . . The name doesn't matter. What matters is that every major development in human communication has begun with a major new development in communication technology. (Schramm, 1988, p. 341)

For 50 years, mass communication research developed in an environment of relatively stable media technologies. Our mass media interfaces—

---

[1] This article began as a set of reflections on the direction of my research during a leave. Back in Silicon Valley where I once worked, I had a chance to return to my earlier interests in communication technology and cognition. With time to think, I critically relistened to the distant echoes of the Marshall McLuhan lectures I heard as an undergraduate in Canada. This article and others emerged as part of my attempt to refocus my questions about communication technologies and my approach to communication research.

books, newspapers, radio, and television—remained relatively unchanged from the late 1940s to the 1990s. Wilbur Schramm, a father of the
academic discipline of communication research, was there in the first
decade, and he was able to sense the changing, turbulent communication
currents in the last decade of his life. During the intervening 50 years,
technological innovation did not radically alter the face of our media interfaces or the experience of mediated communication. Media evolved
slowly. Most research, especially experimental and critical research, focused mostly on those elements of the media that were most pliable and
changeable: media messages.

This is changing. The computer has evolved from a glorified typewriter-
calculator into a major communication medium. From 1980 to 1990 the
annual consumption of personal computers increased by approximately
900 percent. Expenditures on personal computers rose by 1100 percent
during that same decade, and in 1991 the computer industry was a $43.2
billion industry (CBEMA, 1991). The personal computer is only the beginning of a larger wave of smart, interactive communication devices and
systems (Millison & LeGrow, 1993).

The collision and merger of the communication and computer industries is sending a shock wave of change and instability through the
mediated communication environment. This can be characterized as the
restructuring—some say convergence, I say expansion—of (a) communication systems and (b) communication interfaces.

The Expansion of Cyberspace

Communication systems can be defined as follows: *Communication Systems = [Transmission Channel(s), Organizational Infrastructures, Communication Interface(s)].*

Our communication systems have entered a state of flux. Americans use
a number of transmission channels: the broadcasting spectrum, the telephone system's copper wires, cable's coaxial wires, the cellular spectrum,
direct broadcast satellites, and fiber-optic cabling. The organizations that
own each channel are fighting for the opportunity to pump more and
more mediated information into the home, the car, and, ultimately, the
mind. The spread of digital communication methods and information
compression is allowing each one of these channels to offer more information and an expanded but overlapping range of communication
services. The organizational infrastructures associated with these transmission channels—the telephone, broadcast, cable, and computer industries—are entering a period of intense competition and restructuring. It is
a struggle; new structures and alliances are being formed. As a result, the
interfaces and information flows associated with these institutions are in
flux.

The social communication environment is expanding and restructuring;

our mediated communication experiences are certain to take on new features and patterns as we become more strongly interconnected by our communication systems.

> *The protean nature of the computer is such that it can act like a machine or like a language to be shaped and exploited. It is a medium that can dynamically simulate the details of any other medium, including media that cannot exist physically. It is not a tool, although it can act like many tools. It is the first metamedium, and as such it has degrees of freedom for representation and expression never before encountered and as yet barely investigated.* (Kay, 1984)

Alan Kay, master designer at Xerox Park, Apple, and MIT's Media Lab, lays out the challenge for communication scholars very clearly. The computer can give birth to a medium that can simulate any other medium. At the same time it offers a means of communication and expression that has yet to be explored and understood (i.e., virtual reality?). How can communication research take up this challenge?

The interface and the type of communication that it supports is inevitably a central concern of human communication research. By way of definition I offer the following: *Communication Interface = (Physical Media, Codes, Information) + Sensorimotor Channels.*

Like the communication systems to which they are linked, communication interfaces are also changing. The communication interface is becoming more complex and flexible as it addresses more communication needs. For example, advanced communication and simulation interfaces like virtual reality attempt to engage a wider range of sensory channels (Biocca, 1992b; Biocca & Delaney, in press). Within each sensory channel, sophisticated displays are presenting a wider bandwidth of information. For example, projection HDTV (high-definition television) fills more of the visual field at higher levels of resolution. Further developments in virtual reality interfaces have as their explicit logic the attempt to match and, maybe, surpass the sensory variety and information density of the physical environment. *The communication interface is expanding to envelop more of the senses.*

The phrase *interactivity* is often used. It is a misleading term. At its most basic, it simply describes the fact that advanced communication interfaces are increasingly responsive to somatic activity: behaviors like key presses, mouse movements, eye movements, body position, as well as changes in autonomic activity. New input devices connect more and more somatic activity to the interface and to the communication system (Biocca & Delaney, in press). *The body is being immersed into the communication interface.*

Let me summarize the historical juncture I have described. Communication systems and interfaces are not just being merged, they are being expanded: expanded coverage of the sensory channels, expanded range of somatic inputs, expanded geographic coverage, expanded bandwidth of information and services. *Like the physical universe, cyberspace is expanding.*

Responding to Changes in Systems and Interfaces

What are the implications for communication research? History suggests that it is extremely likely that communication theory and research will be influenced by changes in the communication environment. Faced with increased changes in the communication environment, communication researchers could simply be *reactive*: forecast or evaluate developments, trends, and effects. This response remains important. But how might communication research help achieve Mumford's goal and subdue communication systems and interfaces to human purposes? If we are entering a period of creative instability and expansion, communication research might be more *proactive*. Researchers might find it valuable to not just critique or measure, but also participate in the social construction of mediated communication—in communication *design*. Communication researchers might try to creatively engage and influence the evolutionary process of media design—of human–computer interaction.

Upon analysis, we find that not just each medium but each act of communication is an act of design. For example, this sentence is an act of design. If researchers systematically study these acts of communication design—be they psychological or social—then communication science must be, in part, a science of design. We often ask how humans design messages and media, and how those designs somehow redesign their thoughts, behaviors, and social environments. We are increasingly aware that communication is an environment, and that we participate in its design.

To use an existing term, let me suggest what might be called the *human factors approach* to communication research and design. I borrow the term *human factors* from a large and growing interdisciplinary community of researchers. Borrowing a term is often dangerous. The term *human factors* may carry with it some unwanted baggage. I will probably not be able to avoid this problem here. But as with other examples of paradigmatic appropriation (Kuhn, 1970), I hope to begin the process of tailoring the term for communication research in the course of this article.

In the fields of engineering, computer science, and psychology, the words *human factors* refer to important aspects of human performance, behavior, and desire that must be considered in the design of any machine, hardware, program, or information system. For example, interdisciplinary teams of engineers, computer scientists, and psychologists are concerned with human factors when designing a new communication interface, or, more generally, human–computer interaction (Biocca, 1992b;

Card, Moran, & Newell, 1983; Laurel, 1990, 1991; Rubenstein & Hersh, 1984; Shneiderman, 1992; Vassilou, 1984). Like communication researchers, many human factors and human–computer interaction researchers live on the boundaries between traditional fields. Many can be found at the intersection of engineering, computer science, psychology, and anthropology (e.g., Baecker & Buxton, 1987; Hancock, 1987; Rubenstein & Hersh, 1984; Slavendy, 1987; Wickens, 1984; Woodson, 1981). To fully consider various human factors, human–computer interface design teams are often interdisciplinary (e.g., Apple's design group; see Kim, 1990). One can increasingly find communication researchers involved in interface design (e.g., Biocca & Levy, in press; Fish, Kraut, Root, & Rice, 1992; Gay & Grose-Ngate, 1993; Heeter, 1992; Meyer, Applewhite, & Biocca, 1992; Nilan, 1992). In some ways, the human factors approach to communication may signal a return to the creative interdisciplinary synergy that seemed to characterize the spirit of communication research in its early years (e.g., Cherry, 1957).

A concern with human factors has become central to the design of human–computer interfaces (HCI). The design of most technologies, including communication technologies, has moved from what Bullinger (1986) calls a "technocentric approach" (technology controls man) to an "anthropocentric approach" (man controls technology). The technocentric approach is exemplified by the work of industrial engineer Fredrick Taylor (Nelson, 1980) and the Gilbreths (Mandel, 1989), whose time and motion studies sought to make humans conform to the machine processes of industry. By contrast, the goal of many HCI researchers is to have the interface conform to the thinking, the objectives, and the habits of a particular group of users (Shneiderman, 1992). If one could summarize the prevailing attitude of human factors researchers in one sentence, it might read: The user should not have to conform to the machine; the machine should conform to the user.

The intense study of human factors in human–computer interaction emerged from the design of "life-critical" computer systems such as air traffic controller systems, nuclear power plant control systems, advanced fighter cockpits, and medical emergency systems. In such systems, failure of the computer interface to conform to the information and communication needs of the users could lead to death and disaster. The interface between the computer and the human is a communication system, a messaging system for encoding and decoding information. In life-critical systems the communication interface must facilitate the critical exchange of life-and-death information.

Communication Design as Theory and Research

Of all technologies, communication technologies provide the greatest human factors challenge. It is interesting to note that the very word *tech-*

Table 1. Examples of Human Factors Considered in Designing Communication and Human-Computer Interfaces

Cognitive
> *Perception and psychophysics*: capacities and properties of human sensory channels
> *Attention*: reaction time, time sharing, workload, and fatigue
> *Motivation*: individual differences, task motivation, emotional satisfaction
> *Decision-making and judgment*: risk perception, constraint satisfaction
> *Procedural memory*: skill learning, usage, and forgetting
> *Semantic memory*: meaning construction and mental models

Instrumental
> *Individual communication goals*: related to interface usage
> *Organizational communication goals*: related to interface adoption and usage
> *Social communication goals*: economic, public policy, social advocacy
> *Relationship to other communication instruments*: range, access, usage, and perceived utility of other interfaces

Normative
> *Cultural variation in communication expectation, performance, or norms*
> *Contextual and environmental factors*: affecting or defining performance
> *Ethical considerations*: related to interface use or abuse

Expressive
> *Code usage*: extent and usage of existing communication codes
> *Message variability*: likely range of message content and contexts

Somatic
> *Anthropometry*: relationship of human body shape to the interface hardware
> *Biomechanics*: forces, motion, and feedback patterns of the human body during use of the interface

---

nology is a merger of the Greek words *techne,* meaning "art, craft," and *logos,* meaning "word, speech." The origins of the word suggest that communication is at the heart of technology. There may be something essentially different about considering human factors in the design of communication technology—systems that are true to the early Greek sense and improve the craft and art of making signs, words, and speech. More than just architectural or engineering design, communication systems subsume both mind and body into simulations of physical and social reality—to borrow an emerging term, into *virtual realities.*

Communication science has two arms that grapple with the forms and structures of human communication. One arm tends towards basic science. Our basic understanding of the human factors themselves is an extension of the basic sciences. The interdisciplinary human factors research team may consider a number of human factors when designing interfaces, including communication interfaces. Table 1 lists some of these.

Communication science has another arm, one that tries to design and build. Design is a form of human expression; it builds on the essential freedom and indeterminacy of human communication. Its success must be measured not just by the truth of its principles but also by its ability to influence the design of communication tools and environments. Otherwise, communication science is like the architect whose buildings are never built.

It would be narrow-minded to merely dismiss design as the application of communication research. Design principles are theories of communication. Functioning designs, like a new computer interface, can be models of communication theory. An interface design can also be a form of experimentation, especially when systematic evaluation is a part of a design implementation. Designing interfaces as vehicles for theory building and experimentation has a key advantage: The concrete, specific details of implementation force the researcher to specify his or her ideas in detail. Theory, interface design, experimentation, evaluation, and interface redesign can be part of one continuous flow of proactive communication scholarship (see Biocca, 1993).

Habermas (1981) argues that there is a decoupling of what he calls the "system" and the "lifeworld." The lifeworld is increasingly subordinated to the demands of the system. Distrust of technology emerges in part from the feeling that our sphere of freedom is being restricted by the system (i.e., the scientific-technical-industrial apparatus). In the past our reaction to the computer was a reaction to one of the most visible parts of the impersonal industrial system. Computer cards came stamped with the message "do not fold, spindle, or mutilate" from a regime that seemed to "fold, spindle, and mutilate" the users of the system. Critics like Ellul (1964) bemoaned our preoccupation with technique and how it had helped divorce technology from human ends.

As computers and communication technologies begin to merge, a creative, informed, and uncompromising human factors approach to communication research will struggle to regain control over technology and to subordinate aspects of the system to the goals of individual users. Good interface design tries to turn seemingly impenetrable systems into flexible communication tools. We must assume that it is possible to develop a communication research approach that can do more than simply critique practices or assemble lists of possible effects, but can actively help reclaim communication technology for the individual—for the exercise of creative power, enhanced expression, and unfettered communication. Obviously, this is as much a utopian vision as a research program. I am painfully aware that the Janus-like nature of technology does not guarantee positive outcomes.

But a strength and value of a human factors approach to communication design must be a recognition of its limits. Only some aspects of communication systems can be designed by designers; communication design is at best facilitation, not specification. The pen, the printing press, and the telephone are good models. They facilitate communication; they don't specify it. Many key aspects of communication environments are not designed by designers; they are designed by users. Users must be co-designers. Like languages, users finalize the design of communication systems [i.e., Saussure's (1959) distinction between *langue* and *parole*].

Communication environments are ultimately defined by the unpredictable diversity of communication action undertaken by individuals and social groups inside a communication environment. The strength of a human factors approach stems not from a passive observation or measurement of this fact, but from an active engagement with, support of, and amplification of the users' communication actions. The iterative design, evaluation, and redesign of communication environments becomes a natural part of communication theory and research.

References

Baecker, R., & Buxton, W. (1987). *Human–computer interaction: A multidisciplinary approach*. Los Altos, CA: Morgan Kaufmann.

Benedikt, M. (Ed.). (1992). *Cyberspace: First steps*. Cambridge, MA: MIT Press.

Biocca, F. (1987). Sampling from the museum of forms: Photography and visual thinking in the rise of modern statistics. In M. McLaughlin (Ed.), *Communication yearbook 10* (pp. 684–708). Beverly Hills, CA: Sage.

Biocca, F. (1988). The pursuit of sound: Radio, perception, and utopia in the early twentieth century. *Media, Culture, & Society, 10,* 61–80.

Biocca, F. (1992a). Communication within virtual reality: Creating a space for research. *Journal of Communication, 42*(4), 5–22.

Biocca, F. (1992b). Virtual reality technology: A tutorial. *Journal of Communication, 42*(4), 23–72.

Biocca, F. (1993). *Communication technology matrix*. Chapel Hill: Center for Research in Journalism and Mass Communication, University of North Carolina at Chapel Hill.

Biocca, F., & Delaney, B. (in press). The components of virtual reality technology. In F. Biocca & M. Levy (Eds.), *Communication in the age of virtual reality*. Hillsdale, NJ: Lawrence Erlbaum Associates.

Biocca, F., & Levy, M. (Eds.). (in press). *Communication in the age of virtual reality*. Hillsdale, NJ: Lawrence Erlbaum Associates.

Bullinger, H. (1986). Technology trend—A challenge to research and industry. In A. Gerstenfeld, H. Bullinger, & H. Warnecke (Eds.), *Proceedings of the conference on manufacturing research: organizational and institutional issues*. Amsterdam: Elsevier.

Card, S., Moran, T. P., & Newell, A. (1983). *The psychology of human computer interaction*. Hillsdale, NJ: Lawrence Erlbaum Associates.

CBEMA. (1991). *The information technology industry databook*. Washington, DC: Author.

Cherry, C. (1957). *On human communication: A review, a survey, and a criticism*. Cambridge, MA: MIT Press.

Ellul, J. (1964). *The technological society*. New York: Vintage.

Fish, R., Kraut, R. E., Root, R. W., & Rice, R. E. (1992). Evaluating video as a source of informal communication. In P. Bauersfeld & G. Lynch (Eds.), *CHI'92 conference proceedings*. New York: Association for Computing Machinery.

Gay, G., & Grose-Ngate, M. (1993). *Collaborative design in a networked multimedia environment: Emerging communication patterns.* Ithaca, NY: Department of Communication, Cornell University.

Habermas, J. (1981). *The theory of communicative action (Vols. 1 & 2).* Boston: Beacon Press.

Hancock, P. A. (Ed.). (1987). *Human factors psychology.* New York: North Holland.

Heeter, C. (1992). Being there: The subjective experience of presence. *Presence, 1*(2), 262–270.

Kay, A. (1984, September). Computer software. *Scientific American, 251*(3), 52–59.

Kim, S. (1990). Interdisciplinary cooperation. In B. Laurel (Ed.), *The art of human–computer interface design* (pp. 31–44). Reading, MA: Addison-Wesley.

Kuhn, T. (1970). *The structure of scientific revolutions.* Chicago: University of Chicago Press.

Laurel, B. (Ed.). (1990). *The art of human–computer interface design.* Reading, MA: Addison-Wesley.

Laurel, B. (1991). *Computers as theater.* Menlo Park, CA: Addison-Wesley.

Mandel, M. (1989). *Making good time: Scientific management, the Gilbreths, photography and motion, futurism.* Santa Cruz, CA: M. Mandel, California Museum of Photography.

Meyer, K., Applewhite, H., & Biocca, F. (1992). A survey of position trackers. *Presence, 1*(2), 173–200.

Millison, D., & LeGrow, C. 1993. Multimedia "black boxes": Advent of the digital media home. In M. De Sonne (Ed.), *Multimedia 2000* (pp. 103–120). Washington, DC: National Association of Broadcasters.

Mumford, L. (1934). *Technics and civilization.* New York: Harcourt & Brace.

Nelson, D. (1980). *Frederick W. Taylor and the rise of scientific management.* Madison: University of Wisconsin Press.

Nilan, M. S. (1992). Cognitive space: Using virtual reality for large information resource management problems. *Journal of Communication, 42*(4), 115–135.

Rubenstein, R., & Hersh, H. (1984). *The human factor.* Burlington, MA: The Digital Press.

Saussure, F. (1959). Course in general linguistics. New York: Philosophical Library.

Schramm, W. (1988). *The story of human communication: Cave to microchip.* New York: Harper & Row.

Shneiderman, B. (1992). *Designing the user interface: Strategies for effective human–computer interaction.* Reading, MA: Addision-Wesley.

Slavendy, G. (Ed.). (1987). *Handbook of human factors.* New York: John Wiley.

Vassiliou, Y. (1984). *Human factors and interactive computer systems.* Norwood, NJ: Ablex.

Wickens, C. (1984). *Engineering psychology and human performance.* Columbus, OH: Charles E. Merrill.

Winograd, T., & Flores, F. (1987). *Understanding computers and cognition.* Reading, MA: Addison-Wesley.

Woodson, W. (1981). *Human factors design handbook.* New York: McGraw Hill.

The Future of Political Communication Research: A Japanese Perspective

by Ito Youichi, Keio University

While new mass communication theories developed in the West are quickly introduced in Japan, usually by young scholars, they are not always quickly accepted. For one thing, some of their assumptions, such as the *atomized individuals* idea of mass society theory, do not apply to Japan. Most of Japan's internal social, political, and economic problems or international cultural frictions are caused not because the Japanese are atomized, but because they are too cohesive. Further, many empirical surveys indicate that the Japanese tend to avoid political subjects in personal conversations because they do not want to disagree or confront other people (Ito & Kohei, 1990; Nishihira, 1987; Okamoto, Oda, Shiga, Toga, & Hata, 1986).

These cultural characteristics make it difficult to apply Western theories on public opinion formation to Japan or to test the spiral of silence hypotheses (Ikeda, 1989; Iwabuchi, 1989; Okamoto et al., 1986; Tokinoya, 1989). One approach that has been accepted by Japanese scholars is the idea that the mass media have powerful effects because they overwhelm our information environment. According to the Japanese interpretation of the powerful effects model, individuals naturally adapt to information environments around them, which are dominated by the mass media. These information environments have been called "the world [made] of copies" (Shimizu, 1951) or a "pseudo environment" (Fujitake, 1968).

These ideas were well accepted in Japan because the Japanese heavily depend on mass media as information sources. Compared to the United States, Germany, and Hong Kong, Japanese respondents are more dependent on mass media (newspapers, television, magazines, and mass weeklies) and less dependent on interpersonal communication and personal observation (see Edelstein, Ito, & Kepplinger, 1989, p. 206). If people depend heavily upon mass media for information, and if personal influence

Ito Youichi is a professor in the Faculty of Policy Management at Keio University at Shonan Fujisawa and associate director of the Institute for Communications Research at Keio University in Japan. The author wishes to thank Anne Cooper-Chen of Ohio University and Lee B. Becker of the Ohio State University for helpful comments on earlier drafts of this article.

Copyright © 1993 *Journal of Communication* 43(4), Autumn. 0021-9916/93/$5.00

in political communication is weak, then the political influence of mass media may be overwhelming.

The *Joho Kohdo* (Information Behavior) Model

In the early 1970s, a group of scholars emerged who claimed that individuals not only passively received information from their information environment but also *extracted* information from the information environment *and* from the physical environment. These researchers claimed that as long as individuals *extract* information from their physical environment, one cannot easily make a clear distinction between information and physical environments (Kato, 1972; Kitamura, 1970; Nakano, 1980).

Extracted information plays an important role in Japanese interpersonal communication. Because Japanese tend to use many circumlocutions and indirect expressions to avoid personal confrontation and maintain group harmony, people are always expected to extract true meanings (*honne*) out of ambiguous, roundabout expressions or *tatemae* (principles or official stance) (Ito, 1989, 1992; Midooka, 1990; Minami, 1971). Sensitivity to extraction of information, while quite Japanese, is also universal.

Many nations were hostile to Japan after World War II. Naturally, the Japanese were strongly concerned about improving their poor image, which government agencies, mass media companies, private research institutes, and university researchers measured through numerous surveys in foreign countries (e.g., Kawatake, 1988; Tsujimura, Furuhata, & Akuto, 1987; Tsujimura, Kim, & Ikuta, 1982). The surveys pointed out that information from the mass media was only one of many factors contributing to the image of and attitudes toward Japan and the Japanese. Other major factors included the state of political and economic relations, personal contacts, travel to Japan, Japanese industrial products, and exposure to Japanese traditional culture.

Although information from mass media certainly accounted for the largest portion of information about Japan, some researchers were skeptical about its effects. For example, according to a survey conducted in the United Kingdom (Sugiyama & Kimura, 1988), those who mentioned television and newspapers as their major sources of information about Japan tended to reply that they did not know much about Japan, whereas those who mentioned friends and traveling as major sources of information tended to reply that they knew a fair amount about Japan. Even those who mentioned Japanese products as a major source of information felt they knew more about Japan than those who depended upon television and newspapers.

The data in Table 1 seem to suggest that people give higher credibility to information extracted from their own experiences and observations, including exposure to Japanese products, than to mass media information.

Table 1. Major Information Sources and Levels of Knowledge about Japan

Information source	% Naming as major source of information	Self-assessed level of knowledge about Japan[b]	
	All respondents[a] (N = 2,267)	Know well or fairly well	Don't know well or at all
TV	56	7	93
School	16	6	94
Newspapers	16	8	92
Books	15	18	82
Magazines	4	10	90
Friends	4	22	78
Products	4	10	90
Travel	1	55	45

Source: Sugiyama & Kimura (1988, p. 41), Tables 1-5 and 1-6 and additional data provided directly by Sugiyama.

[a] Percentages do not sum to 100% because respondents were allowed to give multiple answers.

[b] Percentage of those respondents who chose that source as their major information source.

In a survey conducted in the United States, Japan, Germany, and Hong Kong, researchers asked *why* the medium that respondents thought most useful *was* useful. Among Japanese respondents, 13 percent mentioned credibility as the reason why they thought direct observation most useful, while 9 percent among those relying on interpersonal communication and only 6 percent of those relying on newspapers and television gave this reason. The results in the other three countries were not so obvious as those in Japan, but similar tendencies were discerned (Edelstein et al., 1989, p. 174).

In fact, Homma (1975) argued that what contributed most to the remarkable improvement in Japan's image in the 1960s and early 1970s was the quality of Japanese industrial products. Foreigners exposed to postwar Japanese industrial products extracted from them information and images such as *hardworking, intelligent, progressive, scientific, rational, clean, sophisticated, reliable,* and *prosperous.*

What happens if mass media information contradicts information that people extract from their own experiences and observations? As the data in Table 2 indicate, when people evaluate foreign countries or other ethnic groups, they tend to give higher credibility to the information gained from their direct experiences and observations, such as personal contact and traveling, than to mass media information.

Makita and Uemura (1985) asked respondents to name the medium most useful for thinking about political and social issues and found that 46 percent mentioned newspapers, 43 percent named television, 2 percent selected radio, and less than 1 percent listed other media. Asked which medium exerted the most influence on public opinion concerning

Table 2. Information Sources and Evaluations of Foreign Countries and Ethnic Groups

Alternate information source	Answers to the following question: "If there is a contradiction between mass media and the following information sources regarding the evaluation of foreign countries or other ethnic groups, which would you believe?"		
	Believe alternate source	Believe mass media	Don't know or no answer
Personal contact with people belonging to that country or ethnic group	73	7	20
Rumor	5	50	45
Travel experience	78	4	18
Products and industrial goods from that country	39	29	32

Note: Respondents ($N = 170$) were students of Keio University. This survey was conducted in 1992 as part of an international collaborative research project on images of foreign countries and other ethnic groups.

political and social issues, 48 percent mentioned television, 43 percent named newspapers, 2 percent selected weekly magazines, 2 percent chose radio, and less than 1 percent listed other media. However, when asked "Do you think that what the mass media report is basically true?" 37 percent of respondents replied "yes" and 43 percent replied "no."

The Makita and Uemure (1985) survey also indicates that although people depend on the mass media for information gathering, they do not necessarily accept its contents uncritically. The studies mentioned above suggest that if the mass media continue to provide information that is contradictory to people's extracted information, the credibility of mass media will be further undermined.

Many scholars in the West have emphasized the importance of interpersonal communication or personal influence vis-à-vis mass communication (Brouwer, 1967; Katz & Lazarsfeld, 1955; Wiebe, 1955). However, they regarded interpersonal communication as a factor that reinforces or weakens mass media influence and did not regard it as something that can undermine the credibility of the mass media.

The long history of credibility research in the West (for an overview, see Edelstein et al., 1989, pp. 155–183) has emphasized credibility as a determinant factor of persuasiveness or has compared the degrees of credibility among different kinds of mass media. But Western scholars have not compared the degrees of credibility between mass media information and non-mass media information, nor have they theorized about the mechanism by which credibility is gradually lost when mass media contradict information that people extract from their own experiences and observations.

Leon Festinger's cognitive dissonance theory (Festinger, 1957) is helpful in understanding the process by which mass media credibility is lost. Contradiction or incongruence between mass media information and extracted information creates psychological dissonance. Because the credibility of extracted information is usually higher than that of mass media information, people resolve the dissonance by conferring less credibility on the mass media. The most important point of the *joho kohdo* model, however, is not this psychological mechanism, but the fact that people extract information from their own experiences and observations and use it to check the credibility of mass media information.

A Tripolar *Kuuki* Model

The second problem with applying the dominant paradigm to Japan (i.e., the notion that the mass media have powerful effects because they dominate people's information environments) is that it fails to account for the direct influence of government on the Japanese public. The classic or ideal model of democracy, in which people first receive information from the mass media, then form public opinion, and finally influence government decisions, has probably never existed. In the 20th-century mass democracy, after constituencies choose their delegates for parliaments or congresses by whatever standards, they leave decisions on "difficult matters" to their delegates and government leaders. This tendency is especially strong in Japan.[1] Therefore, if the mass media cover a prime minister's speech, the content of that coverage has a strong effect on the masses, regardless how the mass media portray the speech—even if it is criticized in editorials.

This does not mean, however, that the Japanese public always obeys the government. It sometimes stubbornly and fiercely revolts against the government when the masses' direct economic interests are at stake (e.g., the long-lasting struggle between the government and farmers on the expansion of Narita Airport).

The Japanese do speak out and express themselves in areas where they think they know best and can make up their own minds. Such areas include government corruption and misconduct among government officials. A powerful coalition of the mass media and the masses forced Pre-

---

[1] Due to a long history of feudalism, the Japanese, especially in rural constituencies, tend to leave decisions on difficult matters to *okami* (people above us) and are not affected by the mass media's editorials as far as those difficult matters are concerned. Those people, suspicious and nervous about political questions, often do not give honest answers in questionnaire surveys or give ambiguous answers, such as "It depends" or "I don't know" (see Ito & Kohei, 1990, p. 107, or Nishihira, 1987, p. 160).

mier Tanaka to resign in 1974, Premier Uno in 1989, Premier Takeshita in 1989, and Vice-Premier Kanemaru in 1992—all in the wake of scandals.

By contrast, sometimes the government and the mass media have formed coalitions that have led the Japanese public in a desired direction despite its initial resistance (e.g., democratization, contraction of working hours, and social participation of women). How do these coalitions function? Nakano (1977) and Tsujimura (1976) posited a *tripolar* relationship between the mass media, the government, and the public, stating that the mass media affected the government and political decision-making only when media were supported by the masses (as in the cases of government officials' misconduct). The mass media, however, cannot affect government decision-making on difficult subjects such as foreign and defense policies, because the masses tend to leave such decisions to government leaders.

Ito (1990) expanded this idea further by suggesting that when two of the three major components of political consensus formation agree with each other, what he called *kuuki* (a Chinese-Japanese-Korean concept meaning an air, atmosphere, or climate of opinion requiring compliance)[2] is created and functions as a social pressure on the third component. When this *kuuki* pressure is strong enough, the third component concedes and a consensus is formed. Actually, however, each of these three components is divided into proponents and opponents. Therefore, Ito (1993, in press) suggested a model of *dominant kuuki* (see Figure 1).

Each box in Figure 1 represents 100 percent of opinion within that sector or the total amount of influence held by that sector (the masses, mass media, and government, assuming that the total amount of influence is the same among these three sectors). The dividing lines in the three boxes are drawn based on the data obtained from (a) public opinion surveys, (b) content analyses of the mass media, and (c) questionnaire surveys of National Diet members (when available). The area created by connecting *x, y,* and *z* represents the mainstream opinion regarding that particular issue and the social pressure *kuuki*. The larger this area is, the stronger the social pressure on those who do not belong to this mainstream area.

Ito (1993, in press) applied this model to explain the political process in Japan in the 1930s, the amendment of the Consumption Tax Law in 1989, and the withdrawal of the proposed U.N. Peace Cooperation Bill in 1990. In the first case, the mass media carried chauvinistic articles at the time of the Manchurian Incident in 1931. This helped to intensify chauvinistic *kuuki,* and aided in the militarists' takeover of the Japanese government and start of the war with the United States and Great Britain. In

---

[2] *Kuuki* in Japanese is similar and related to *qi feng* in Chinese and *kong ki* in Korean. In applying the concept of *kuuki* to this model, I was inspired by an excellent book, *Kuuki no Kenkyu* [A Study of *Kuuki*], written by social critic Yamamoto Shichihei (Yamamoto, 1977)

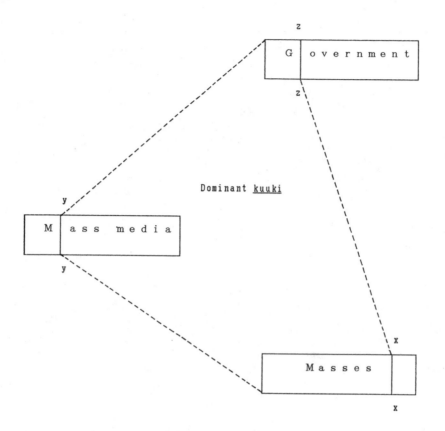

Figure 1. A tripolar *kuuki* model of political consensus formation.

the second case, the general public's strong antagonism against the newly introduced Consumption Tax affected the mass media and intensified the "anti-Consumption Tax *kuuki*." The government finally agreed to amend the Consumption Tax Law in December 1989. In the third case, the public's strong antiwar feelings (a result of the public's memory of its World War II experience) created "antiwar *kuuki*" during the Gulf Crisis in 1990. This forced the government to withdraw the U.N. Peace Cooperation Bill that would have dispatched Japanese noncombat troops to the Middle East. In each case, Ito (1993, in press) used public opinion survey data, content analyses of newspapers, elections results, support and nonsupport percentages of cabinets, and opinion survey results of National Diet members (when available) to show the functions and effects of *kuuki*.

These case studies indicate that it is too simplistic to claim that the government always dominates and manipulates the mass media or that the mass media always have powerful effects. Mass media effects differ from one case to another. The mass media have effects only when they stand on the majority side or the mainstream in a triadic relationship that creates and supports the *kuuki* that functions as a social pressure on the minority side. Actually, however, the masses, the mass media, and even the government can be divided between proponents and opponents, and only the mass media that join the formation of the mainstream have effects.

This model is similar to the spiral of silence theory (see Noelle-Neumann, 1984). However, unlike the spiral of silence theory or other conventional bipolar models (consisting of only the sender, the receiver, and intermediate variables), the *kuuki* model is tripolar. In the spiral of silence theory, only the mass media create social pressure for compliance. Therefore, changes in mass media content or government policies can never be explained or predicted. The *kuuki* model can explain or predict changes of mass media content (or government policies) in terms of compliance to *kuuki* created and maintained by a coalition of the government and the masses (or the mass media and the masses).

Testing the Models

This article introduced two Japanese models of mass media effects. The first, the *joho kohdo (information behavior) model,* suggests that people use extracted information to check the credibility of mass media information. What results would the loss of mass media credibility bring? In societies where mass media are operated as private businesses independent of the government, loss of credibility would result in a decrease in circulation or ratings, financial difficulties, and bankruptcy of mass media companies. Some critical scholars have claimed that the mass media in capitalist societies manipulate the masses to profit the elite (capitalist) class (Hall, 1982, pp. 87–88; Nordenstreng & Varis, 1973, p. 403). The masses in the 20th century, however, are not like 18th-century peasants. They travel overseas and have friends in foreign countries. If the mass media "cheat" the masses, they will lose mass support, and alternative media—such as "mini media," community or local papers and magazines, and participatory, two-way cable television—should emerge, flourish, and replace the mass media.

In societies where the mass media are directly controlled and operated by the state, and where free entry by private businesses is not allowed (as was the case in the former Soviet Union and many Eastern European countries), the loss of the mass media's credibility could lead to the loss of the government's, and then the entire regime's, credibility.

Promising studies testing the *joho kohdo* (information behavior) model include case studies in the former Soviet Union and Eastern European countries where mass media credibility has been lost gradually. In these cases, broadcasting from foreign countries and information extracted from imported foreign commodities, foreign travelers, letters from foreign countries, and so forth contributed to the erosion of the credibility of state-controlled mass media, the government, and then the entire regime.

Mass communication scholars who believe in powerful mass media effects have assumed that mass media credibility is always high and stable. But this assumption may not be appropriate. Scholars should pay more attention to the conditions under which mass media credibility is or is not maintained.

The second model, the *tripolar kuuki model,* suggests that the mass media have effects only when they are in the mainstream in the triadic relationship among the masses, the mass media, and the government. So far, only three case studies have tested this model, and all of them are cases in which government policies were influenced by *kuuki* created by the mass media and the masses. More case studies are needed to test this model—especially cases in which the mass media (or the masses) change under the influence of *kuuki* created by the government and the masses (or the government and the mass media).

Finally, what kind of future predictions can be made using these models? A prediction based on the *joho kohdo* model proposes that what happened in Eastern Europe is likely to happen eventually in other socialist countries unless they drastically change their propaganda strategies. The reason is that propaganda campaigns by state-controlled mass media that claim socialism is superior in raising people's standards of living contradicts information that the masses extract from their foreign friends, foreign products, overseas travel, and other sources.

A modest future prediction based on the tri-polar *kuuki* model would read something like this: The Japanese themselves have described contemporary Japan as "a country with a first-rate economy and third-rate politics." Since at least the early 1970s, however, the coalition between the mass media and the masses has been powerful in exposing political corruption and misconduct by high government officials. Each time scandal-tainted high government officials were ousted from their positions, some political reform took place. Therefore, as long as the tripolar *kuuki* model functions as it does at present, Japan will eventually evolve from being a country with third-rate politics to being a country with second-rate politics.

References

Brouwer, M. (1967). Prolegomena to a theory of mass communication. In L. Thayer (Ed.), *Communication: Concepts and perspectives* (pp. 227–239). Washington, DC: Spartan.

Edelstein, A. S., Ito, Y., & Kepplinger, H. M. (1989). *Communication & culture: A comparative approach.* New York: Longman.

Festinger, L. (1957). *A theory of cognitive dissonance.* Stanford, CA: Stanford University Press.

Fujitake, A. (1968). *Gendai masu komyunikeishon no riron* [Contemporary mass communication theories]. Tokyo: Nihon Hoso Shuppan Kyokai.

Hall, S. (1982). The rediscovery of "ideology": Return of the repressed in media studies. In M. Gurevitch, T. Bennett, J. Curran, & J. Woollacott (Eds.), *Culture, society and the media* (pp. 56–90). London: Methuen.

Homma, N. (1975). Kokusai komyunikeishon to shiteno nichibei kankei [U.S.–Japan relationship from an international communication perspective]. In Institute for Communications Research, Keio University (Ed.), *Nihon, Ajia, Amerika: Kokusai komyunikeishon no kenkyu* [Japan, Asia, America: A study of international communication] (pp. 25–38). Tokyo: Institute for Communications Research, Keio University.

Ikeda, K. (1989). "Spiral of silence" hypothesis and voting intention: A test in the 1986 Japanese national election. *Keio Communication Review, 10,* 51–62.

Ito, Y. (1989). Socio-cultural backgrounds of Japanese interpersonal communication style. *Civilisations, 39*(1), 101–127.

Ito, Y. (1990). Mass communication theories from a Japanese perspective. *Media, Culture and Society, 12*(4), 423–464.

Ito, Y. (1992). Theories on interpersonal communication styles from a Japanese perspective: A sociological approach. In J. Blumler, J. McLeod, & K. E. Rosengren (Eds.), *Comparatively speaking: Communication and culture across space and time* (pp. 238–268). Newbury Park, CA: Sage.

Ito, Y. (1993). New directions in communication research from a Japanese perspective. In P. Gaunt (Ed.), *Beyond agendas: New directions in communication research* (pp. 119–135). Westport, CT: Greenwood.

Ito, Y. (in press). From bipolar models of mass media influence to a tri-polar model of social consensus formation. In W. B. Gudykunst (Ed.), *Communication in Japan and the United States*. Albany: State University of New York Press.

Ito, Y., & Kohei, S. (1990). Practical problems in field research in Japan. In U. Narula & W. B. Pearce (Eds.), *Cultures, politics, and research programs: An international assessment of practical problems in field research* (pp. 89–121). Hillsdale, NJ: Lawrence Erlbaum.

Iwabuchi, Y. (1989). Seijiteki ronso to yoron keisei katei: Chimmoku no rasen riron no jisshouteki kenkyu [Political issues and the process of public opinion formation: An empirical study of the spiral of silence theory]. *Seigakuin Daigaku Ronso, 2,* 55–79.

Kato, H. (1972). *Joho kohdo* [Information behavior]. Tokyo: Kohdan-sha.

Katz, E., & Lazarsfeld, P. F. (1955). *Personal influence*. Glencoe, IL: The Free Press.

Kawatake, K. (1988). *Nippon no imehji: Masu media no kohka* [Images of Japan: Mass media effects]. Tokyo: Nihon Hoso Shuppan Kyokai.

Kitamura, H. (1970). *Joho kohdo ron* [Information behavior]. Tokyo: Seibundo Shinkoh-sha.

Makita, T., & Uemura, S. (1985). Nihonjin to terebi, 1985 [Japanese and television, 1985]. *Hoso Kenkyu to Chohsa, 35*(8), 2–17, 70–77.

Midooka, K. (1990). Characteristics of Japanese-style communication. *Media, Culture and Society, 12*(4), 477–489.

Minami, H. (1971). *Psychology of the Japanese people*. Toronto: University of Toronto Press.

Nakano, O. (1977). Shimbun to yoron [Newspapers and public opinion]. In M. Inaba & N. Arai (Eds.), *Shimbungaku* (pp. 262–271). Tokyo: Nihon Hyoron-sha.

Nakano, O. (1980). *Gendaijin no joho kohdo* [Information behavior of modern individuals]. Tokyo: Nihon Hoso Shuppan Kyokai.

Nishihira, S. (1987). *Yoron chohsa ni yoru dohjidaishi* [Chronology by public opinion polls]. Tokyo: Brehn Shuppan.

Noelle-Neumann, E. (1984). *The spiral of silence: Public opinion—our social skin*. Chicago: University of Chicago Press.

Nordenstreng, K., & Varis, T. (1973). The nonhomogeneity of the national state and the international flow of communication. In G. Gerbner, L. P. Gross, & W. H. Melody (Eds.), *Communications Technology and Social Policy* (pp. 393–412). New York: John Wiley & Sons.

Okamoto, S., Oda, M., Shiga, T., Toga, T., & Hata, E. (1986). Nihon ni okeru supairaru obu sairensu no jissho [A test of the spiral of silence theory in Japan], *Seijigaku Kenkyu, 16,* 199–210.

Shimizu, I. (1951). *Shakai shinrigaku* (Social psychology). Tokyo: Iwanami Shoten.

Sugiyama, M., & Kimura, Y. (1988). Igirisu no terebi ni egakareta "Nippon" ["Nippon" as described in British television]. *Hoso Kenkyu to Chohsa, 38*(4), 36–53.

Tokinoya, H. (1989). Testing the spiral of silence theory in East Asia. *Keio Communication Review, 10,* 35–49.

Tsujimura, A. (1976). Yoron to seiji rikigaku [Public opinion and political dynamics]. In Nihonjin Kenkyu Kai (Ed.), *Nihonjin kenkyu, No. 4: Yoron towa nanika* (pp. 173–238). Tokyo: Shiseido.

Tsujimura, A., Furuhata, K., & Akuto, H. (Eds.). (1987). *Sekai wa nihon wo do mite iruka: Tainichi imehji no kenkyu* [How does the world view Japan?: A study of the images of Japan]. Tokyo: Nihon Hyoron-sha.

Tsujimura, A., Kim, K., & Ikuta, M. (Eds.). (1982). *Nihon to kankoku no bunka masatsu* [Cultural frictions between Japan and Korea]. Tokyo: Idemitsu Shoten.

Wiebe, G. D. (1955). Mass communications. In E. L. Hartley & R. E. Hartley (Eds.), *Fundamentals of social psychology* (pp. 159–195). New York: Knopf.

Yamamoto, S. (1977) *"Kuuki" no kenkyu* [A study of *"kuuki"*]. Tokyo: Bungei Shunju-sha.

Has Communication Explained Journalism?

by Barbie Zelizer, Temple University

Regardless of one's view of the discipline of communication, journalism has occupied a central place in it. Communication researchers have long used journalism to explain how communication works, and journalists' visibility in mediated discourse has made them a target for scholars seeking to understand the work of communication practitioners and the communication process. But has communication done its job? Has communication scholarship provided the tools necessary to explain how and why journalism works? Has it explained why publics let reporters present themselves as cultural authorities for events of the "real world"? In short, has communication adequately explained journalism and journalistic authority?

This article argues that it has not. Journalism researchers, I contend, have allowed media power to flourish by not addressing the ritual and collective functions it fulfills for journalists themselves. The paper argues for a more interdisciplinary approach to journalism scholarship in order to provide a fuller account of media power. It also briefly considers the notions of *performance, narrative, ritual,* and *interpretive community* as alternative frames through which to consider journalism.

Power of the Fourth Estate

Media power is one of the outstanding conundrums of contemporary public discourse, in that we still cannot account for the media's persistent presence as arbiters of events of the real world. Audiences tend to question journalistic authority only when journalists' versions of events conflict with the audience's view of the same events. And while critical appraisals of the media should be part of everyday life, journalistic power burgeons largely due to the public's general acquiescence and its reluctance to question journalism's parameters and fundamental legitimacy.

Barbie Zelizer is an assistant professor in the Department of Rhetoric and Communication at Temple University. She thanks all those in communication and beyond who have alerted her to the value of interdisciplinary study.

In part, this has had to do with the rather basic fact that journalists do not invite or appreciate criticism. The media, Lule recently argued, "engage in critical evaluation of every institution in society except [themselves]" (1992, p. 92). Journalists ignore criticism leveled at them in journalism reviews, academic conferences, books, and the alternative press, trying to maintain a stance of autonomous indifference both vis-à-vis the events of the real world and that world's most vocal inhabitants, their critics.

Yet scholars studying journalism have also been partly responsible for the public's inability to grasp fully the power of journalists. News has been approached primarily by communication researchers "as a sociological problem" (Roshco, 1975, p. 2). Inquiry has favored examining the dominant rather than deviant form of practice. It has addressed the identifiable and relatively finished products—news text (Fowler, 1991; Glasgow University Media Group, 1980; van Dijk, 1988), news-gathering setting (Fishman, 1980; Gans, 1979; Gitlin, 1980; Tuchman, 1978), or news audience (Graber, 1988; Robinson & Levy, 1986)—rather than the continuous negotiation toward such products. It has generated linear notions of media power that have explained it as a dominance over the weak by the strong, supporting the view, recently voiced by sociologist Dennis Wrong, that power "is the capacity of some persons to produce intended and foreseen effects on others" (1979/1988, p. 2). In adopting a sociological tenor in their scholarship, journalism researchers have thereby fashioned a view of journalism that fits neatly within sociology, but perhaps nowhere else in the academy. Given journalism's complex and multifaceted dimensions, this may mean we have missed much of what constitutes journalism. Reporters' burgeoning authority underscores the degree to which we have understated journalists' consolidation of the power derived from reporting any given event. We thereby need to explore other lenses for examining the trappings of journalism, and to consider how authority and power function as a collective code of knowledge for journalists.

Humanistic Inquiry Into Journalism

An alternative view of media power and authority is supported by humanistic inquiry. Rather than conceptualize power only as the influence of one group over another, humanistic inquiry supports a view of power as also having ritual or communal dimensions. These communal dimensions have prompted certain communication scholars to consider the limitations imposed by our preference for sociological inquiry (Carey, 1975; Hirsch & Carey, 1978; Park, 1940; Schlesinger, 1989; Schudson, 1991). These scholars, and others, have produced work that addresses the collective codes by which journalism constitutes itself (Carey, 1974; Hardt, 1990; Nord, 1988; O'Brien, 1983; Schudson, 1988, 1992). But humanistic inquiry has not made sufficient inroads into journalism scholarship, generating concerns about how to capitalize on what the inquiry offers—a way to examine journalism through its own collectivity in addition to its influence on others. In such a view, authority and power become a construct of community, functioning as the stuff that keeps a community together.

Four notions employed in humanistic fields of inquiry suggest this explicit address to journalism as a collectivity: performance, narrative, ritual, and interpretive community. Each provides an alternative frame for explaining journalism, a frame that regards journalistic practice as a way to connect journalists with each other as well as to relay news. Thus in different ways each frame addresses the basic yet overlooked fact that journalists use news to achieve pragmatic aims of community.

Journalism as Performance

Within folklore, anthropology, and particularly performance studies, the notion of performance has generated widespread interest as a way of explaining practice (Bauman, 1986; Schechner, 1985). The performance frame signifies both accomplishment and artfulness, and it has been used to reference both small theatrical productions and large-scale public events (Abrahams, 1977; Bauman, 1989; MacAloon, 1984; Turner, 1982; Zelizer, 1989). In anthropologist Victor Turner's words, performance is a circumstance of "making, not faking" (1982, p. 93). It offers group members a way to negotiate their internal group authentication (Abrahams, 1986).

Generally contrasted with *text,* performance suggests a fluid, less fixed frame for understanding practice (Bauman, 1986; Fine, 1984). In its most fundamental form, text refers to the actual script of a play, performance to how it is acted out. Within this frame, journalism scholars have begun to examine the unfolding of news rather than the finished product, an aim called for by Bauman (1989). Ettema (1990) and Wagner-Pacifici (1986) used the performative dimensions of the social drama to consider media coverage of, respectively, Chicago's Cokely affair and the Aldo Moro kidnapping. Dayan and Katz (1992) analyzed the media event as a type of performance. When seen as performance, news is understood as a situationally variant process that is neither static nor fixed. Reporters negotiate their power across a variety of situations, allowing analysts to map out the patterns of cultural argumentation by which an event becomes news. Given a news event's inherently unstable meaning, this frame may turn out to be particularly useful for considering journalism.

Journalism as Narrative

Journalism as narrative is another way to account for journalism's commonality. Borrowed from literary studies, this frame indexes a group's ability to consolidate around codes of knowledge by examining which narratives are upheld, repeated, and altered. In this view, narrative helps us construct our view of the world, by allowing us to share stories within culturally and socially explicit codes of meaning (Barthes, 1977, 1979). Narrative gives us a way to challenge or interrupt a community's so-called "master discourses" (Bhabha, 1990). In social history and historiography, narrative is seen as a strategic way of remembering and representing the past (Halbwachs, 1950/1980; Kammen, 1991). Hayden White (1981) in particular has argued that over time narratives offer narrators ways of reconfiguring their authority for events.

Narrative has received considerable attention from communication scholars during the last decade (Fisher, 1987; Lucaites & Condit, 1985). News scholarship has produced discussions of common narrative frames and themes (Barkin & Gurevitch, 1987; Bennett & Edelman, 1985; Campbell & Reeves, 1989; Carragee, 1990) and narrativizing strategies (Bird & Dardenne, 1988; Mander, 1987; Schudson, 1982). An issue of the journal *Communication* focused on the theme of "news as social narrative" (Carey & Fritzler, 1989). Recent interest in collective memory has also generated interest in narrative, whereby oft-repeated narratives are seen as offering speakers a way to compete for a place in public discourse about the past (Schudson, 1992; Zelizer, 1992b).

Narrative helps us explain journalism by stressing elements that are formulaic, patterned, finite, yet mutable over time. In this sense, news as narrative offers analysts a way to account for change within predictable and defined patterns of news presentation. Not only does this focus on how narratives are repeated in conjunction with certain events, but it also suggests that narratives change and thereby affect the power of their narrators, the journalists.

Journalism as Ritual

Yet another way of viewing journalism is through the frame of ritual. Taken from anthropology and folklore, rituals are seen as offering "a periodic restatement of the terms in which [people] of a particular culture must interact if there is to be any kind of a coherent social life" (Turner, 1968, p. 6). Rituals provide moments for individuals to question authority and consolidate themselves into communities (Turner, 1969).

Certain work within communication and journalism scholarship has called for an examination of ritual (e.g., Glasser & Ettema, 1989; Rothenbuhler, n.d.). Rothenbuhler provided an extended discussion of the efficacy of ritual as a communication concept; he characterized ritual as "communication without information" or communication that had more to do with meaning-making than with informing. Glasser and Ettema (1989) examined the way in which rituals in journalistic work helped promote cultural change. Press rites were explored for the ways in which they allowed reporters to flex their muscles (Elliot, 1982; Manoff & Schudson, 1986; Tuchman, 1972). Press criticism was found to function symbolically for reporters, who use media criticism not so much to generate real debate over appropriate journalistic practice as to uphold and maintain the larger canon of objective journalism (Lule, 1992). Such "allowed" cri-

tiques constitute journalism's way of underscoring its own apparent reflexivity. Looking at news as ritual suggests a way of examining patterned
behavior that emerges through collectivity, for rituals do not work without solid support behind them. This again allows us to examine journalism via its own commonality, rather than merely accounting for its influence on others.

Journalism as Interpretive Community
The idea of the interpretive community has long been of interest in literary studies, anthropology, and folklore. Hymes defined "speech communities" as groups united by shared interpretations of reality (1980, p. 2). In
literary studies, Fish defined interpretive communities as those that produce texts and "determine the shape of what is read" (1980, p. 171).
Scholars have examined "communities of memory," where group members create shared interpretations over time (Bellah, Madsen, Sullivan,
Swidler, & Tipton, 1985). Each view suggests that communities arise
through patterns of association derived from the communication of
shared interpretation.

   In communication, the idea of the interpretive community has been invoked most avidly in audience studies (Lindlof, 1987; Radway, 1984). But
it is possible to examine communicators themselves as an interpretive
community (Zelizer, 1993). Journalists, in this view, create community
through discourse that proliferates in informal talks, professional meetings and trade reviews, memoirs, interviews on talk shows, and media
retrospectives. Through discourse, journalists create shared interpretations that make their professional lives meaningful; that is, they use stories about the past to address dilemmas that present themselves while
covering news.

   The idea of the interpretive community has been the thrust of certain
journalism scholarship. It is implied in Schudson's (1991) recent discussion of the "culturological" studies of news, which chart the flow of informal discourse among reporters; also in Pauly's (1988) examination of
journalistic rhetoric about the tabloid, used by reporters to establish standards of "good" journalism; and it can be observed in journalists' reliance
on shared interpretive strategies to address the Janet Cooke affair (Eason,
1986). The prescriptive parameters that underlie journalistic textbooks
similarly reflect consensual assumptions about the kinds of gendered
voices that have been permitted and barred from news (Steiner, 1992).
Journalists functioned as an interpretive community in shaping the tale of
John F. Kennedy's assassination into a story about the legitimation of television news, in fashioning stories of the Gulf War into celebrations of
CNN, or in recasting stories about Watergate and McCarthyism into moral
tales about appropriate journalistic practice (Zelizer, 1992a, 1992b, 1993).
In examining journalism as an interpretive community, we see a group
united by its shared discourse and collective interpretations of key public

events. The shared discourse that they produce thus offers a marker of how journalists see themselves as journalists.

Alternative Explanations

The principal thrust of this article has been to argue for a more interdisciplinary approach to journalism. This is a necessary corrective to our commonly held view of journalism—that it is foremost a sociological problem—for that view has prompted us to examine journalism in narrowly defined ways. In considering these alternative frames for examining journalism, this article suggests a broader understanding of journalism that addresses both how news functions *for journalists* as well as for audiences. Adopting such a focus may help us achieve one of communication scholarship's most pragmatic aims: how to predict and control for the work of communication practitioners, by understanding their collective aims in practice. By recognizing our dependence on sociologically motivated inquiry, we may find we have thus far missed much of journalism's essence. And considering journalism's central role in explaining general communicative practice, this may mean we have missed much of the essence of communication as well.

References

Abrahams, R. (1977). Toward an enactment-centered theory of folklore. In W. R. Bascom (Ed.), *Frontiers of folklore* (pp. 79–120). Boulder, CO: Westview Press.

Abrahams, R. (1986). Ordinary and extraordinary experience. In V. Turner & E. Bruner (Eds.), *The anthropology of experience* (pp. 45–72). Urbana: University of Illinois Press.

Barkin, S., & Gurevitch, M. (1987). Out of work and on the air: Television news of unemployment. *Critical Studies in Mass Communication 4,* 1–20.

Barthes, R. (1977). Introduction to the structural analysis of narrative. In *Image, music, text* (pp. 79–124). New York: Hill and Wang.

Barthes, R. (1979). From work to text. In J. Harari (Ed.), *Textual strategies* (pp. 73–81). Ithaca, NY: Cornell University Press.

Bauman, R. (1986). *Story, performance, and event.* Cambridge, England: Cambridge University Press.

Bauman, R. (1989). American folklore studies and social transformation: A performance-centered perspective. *Text and Performance Quarterly, 9*(3), 175–184.

Bellah, R., Madsen, R., Sullivan, W., Swidler, A., & Tipton, S. (1985). *Habits of the heart.* Berkeley: University of California Press.

Bennett, W. L., & Edelman, M. (1985). Toward a new political narrative. *Journal of Communication, 35*(4), 156–171.

Bhabha, H. (1990). *Nation and narration.* London: Routledge.

Bird, S. E., & Dardenne, R. W. (1988). Myth, chronicle and story: Exploring the narrative qualities of news. In J. W. Carey (Ed.), *Media, myths and narratives: Television and the press* (pp. 67–87). Beverly Hills, CA: Sage.

Campbell, R., & Reeves, J. L. (1989). Covering the homeless: The Joyce Brown story. *Critical Studies in Mass Communication, 6*(1), 21–42.

Carey, J. W. (1974). The problem of journalism history. *Journalism History, 1*(1), 3–5, 27.

Carey, J. W. (1975). A cultural approach to communication. *Communication, 2,* 1–22.

Carey, J. W., & Fritzler, M. (Eds.). (1989). News as social narrative. *Communication, 10*(1), 1–92.

Carragee, K. M. (1990, February). Defining solidarity: Themes and omissions in coverage of the Solidarity trade union movement by ABC news. *Journalism Monographs, 119.*

Dayan, D., & Katz, E. (1992). *Media events: The live broadcasting of history.* Cambridge, MA: Harvard University Press.

Eason, D. (1986). On journalistic authority: The Janet Cooke scandal. *Critical Studies in Mass Communication 3,* 429–447.

Elliot, P. (1982). Press performance as political ritual. In D. C. Whitney, E. Wartella, & S. Windahl (Eds.), *Mass communication review yearbook* (Vol. 3, pp. 583–619). Beverly Hills, CA: Sage.

Ettema, J. S. (1990). Press rites and race relations: A study of mass-mediated ritual. *Critical Studies in Mass Communication 7*(4), 309–331.

Fine, E. (1984). *The folklore text: From performance to print.* Bloomington: Indiana University Press.

Fish, S. (1980). *Is there a text in this class?* Cambridge, MA: Harvard University Press.

Fisher, W. R. (1987). *Human communication as narration: Toward a philosophy of reason, value, and action.* Columbia: University of South Carolina Press.

Fishman, M. (1980). *Manufacturing the news.* Austin: University of Texas Press.

Fowler, R. (1991). *Language in the news: Discourse and ideology in the press.* London: Routledge.

Gans, H. (1979). *Deciding what's news.* New York: Pantheon.

Gitlin, T. (1980). *The whole world is watching.* Berkeley: University of California Press.

Glasgow University Media Group. (1980). *More bad news.* London: Routledge and Kegan Paul.

Glasser, T. L., & Ettema, J. S. (1989). Investigative journalism and the moral order. *Critical Studies in Mass Communication, 6*(1), 1–20.

Graber, D. (1988). *Processing the news.* White Plains, NY: Longman.

Halbwachs, M. (1980). *The collective memory.* New York: Harper and Row. (Original work published 1950)

Hardt, H. (1990). Newsworkers, technology, and journalism history. *Critical Studies in Mass Communication, 7,* 346–365.

Hirsch, P., & Carey, J. W. (Eds.). (1978). Communication and culture: Humanistic models in research. *Communication Research, 5*(3).

Hymes, D. (1980). Functions of speech. In *Language in education: Ethnolinguistic essays* (pp. 1–18). Washington, D.C.: Center for Applied Linguistics.

Kammen, M. (1991). *Mystic chords of memory*. New York: Alfred A. Knopf.

Lindlof, T. (Ed.). (1987). *Natural audiences*. Norwood, NJ: Ablex.

Lucaites, J. L., & Condit, C. (1985). Reconstructing narrative theory: A functional perspective. *Journal of Communication, 35*(4), 90–108.

Lule, J. (1992). Journalism and criticism: The *Philadelphia Inquirer* Norplant editorial. *Critical Studies in Mass Communication, 9*, 91–109.

MacAloon, J. (1984). *Rite, drama, festival, spectacle: Rehearsals toward a theory of cultural performance*. Philadelphia: ISHI Publications.

Mander, M. S. (1987). Narrative dimensions of the news: Omniscience, prophecy and morality. *Communication, 10*, 51–70.

Manoff, R., & Schudson, M. (1986). *Reading the news*. New York: Pantheon.

Nord, D. P. (1988). A plea for journalism history. *Journalism History, 15*(1), 8–15.

O'Brien, D. (1983, September). The news as environment. *Journalism Monographs, 85*.

Park, R. (1940). News as a form of knowledge. *American Journal of Sociology, 45*, 669–686.

Pauly, J. P. (1988). Rupert Murdoch and the demonology of professional journalism. In J. W. Carey (Ed.), *Media, myths, and narratives: Television and the press* (pp. 246–261). Beverly Hills, CA: Sage.

Radway, J. (1984). *Reading the romance*. Chapel Hill: University of North Carolina Press.

Robinson, J. P., & Levy, M. R. (1986). *The main source: Learning from television news*. Beverly Hills, CA: Sage.

Roshco, B. (1975). *Newsmaking*. Chicago: University of Chicago Press.

Rothenbuhler, E. (n.d.). *Ritual as a communication concept*. University of Iowa. Unpublished manuscript.

Schechner, R. (1985). *Between theater and anthropology*. Philadelphia: University of Pennsylvania Press.

Schlesinger, P. (1989). Rethinking the sociology of journalism: Source strategies and the limits of media-centrism. In M. Ferguson (Ed.), *Public Communication: The new imperatives* (pp. 61–83). London: Sage.

Schudson, M. (1982). The politics of narrative form: The emergence of news conventions in print and television. *Daedalus, 3*(4), 97–112.

Schudson, M. (1988). What is a reporter: The private face of public journalism. In J. W. Carey (Ed.), *Media, myths and narratives: Television and the press* (pp. 228–45). Beverly Hills, CA: Sage.

Schudson, M. (1991). The sociology of news production revisited. In J. Curran & M. Gurevitch (Eds.), *Mass media and society* (pp. 141–159). London: Edward Arnold.

Schudson, M. (1992). *Watergate in American memory: How we remember, forget and reconstruct the past*. New York: Basic Books.

Steiner, L. (1992, October). Construction of gender in newsreporting textbooks: 1890–1990. *Journalism Monographs, 135*.

Tuchman, G. (1972). Objectivity as a strategic ritual. *American Journal of Sociology, 77*, 660–679.

Tuchman, G. (1978). *Making news*. Glencoe, IL: Free Press.

Turner, V. (1968). *The drums of affliction*. Oxford, England: Clarendon Press.

Turner, V. (1969). *The ritual process*. Ithaca, NY: Cornell University Press.

Turner, V. (1982). *From ritual to theatre*. New York: Performing Arts Journal Publication.

van Dijk, T. A. (1988). *News as discourse*. Hillsdale, NJ: Lawrence Erlbaum.

Wagner-Pacifici, R. (1986). *The Moro morality play*. Chicago: University of Chicago Press.

White, H. (1981). The value of narrativity in the representation of reality. In W. J. T. Mitchell (Ed.), *On narrative* (pp. 1–23). Chicago: University of Chicago Press.

Wrong, D. H. (1988). *Power: Its forms, bases, and uses*. Chicago: University of Chicago Press. (Original work published 1979)

Zelizer, B. (1989, May). "All the world's a stage": "Performance" as interdisciplinary tool. Paper presented at annual meeting of International Communication Association in San Francisco, CA.

Zelizer, B. (1992a). CNN, the Gulf War, and journalistic practice. *Journal of Communication 42*(1), 68–81.

Zelizer, B. (1992b). *Covering the body: The Kennedy assassination, the media, and the shaping of collective memory*. Chicago: University of Chicago Press.

Zelizer, B. (1993). Journalists as interpretive communities. *Critical Studies in Mass Communication, 10*(3), 219–237.

Can Cultural Studies Find True Happiness in Communication?

by Lawrence Grossberg, University of Illinois at Urbana–Champaign

When the history of cultural studies in the United States is written, I hope the authors will acknowledge the importance of the discipline of communication. As much as (the project if not the reality of) American studies, communication studies has provided key resources for the construction of an American cultural studies. Equally important, along with the field of education, communication studies was one of the first disciplines to take up and provide a space in the U.S. academy for the developing work of British cultural studies. And it has, over the decades, continued to provide a place—albeit not without resistance and, initially, often a minoritarian and marginal place—for emergent trends within the expanding field of cultural studies (e.g., popular culture, global culture, postmodernism).

Cultural studies encompasses a set of approaches that attempt to understand and intervene into the relations of culture and power. Without equating cultural studies with the diverse body of work that is often referred to as British cultural studies, it is still the case that as cultural studies has become more successful, as it has appeared in different national and disciplinary sites, and as alternative traditions have emerged, British cultural studies has provided a common vocabulary and iconography through which these diverse voices have been able to come together (Grossberg, Nelson, & Treichler, 1992).

In part, what British cultural studies has provided is an exemplar of a particular vision of critical practice, a particular operational assumption about the relationship between theory and context. Cultural studies rejects the application of a theory known in advance as much as it rejects the possibility of an empiricism without theory. It is committed to the necessary "detour through theory" while at the same time refusing to be driven by its theoretical commitments. It is driven instead by its own sense of history and politics (Hall, 1991).

There are three consequences of this particular practice worth identifying. First, cultural studies is committed to the fact that reality is continual-

Lawrence Grossberg is a professor in the Department of Speech Communication, the Institute of Communications Research, and the Unit for Criticism and Interpretive Theory at the University of Illinois in Urbana–Champaign. An early version of this paper was presented at the Speech Communication Association meeting, Chicago, October 1992.

ly being made through human action and that, therefore, there are no guarantees in history. As a result, contestation—both as a fact of reality (although not necessarily in every instance) and as a strategic critical practice—is a basic category. Second, cultural studies is continuously drawn to the "popular," not as a sociological category purporting to differentiate among cultural practices but as the terrain on which people live and political struggle must be carried out in the contemporary world.

Finally, cultural studies is committed to a radical contextualism, a contextualism that precludes defining culture, or the relations between culture and power, outside of the particular context into which cultural studies imagines itself to intervene. Consequently, cultural practices cannot be treated as simply texts, as microcosmic representations (through the mediating structures of meaning) of some social other (whether a totality or a specific set of relations). Cultural practices are places where a multiplicity of forces (determinations and effects) are articulated, where different things can and do happen, where different possibilities of deployment and effects intersect. A cultural practice is a complex and conflictual place that cannot be separated from the context of its articulation since it has no existence outside of that context. And if this is the case, then the study of culture can be no less complex, conflictual, and contextual (Frow & Morris, 1993).

The practice of cultural studies then is an attempt to map out the particular relations, the context, within which both the identity and the effects of any particular practice are determined. This is the import of the concept of *articulation,* which describes both the practice by which human reality is made and the practice of cultural studies. Articulation implies that effects (including identities) are the product of contextually defined relations. The question of the politics of culture then involves the work of placing particular practices into particular relations or contexts, and of transforming one set of relations, one context, into another. The identity and effects of a practice are not given in advance; they are not determined by its origin or by some intrinsic feature of the practice itself. Hence, no theory defined independently of the context of its intervention can predefine the relations surrounding a practice, or its specific concrete effects. Articulation guarantees that such relations and effects are real, but also that they are not guaranteed in advance. Some may find it ironic that cultural studies, which has always criticized the tradition of effects research in communications, takes up the language of effects. But the notion of effects here is significantly different—both broader and less determined, and more contextual (Grossberg, 1992).

At the present moment, there seems to be a feeling, among both the general community of (mass) communications scholars and among those committed to cultural studies within that community, that a rapprochement of sorts has been reached. After all, the field has made a space for, and granted the legitimacy of, cultural studies. There may still be some hostility between some individuals or paradigms, but that is relatively insignificant in light of the history of the discipline. Of course, part of this new tolerance has resulted from transformations within communications research as it has moved beyond positivism and taken up issues of meaning, cognition, affect, etc.

Communicational Cultural Studies—Not

I want to contest this illusory harmony, not so much on the basis of the failures of communication studies from a cultural studies perspective, but on the basis of the failures of cultural studies within a communication perspective. In a sense, I want to challenge the increasingly comfortable relationship between cultural studies and communication, a relationship in which cultural studies has too readily compromised itself by remaking itself within the image of communication studies. Here it may be worthwhile to quote from Raymond Williams's discussion of the history of cultural studies and the pressures toward disciplinization.

> *In its most general bearings, this work remained a kind of intellectual analysis which wanted to change the actual developments of society, but then locally, within the institution, there were all the time those pressures that had changed so much in earlier phases: from other disciplines, from other competitive departments, the need to define your discipline, justify its importance, demonstrate its rigour; and these pressures were precisely the opposite of those of the original project. . . . As you separate these disciplines out, and say "Well, it's a vague and baggy monster, Cultural Studies, but we can define it more closely—as media studies, community sociology, popular fiction or popular music; so you create defensible disciplines, and there are people in other departments who can see that these are defensible disciplines, that here is properly referenced and presented work. But the question of what is then happening to the project remains.* (1989, p. 158)

Cultural studies practitioners have too often agreed to reduce cultural studies to a theory of communication by too quickly equating *culture* and *communication*. They have constructed cultural studies itself as *communicational cultural studies*. This reduction can, in part, be laid at the feet of a certain misreading, followed by a rather uncritical adoption, of Raymond Williams's (1961) and James Carey's (1989) identifying of communication with culture. The result has been that the project of cultural studies has been submerged under a model of communication, rendering it an alternative paradigm of, rather than a serious challenge to, the existing forms of communication studies. And equally important, political struggles have been increasingly reduced to struggles over communication and culture, which can be magically solved by the proliferation of communicative and cultural practices.

There is a certain historical rationale to this identification. Writing about the reception of Richard Hoggart's (1957) foundational book, *The Uses of Literacy,* Stuart Hall (1970) acknowledged that it was read "—such were the imperatives of the moment—essentially as a text about the mass media" (p. 2). Consequently, cultural studies was framed, both within and outside of the Centre for Contemporary Cultural Studies, as a literary-based alternative to the existing work on mass communication. Hall continued:

> *The notion that the Centre, in directing its attention to the critical study of "contemporary culture" was, essentially, to be a centre for the study of television, the mass media and popular arts . . . though never meeting our sense of the situation . . . nevertheless came by default, to define us and our work.* (1970, p. 2)

What is at least hinted at here is that cultural studies is a radically alternative practice to the very practice of constituting the mass media and popular culture as objects of study that can be analyzed in isolation from their place within the material contexts of social reality. I want to begin to explore that alternativeness, to suggest a different view of the relationship of cultural studies and communication, by looking at the consequences of this too narrow vision of cultural studies, a vision that abandons the radical contextualism of cultural studies.

First, by identifying communication and culture, communicational cultural studies legitimates its reification of Stuart Hall's (1980) encoding/decoding model. What was originally offered as a theoretical semiotic solution to a particular contextually defined set of empirical problems has become instead *the* general model of cultural studies, an analogy transformed into an identity. In fact, this model has become little more than a recycling of the old, theoretically discredited, linear model of communication: sender–message–receiver. This is justified by presenting the model in the more critical terms of production–practice–consumption. But in fact, by reinterpreting the latter relationship in terms of the linear model of communication, the power and originality of Hall's argument—as well as of Marx's analysis of the relations among these three moments of the cycle of production (see Hall, 1974)—are seriously distorted. The equation of culture, communication, and production/consumption (although usually in communicational cultural studies, it is only consumption that appears relevant) not only effaces the specificity and complexity of culture, but of communication and consumption as well. Consequently, discussions of consumption in communications tend to conflate a number of issues: consumption as a necessary and productive practice, the nature of the commodity in the contemporary world, the nature of and alternatives to a consumer economy, the means by which a consumer culture enables people to construct identities through the market, and the various ideologies of the consumer (late capitalist?) society.

The encoding/decoding model is generally used to frame a tripartite approach to the study of media: The researcher is compelled to study the institutions and practices by which a particular text is produced, followed by a literary-critical analysis of the "encoded meaning" of the text (which is presumably the meaning that the producers "intended"), completed by an "ethnographic" study (usually in the form of interviews, question-naires, or selected observational studies) of the uses to which different audiences put the text and/or the different ways in which they interpret it. In some cases, by selectively deploying the notion of articulation, par-ticular studies may argue against the need for a consideration of the con-text of production since origins do not guarantee effects. And in other cases, even the analysis of the text itself may be considered unnecessary since the intrinsic properties of the text do not guarantee effects. Instead, specific audiences are granted the full weight and power of determining the actual effects (meanings) of cultural practices.

Contexts and Complexities

But the same reasoning that has enabled some analysts to jettison the mo-ment of production and textuality also justifies jettisoning any ethnogra-phy of the audience's uses or self-conscious interpretations. The point is not which of the three moments is actually determining of the effects of communication. Rather, the point is that in understanding culture as com-munication, as divisible into three separable moments, communicational cultural studies fails to locate specific cultural practices within their com-plexly determined and determining contexts. It is not that such ques-tions—of production, textuality and consumption—are irrelevant. On the contrary, they may well be crucial parts of the context within which cul-tural studies must locate specific cultural practices. But their identity and power cannot be identified apart from that context.

In that sense, the study of producers, texts, and audiences merely pro-vides some of the material with which cultural studies must grapple in its attempt to understand specific contextual relations of culture and power. And since such material does not directly reveal how it is located within the context, we cannot know in advance what knowledge it will provide, what we might learn as a result of such studies. At the same time, it may well be the case that such material does not provide the most important determinations. It follows then that cultural studies is not to be modeled on the linear notion of communication, that cultural studies cannot ap-proach communication as the discrete existence of three moments. But even more to the point, theory in cultural studies cannot be identified with—and cultural studies does not need—theories of authors, texts, or audiences. Cultural studies needs theories of contexts and of the com-plexity of cultural effects and relations of power.

Second, communicational cultural studies has confused a particular

contextual struggle to put questions of ideology (signification, representation, and identity) on the agenda of critical studies with the project of
cultural studies itself. As a result, it has reduced the politics of culture to
questions ⌐f meaning and identity (with an occasional nod to pleasure
and desire). It has assumed that power (domination and subordination) is
always hierarchical, understood on the model of either the oppressor and
the oppressed or oppression and transgression. And following the idealist
tradition of modern philosophy and social thought, it has erased "the
real"—the material conditions of possibility and of effectivity, the material
organization and consequences of life—from its agenda, reducing social
life to the experience of everyday life.

Consequently, it is often assumed that research in cultural studies involves a hermeneutical project of uncovering a relationship, a structure of
meaning, that already and necessarily exists. Whether embodied within
the text or within the lives of some fraction of the audience, whether carried out through some form of textual interpretation or ethnographic
practice, communicational cultural studies attempts to open for critical
scrutiny a dimension of human existence—meaning—that cannot be reduced to materiality. Communicational cultural studies is about the ways
people make sense of the world and their lives. Moreover, this dimension
must be relatively independent of the real material and economic conditions of the world and people's lives. Consequently, communicational
cultural studies can never actually confront the question of effects, because it cannot theorize the relationship of meaning to anything else. It
cannot even decide when meaning (signification) becomes representation (ideology), for that involves its articulation to material practices and
social relationships.

Third, communicational cultural studies has followed the practice of
the field of communication in isolating and disciplining cultural practices
according to some predefined internal schema of domains (e.g., interpersonal, organizational, mass), of media (e.g., television, film, music), or of
genres (e.g., romance, comedy, melodrama). Of course, as any scholar of
communication knows, this has little to do with either the context of production or of consumption. How can effects be mapped if we begin by
artificially circumscribing the context of cultural practices according to
the predefined schema of our own critical positions?

Finally, communicational cultural studies has become too comfortable
with the disciplinary boundaries of the field, and has too often abandoned the interdisciplinarity that is at the heart of cultural studies. For
cultural studies, unlike communicational cultural studies, starts with the
assumption that an account of culture will always require an appeal to
things and domains other than the cultural. It emphatically refuses the
claim that everything, including culture, can be explained in cultural
terms. Consequently, cultural studies is always stepping on the toes of
other disciplines, not merely because its questions overlap, but because it
attempts to find its own answers as well as its own questions. The inter-

disciplinarity of cultural studies, then, is never merely a matter of references and footnotes. It is of the very essence of the work of cultural studies. But communicational cultural studies, on the contrary, is usually quite content to remain within the discipline; its questions as well as its answers remain firmly entrenched within the boundaries of culture. Its interdisciplinarity is confined to theoretical borrowings and the occasional reference to other sources, which are given responsibility for describing the context within which its communicative practices are assumed to be located.

Communicational cultural studies, in the end, reduces culture to the symbolic representation of power and grants it a certain apparent autonomy. As a result, communicational cultural studies finds itself constantly rediscovering what it already knew: Regarding domination, that particular cultural practices reproduce the structures of domination and subordination, and that they reinscribe relations of identity, difference, and inequality; and regarding subordination, communicational cultural studies seems satisfied with finding the cracks in the processes of the reproduction and reinscription. The assumption that people are active and capable of struggle and resistance becomes an apparent discovery, and the important empirical questions, of the concrete contextual effects of such practices, are left unanswered.

But without a theory of contexts, communicational cultural studies cannot begin to answer these questions of cultural studies. Questions about the specific forms in which domination and subordination are organized, about the ways they operate, about how they are lived, mobilized, and empowered—that is, questions about the actual ways in which cultural practices are deployed in relations of power and how they themselves deploy power, questions about the actual effects of culture within specific contexts, questions about the relations of culture to practices of governance—remain largely unasked and unanswered. Domination and resistance are assumed to be understood in advance, to operate always and everywhere in the same ways. By identifying culture with communication, and identifying cultural studies with a taken-for-granted model of communication, communicational cultural studies has not only limited what it can say about specific cultural practices and about the place of culture in relations of power, it has also closed off the possibility of a more radical critique of the concept of communication and the discipline of communication studies.

The Place of Communication

I want then to advocate a different notion of cultural studies, one that enables a more radical critique of the discipline of communication. Paradoxically, precisely by setting itself in opposition to notions of communication, a cultural studies built on the concept of articulation has, I believe, more to contribute to the study of contemporary communication and the roles of communication in contemporary life. For such a cultural studies would enable us to rearticulate communication as a practice and a concept by questioning its complicity with specific institutions and technologies of power. What I mean by communication here is quite simple. As a concept, it was the 19th-century invention of the semantic plane as the site of the mediation between the individual and reality (resulting in experience as the basis of modern epistemologies) and of the mediation between the individual and society (resulting in communication as the site of the constitution of intersubjectivity).

Communication—as technology and concept—was a key figure in the construction of modern thought and modern societies, and it is closely related to the celebration of history, temporality, consciousness, and experience as the constitutive features of human life. It is a crucial link in the conceptual chain, the intellectual and discursive structure, by which modern thought and society were constructed and have been kept in place. Cultural studies can and must problematize the place of "communication" (and culture, of course) within broader discursive spaces. It can and must force communication scholars to recognize that the historical and intellectual conditions that both enabled and demanded that we talk about communication as a specific set of practices and within a specific set of discourses also defined the limits of our ability to talk about these practices and the silences that have too often linked communication, and the study of communication, to particular forms, technologies, and practices of power. Sometimes with the complicity of communication, but always in its name, modern forms of global organization (e.g., nationhood, capitalism, colonialism, and imperialism), modern forms of alterity (in which the other is reduced to the different—e.g., racisms and sexisms), and modern forms of control (e.g., disciplinization and normalization) were put into and maintained in place (Grossberg, 1993).

Such claims are not meant as an attack on those who work within the discourses of communication, for in one way or another, in the contemporary world, we all do. Rather, it is meant as a call for a more reflective and critical contextualization of the power of our own discourses as communication scholars. My claim, or at least my hope, is that a more radical understanding of cultural studies, even as it is located within the discipline of communication, can contribute to such a project, as well as to a better understanding of the imbrication of cultural practices in the material organization and deployment of power in the contemporary world. Nor is my argument meant as a call for cultural studies to abandon the discipline of communication. In the first place, there is no more comfortable home and, in the second place, cultural studies must often accommodate itself to a disciplinary home. Rather, I want to suggest that if cultural studies is to work, it must always make such (albeit necessary) disciplinary relations uncomfortable. It must always remind us that the boundaries of our disciplines can never be allowed to define the boundaries of our questions.

References

Carey, J. W. (1989). *Communication as culture: Essays on media and society*. Boston: Unwin Hyman.

Frow, J., & Morris, M. (1993). Introduction. In M. Morris & J. Frow (Eds.), *Australian cultural studies: A reader*. Urbana: University of Illinois Press.

Grossberg, L. (1992). *We gotta get out of this place: Popular conservatism and postmodern culture*. New York: Routledge.

Grossberg, L. (1993). Cultural studies in/and new worlds. *Critical Studies in Mass Communication, 10,* 1–22.

Grossberg, L., Nelson, C., & Treichler, P. (Eds.). (1992). *Cultural studies*. New York: Routledge.

Hall, S. (1970). Introduction to *The Annual Report of the Centre for Contemporary Cultural Studies*. Birmingham, England: Centre for Contemporary Cultural Studies.

Hall, S. (1974). Marx's notes on method. *Working Papers in Cultural Studies, 6,* 132–170.

Hall, S. (1980). Encoding/decoding. In S. Hall et al. (Eds.), *Culture, media, language* (pp. 128–138). London: Hutchinson.

Hall, S. (1991). The local and the global, and Old and new identities. In A. D. King (Ed.), *Culture globalization and the world-system* (pp. 19–68). London: Macmillan.

Hoggart, R. (1957). *The uses of literacy*. New York: Oxford University Press.

Williams, R. (1961). *The long revolution*. London: Chatto & Windus.

Williams, R. (1989). *The politics of modernism*. London: Verso.

Critical Communication Research at the Crossroads

by Robert W. McChesney, University of Wisconsin

I began my graduate work in communication in the autumn of 1983, short-ly after the publication of the "Ferment in the Field" issue. This was a tur-bulent and dramatic period of change for communication research. The recent emergence in the United States of critical studies in communication had torn asunder the guise of scientific neutrality that had masked the ideological premises of dominant communication research for decades. As one who had come to communication from history and economics with an explicit commitment to radical intellectual and political work, I regarded these developments as most heartening.

Indeed, in the past decade the number of tenured and tenure-track communication faculty members with critical perspectives has mush-roomed, and most major research programs have at least one or two left-ists on board. Moreover, a significant proportion of the brightest students are now pursuing critical approaches. The times, it seems, they are a-changing and a-changing for the better.

Over the past decade, some important scholars and significant works have come out of critical communication. In cultural studies, social theo-ry, political economy, law and policy, history, critical communication scholars have produced some good work—even using the frequently dis-paraged quantitative methodologies. I refrain from naming names only for fear that I might neglect someone who deserves mention. Neverthe-less, the amount of quality work has been nowhere near as considerable as it should have been. On balance, I regard the legacy of critical commu-nication over the past decade as one of promise unfulfilled. Moreover, if by *critical* we mean an explicit skepticism toward dominant institutions (including the university), ideologies, and social relations and an implicit commitment to a more democratic, egalitarian, and humane social order, then the trajectory for critical scholarship in communication may be in the wrong direction altogether. Arguably, we are heading toward the point where the term *critical* may lose its bite and simply refer to bounded

Robert W. McChesney is an assistant professor in the School of Journalism and Mass Com-munication at the University of Wisconsin–Madison. The author thanks Inger L. Stole and Vivek A. Chibber for their helpful comments on an earlier draft of this article.

areas of study or methodologies, its link to movements for radical social change buried beneath so much academic rigmarole.

In this article I discuss what I see as several weak spots in critical communication research. Since *cultural studies* is currently the dominant component of critical communication research in the United States, it may seem that most of the criticism that follows is directed at it. That is not my intent. In fact, much of the criticism that follows could be extended to left scholarship in general. I must concede at the outset that I am guilty of some of the wrongs I am about to describe. The point of this exercise, then, is not to mount a self-appointed tribunal to locate and demonize poseurs and backsliders as much as it is an attempt to initiate a frank and comradely internal discussion among critical communication scholars about what we are doing and what we should be doing.

Critical Reading

The most striking weak spot in critical communication scholars' research is its insularity, which manifests itself in three distinct ways. First, many critical communication scholars reveal a shocking ignorance of the entire tradition of critical social and political thought over the past 200 years. For scholars who define themselves as part of the critical or radical tradition, this ignorance of their own history is absolutely unconscionable. Many critical scholars appear to have learned most of their theory from reading people like Foucault or Stuart Hall or, God help us, Baudrillard. The more adventurous may have taken a dive into the turbulent waters of Gramsci or Althusser. Few, however, have conducted systematic readings in Marx, Engels, Luxemburg, Lenin, Lukács, Adorno, Habermas, Veblen, DuBois, Sweezy, or Mills, or the broader traditions that they were responding to and influencing. Consequently, in the early 1980s critical communication scholars often acted as if they were making some enormous revolution in philosophy of science as they attacked mainstream research, when in fact they were merely acting out another episode—a small one at that—in a two-century intellectual battle. The lesson for critical communication scholars is clear: We must make sure that our students, regardless of their eventual area of specialization, receive a rigorous grounding in political and social theory, radical and mainstream. And we might want to take a good dose of the medicine ourselves. This should be regarded as nonnegotiable, the ante for admission to critical communication scholarship.

Second, many critical communication scholars have only the most hackneyed notions about the histories of the various social movements—labor, socialist, civil rights, feminist, and so forth—of the past 200 years. If the entire purpose of critical work is to not merely interpret the world, but to change it and change it radically, this ignorance defies comprehension. I cannot help but roll my eyes when I hear some "critical" academic dismiss all previous labor and socialist movements as racist or sexist, and therefore insufficiently pure, based at best on anecdotal evidence, when that person has never lifted a finger for social change and probably never will.

Third, and in a less historical vein, critical communication research is also far too isolated from critical scholarship in other disciplines and from the broader left intellectual community. This insularity is revealed in all too many papers by critical scholars that have predictable reference lists of the usual suspects. We simply need to push ourselves harder to expand our vistas (as much as our vitae) and our audiences. In view of the inter-disciplinary nature of both critical work *and* communication, this insulari-ty is indefensible. As much as anyone, critical communication scholars should be the thread that links the various patches of critical scholarship together. Moreover, this isolation partially explains why so few critical communication scholars are ever cited by scholars in other disciplines. Sadly, much like our mainstream colleagues, too many of us are simply producing inbred and unimportant work for a handful of colleagues who are doing the same.

Barriers and Barricades

Yet I do not advocate that critical communication scholars merely ape their compatriots in the more established academic disciplines. We are justifiably sensitive about the relative lack of importance accorded com-munication research—mainstream and critical—at major U.S. research universities. On my campus, with two of the most productive and presti-gious communication programs in the nation, communication is seeming-ly regarded by the pooh-bahs in history, political science, and sociology as having roughly the same intellectual merit as, say, driver's education.

Some communication scholars—mainstream and critical—have reacted to this by erecting the same type of intellectual barriers-to-entry as main-stream social scientists in order to justify and exalt the field's existence. To the extent that critical communication scholars engage in this exer-cise—for example, adopting trendy but mostly incomprehensible jar-gon—is the extent to which critical communication scholars insure their irrelevance, even asininity. In short, we cannot succeed in academia by imitating the established fields. We have to boldly strike out in a popular and interdisciplinary manner that runs directly counter to the dominant trends in the academy.

Moreover, the present intellectual climate, typified by the rightward re-treat of formerly left intellectuals, is one that needs to be examined though not embraced by critical communication scholars (Geras, 1990; Miliband & Panitch, 1990; Wood, 1986). Without a deep theoretical or his-torical perspective, or any experience as political activists, young critical communication scholars can get swept up in this current and come to regard the radical overthrow of capitalism as impossible, or irrelevant, or perhaps even dangerous.

Severed from an explicit commitment to an anticapitalist future, critical communications research becomes a caricature of radical intellectual work. Cultural studies, with its de-emphasis of the institutional relations of cultural production and emphasis on cultural consumption (at worst, the capacity of audiences to decode messages "oppositionally"), has been particularly guilty in this regard (Savage & Frith, 1993; Seaman, 1992). I have been to two conferences where papers were presented that employed cultural studies "methodologies" to help advertisers "improve" the effectiveness of their messages. In grand irony, a major advertising agency in Chicago now uses a seminal work in cultural studies in its training sessions, as a text in how advertisers might better influence consumers.

Divorced from any notion that radical social change is possible, the relativism of the early 1980s has gradually degenerated into a politically fashionable antirationalism. Some of this has been fueled by the influence of postmodernism and poststructuralism. The argument that rejected the universality and neutrality of mainstream social science has evolved into a rejection of the very notions of truth, rationality, reason, science, logic, or evidence. This is a phenomenon across the academic left, but it is most pronounced in literature and communication, where symbolic interpretation has traditionally been more influential (Lubiano, 1992). This is a striking rejection of the enlightenment project, of the notion that all humans share the capacity to reason and that reason could be used to liberate the species (Chomsky, 1992b). To the contrary, we are informed that science is a white male invention to maintain hegemonic rule over the "other," and that there is no such thing as truth. Linked to antiracist and antisexist sentiments, these irrationalist claims can even sound progressive, but critical communication scholars need to evaluate them with the greatest care (Epstein, 1991). Historically, reason has been the weapon of the oppressed to do battle with their oppressors. Moreover, the last great wave of irrationalism in the world led directly to the emergence of fascist political theory. If we abandon the notion that all people share certain fundamental qualities and capacities and are capable of reason, we open the door to justifications for treating people fundamentally differently. And when that happens, it is rarely the dispossessed who call the shots (Ehrenreich, 1992).

Privileging Journalism

In communication this irrationalism is perhaps most apparent in the low regard accorded the practice of journalism. Now I do not mean to defend the manner by which journalism is practiced in the United States; for the most part David Barsamian (1992) is on the mark when he characterizes journalists as "stenographers to power." However, the alternative is not to dismiss journalism per se as simply a series of stories or myths holding no more intrinsic claim to truth or importance than advertisements, baseball cards, pulp novels, or Nazi propaganda tracts. Indeed, critical communication scholars need to frankly *privilege* journalism as a form of discourse. It must be privileged because it is virtually impossible to conceive of a democratic society without journalism playing a central role in the political process. The same cannot be said about many other forms of communication. Hence, if we are serious about participatory democracy, we must be serious about journalism and its relation to democracy. We must examine why journalism does not serve democratic ends at present and what needs to be done so we may have a more democratic journalism in the future. In fact, in a more sane world, journalism would be among the most prestigious fields in the academy.

This commitment to democratic communication and to journalism also provides critical communication scholars with an entrée to political activism that is unique among left academics. We are poised to work with activist groups and the public at large to help them decode the media critically, use the existing media systems to best effect, and produce their own communications. There is an enormous need for this type of expertise and it is the special mandate of critical communication scholars to fill this role (Ryan, 1991). In addition, this type of practice also keeps critical communication scholars from getting too wrapped up in their own academic verbiage and too concerned with only academic feedback to their work.

Granted, helping a labor union or an environmental or feminist group work with the media might not generate many brownie points at the moment of tenure review, but what is the point of being a critical scholar after all? We all have to take care of professional business—one glance at my curriculum vitae will indicate that I have internalized this maxim—but we have to remember never to put the proverbial cart before the horse. The Union for Democratic Communication is an organization explicitly chartered to assist this marriage between communication scholars and activists. It is a group all critical communication scholars should support.

Critiquing Capitalism

Privileging journalism and engaging in activism, however, are not enough for critical communication scholars. We also need to incorporate capitalism into the heart of our calculations. With the collapse of the communist regimes in Eastern Europe, a new generation of liberals is informing us that the term *capitalism* is meaningless and should be dispensed with, as capitalism is the only manner by which a modern industrial economy can possibly be organized (Fukuyama, 1992; Herman, 1992, pp. 111–112; Rorty, 1992). This strikes me as an extraordinary abrogation of intellectual responsibility, conceding the most important political and intellectual question of our era before we have even subjected it to scrutiny. On a worldwide scale we presently see the collapse of living standards, the demise of the public sector, and the deterioration of the environment, all of which are contributing to immense human suffering and staggering nationalist and racist tensions. None of this makes much sense if ascribed to the vague workings of an industrial economy; it begins to come into focus if we understand these phenomena as, at the very least, the partial consequence of the limitations of a profit-driven economic system. Hence, it seems that a hard and ruthless examination of capitalism is as necessary now as ever.

Moreover, for critical communication scholars it seems that any attempt to generate a notion of a more democratic communication system, and therefore a more democratic society, must deal immediately and constantly with the inherent conflicts between a class-divided capitalist economic system and the egalitarian components of a democratic political system (MacPherson, 1977). We stand on the threshold of an era in which society finally has the communication technologies to democratize societies in a manner unfathomable only a few years ago. The great barrier to the democratic application of these technologies is the corporate control on communication and the relative powerlessness of the public in capitalist societies. I realize that class and capitalism are unfashionable terms for critical analysis at present, and have few defenders in the academy. Nevertheless, I have yet to see any evidence that a progressive commitment to smash racism and sexism or to protect the environment can ever amount to much unless part and parcel of a broader social movement to reconstruct the political economy and address the extraordinary class imbalances of capitalist societies. Accordingly, I have yet to see any evidence that critical scholarship can be meaningful unless it is at least implicitly anticapitalist.

In sum, what critical communication scholars need to do is regard themselves, first and foremost, as public intellectuals (Jacoby, 1987). We have to remember that as recently as the 1950s it was virtually impossible for a radical intellectual to get a tenured position in a U.S. university; we have to realize that we have a special obligation to utilize our privileges in a selfless manner. The immediate future is not necessarily going to be one of great victories for the political left or for critical scholarship. In the short term, we need to keep our historical perspective and struggle to maintain our intellectual and spiritual sanity. More broadly, we have to dig in for the long haul, recognizing that if we conceive of social change in terms of only a few years or a decade, or even our own lifetimes, then our time frame of reference is too short (Chomsky, 1992a, pp. 139–161).

> *If you should think this is Utopian, then I would ask you to consider why it is Utopian.*
>
> —Bertolt Brecht (quoted in Raboy, 1990, p. 357)

References

Barsamian, D. (1992). *Stenographers to power: Media & propaganda*. Monroe, ME: Common Courage Press.

Chomsky, N. (1992a). *Chronicles of dissent: Interviews with David Barsamian*. Monroe, ME: Common Courage Press.

Chomsky, N. (1992b). Rationality/science. *Z Papers, 1*(4), 52–57.

Ehrenreich, B. (1992). Truth, justice, and the left. *Z Papers, 1*(4), 58–64.

Epstein, B. (1991). "Political correctness" and collective powerlessness. *Socialist Review, 21*(3), 13–35.

Fukuyama, F. (1992). *The end of history and the last man*. New York: The Free Press.

Geras, N. (1990). *Discourses of extremity: Radical ethics and post-Marxist extravagances*. London: Verso.

Herman, E. S. (1992). *Beyond hypocrisy: Decoding the news in an age of propaganda*. Boston: South End Press.

Jacoby, R. (1987). *The last intellectuals: American culture in the age of the academe*. New York: Basic Books.

Lubiano, W. (1992). To take dancing seriously is to redo politics. *Z Papers, 1*(4), 17–23.

MacPherson, C. B. (1977). *The life and times of liberal democracy*. New York: Oxford University Press.

Miliband, R., & Panitch, L. (Eds.). (1990). *The retreat of the intellectuals: Socialist register 1990*. London: Merlin Press.

Raboy, M. (1990). *Missed opportunities: The story of Canada's Broadcasting Policy*. Montreal: McGill-Queen's University Press.

Rorty, R. (1992). The intellectuals at the end of socialism. *The Yale Review, 80*(1), 1–16.

Ryan, C. (1991). *Prime time activism: Media strategies for grassroots organizing*. Boston: South End Press.

Savage, J., & Frith, S. (1993). Pearls and swine: The intellectuals and the mass media. *New Left Review, 198,* 107–116.

Seaman, W. R. (1992). Active audience theory: Pointless populism. *Media, Culture & Society, 14*(2), 301.

Wood, E. M. (1986). *The retreat from class: A new "true" socialism*. London: Verso.

Rethinking Political Economy: Change and Continuity

by Eileen R. Meehan, University of Arizona, Vincent Mosco, Carleton University, and Janet Wasko, University of Oregon

The explanatory strength of materialist scholarship generally, and of Marxist scholarship particularly, is rooted in its ability to provide compelling accounts of economic structures and their effects on politics and ideology. For the left academy, political economy has been the starting point of scholarship, even for the Western Marxists who ritually ascribe the worst abuses of vulgar Marxism to political economists. In communications, political economy has long been a mainstay of the critical approach. Political economists have produced considerable research examining telecommunications and media as industries. This research uncovers connections between ownership, corporate structure, finance capital, and market structures to show how economics affects technologies, politics, cultures, and information. In the U.S., where media are always and primarily businesses, the approach has been crucial to understanding why we get what we get.

While the cliché reminds us that the world is ever changing, recent events suggest that even clichés can be meaningful. The dramatic changes in Eastern Europe and in the former Soviet Union have been widely reported as the demise of communism, socialism, and Marxism—as the ultimate proof that Marx, Lenin, and Mao were wrong. Some conservative colleagues argue that the collapse automatically invalidates our research, leaving us with no empirical or ethical ground from which to argue. The new world order, we are told, is firmly rooted in democracy, which naturally implies capitalism. Positing a new "positive" political economy, they apply neoclassical economic models to political and social behavior. From this point of view, buttressed by a few Nobels in economics, we are passé.

The collapse of communism is but one challenge; political economists also face two other "real world" challenges. Two decades of economic stagnation in the capitalist world have left Keynesians and monetarists

Eileen R. Meehan is an associate professor in the Department of Media Arts at the University of Arizona. Vincent Mosco is a professor in the School of Journalism and Mass Communications at Carleton University. Janet Wasko is an associate professor at the School of Journalism and Communication at the University of Oregon.

completely perplexed. In the last decade, financial markets boomed, the industrial sector rusted, and the agricultural sector withered. In the state-sanctioned speculative frenzy of the 1980s, Western capitalism appeared to be eating itself. Attracting much attention from critical political economists, these contradictions of Western capitalism remained untheorized by the conservatives.

The second challenge rises from the transformation of the former third world into a new set of haves and have-nots. In the newly industrialized nations of Asia (NICs), the "economic miracle" achieved by local capitalist elites has eroded the economic dominance of Western nations in both the world market and Western domestic markets. Within the NICs, the gap between the elite and the majority continues to widen. In other regions of the old third world, the majority of people are excluded from the world economy such that their national relations to the West no longer merit the term *dependent*. These developments generally, and their implications for communications specifically, make it essential that we press forward with the process of rethinking political economy.

But our challenges do not issue solely from changing world conditions; from within the academy, postmodernists have mounted a serious challenge. From their perspective, both conservative and critical academics are equally passé. Put simply, communism's collapse and capitalism's contradictions sound the death knell of modernism: The monolithic states, economies, cultures, and identities of modernism are swept aside by the new age. This new postmodernist age fractures, deconstructs, and recycles everything that modernism posits as stable. Why fuss over the financing of the latest Disney film? What really matters is one's personal, whimsical appropriation of that film as a signifier shifting over no signifieds.

We agree with both camps on one point: Things have changed. We disagree with their interpretations of those changes. In this essay, we argue that an adequate analysis of current political, economic, and cultural changes requires a firm grounding in political economy. We note, as Marx noted, that political economy is but one approach and that it is best adapted to the study of economic structures, relations of production, and political systems that protect economic structures. In communication industries where culture and audiences are manufactured, political economy provides an essential first step by analyzing the conditions under which manufacture and distribution occur. But political economy is ill adapted to analyzing cultural commodities as artifacts or audience commodities as cultural collectivities; materialist cultural studies has developed the appropriate methodologies and theories for such research. Recognizing the need for these two branches of critical communications research to work more closely together, we reflect on methodological difficulties that separate the two and problematize the general research project undertaken by political economists of communications. We conclude by arguing for a holistic approach to the study of culture and communication—an approach firmly rooted in political economy. Let us begin, then, by clarifying our terms.

What Is Political Economy?

Definitions of political economy range from the most general to the concrete. The former is exemplified in the idea that political economy is about survival—how societies organize themselves to produce what they need to reproduce themselves—and control—how societies maintain order to meet economic, political, social, and cultural goals. Specifically, political economy examines the production, distribution, and consumption of resources, including communication and information resources. These definitions and their variations provide a starting point for understanding political economy. One dimension of rethinking the approach involves broadening these conceptions by considering four essential dimensions of political economy: history, the social totality, moral philosophy, and praxis.

Political economy has consistently focused on the process of *social change*. For classical theorists like Smith, Ricardo, and Mill, this meant understanding the transformation from feudalism to capitalism. For their Marxist and institutional critics, it meant uncovering the dynamics of capitalism, including its cyclical reconstitution to fend off crises, class upheavals, and the resulting growth of monopoly capital and of a massive state apparatus. Political economists debated the wider significance of these developments: Did they portend a transition to socialism, full democracy, authoritarianism? In contrast, the positivist economists jettisoned history and concentrated on static, descriptive models. The political economy of communication is crucial to our current understanding because a wide range of analysts have identified the production, distribution, and use of communication and information as a central dynamic force in the global political economy. The challenge for political economy is to examine unflinchingly the range of historical trajectories—including globalization, postindustrialism, post-Fordism, and postmodernism—that challenge some of our accepted notions.

Starting with the classical political economists, the approach has been concerned with a holistic analysis, with comprehending the *social totality*. In concrete terms, this meant understanding the relationship among commodities, institutions, social relations, and hegemony. This characteristic of the political economic approach has faced and continues to face a range of broad and specific challenges. The broadest include the questions of determination among these elements. Furthermore, political economists reaffirm the need to examine totalities as they broaden their sense of what comprises them.

These points are carried forward in the consideration of specific challenges. Traditionally, the *commodity* form has been defined by the use of wage labor to produce goods (from sitcoms to data bases), which are valued for what they bring in the marketplace. This notion has been broadened in the debate over how to think about the audience (Smythe, 1977) and about symbolic commodities (Baudrillard, 1981). Moreover, the growth of spatial forms of analysis leads to reflections on the interpenetration of commodified, public, and personal space. Capital, labor, and the state have traditionally constituted the *institutional* base of political economic analysis. Here challenges arise from the deepening tensions brought about by globalization, which substantially reconfigures the relationship between capital, labor, and the state. Moreover, the integration of once distinct industry sectors—such as publishing, broadcasting, telecommunications, financial services, marketing, etc.—blurs the boundary between those institutions included and excluded from the political economy of communication. Similarly, the *social relations* of communication—which, in political economy, have been taken up largely along with the class question—now face challenges ranging from those who press for systematically incorporating race and gender into the analysis to those who question the value of any central categories, opting instead for a shifting spectrum of subjective characteristics. Finally, *hegemony* comprises the common sense, taken-for-granted reality that Gramsci used to build a case for going beyond coercion in the analysis of social control. Political economists generally accept Gramsci's point but have yet to take a close look at what it means for understanding power as intersubjective and for our methods—that is, how to link political economy to ethnography.

Political economy embedded its historical and institutional analysis in a *moral philosophical* outlook. Though Marx, particularly in his early work, exemplified this integrative thinking, neither his work nor the research inspired by him are alone in holding this view. Adam Smith held a professorship in moral philosophy and his book *The Theory of Moral Sentiments* (1759/1986) established a program for examining values as forces for social change, in this case the moral code of individual acquisitiveness, and for taking up values as goals and guides for social action. His analysis of the need for state protection of workers abused by the factory system and for consumers abused by unregulated bank speculation holds up well today. The assumptions embodied in a moral perspective have inspired considerable debate about the origins, originators, and standpoints of a moral position—setting aside the seeming irony of how any moral sensibility can come out of this bloody century. Despite these challenges, political economists of communication have maintained a firm commitment. For Golding and Murdock, it is the distinguishing characteristic of political economy: "perhaps most importantly of all, it goes beyond technical issues of efficiency to engage with basic moral questions of justice, equity, and the public good" (1991, pp. 18–19).

Building on this moral concern, political economy is committed to *praxis,* that is, it seeks to transcend the distinction between research and social intervention. Challenges to this view range from the traditional positivist separation of fact and value to the Weberian compromise that would see values as influential but distinguishable from the reality that research seeks to represent. Political economy starts from the view that research is a form of both labor and social intervention. Therefore, research is enmeshed in the very social totality that it aims to examine (we produce books under a specific set of industry conditions) and thus cannot avoid, even if it were to try, the value questions that saturate this totality. The goal is therefore more than a simple reflection of social reality but a self-reflexive process of questioning and acting on the object of analysis.

With these traditional and strong commitments to history, social totality, moral philosophy, and praxis, political economy is well prepared to rise to the challenges posed by economic crisis, national transformations, and reorganizations of the global order. Political economists of communication are well positioned to research the deepening divisions between communication haves and have-nots, the growth of the panopticon, and the role played by entertainment in the creation of hegemony. Driving such work is the position's strong commitment to the self-reflexive production of knowledge and to the progressive use of knowledge by people acting in the world. Perhaps because of political economy's four-square assumptive base, political economists initially dismissed the challenge issued by postmodernist academics as frivolous. For postmodernists, history was dead; the social totality was a fiction shattered by the advent of the new age; moral philosophy was only worthy of deconstruction; and praxis was replaced by posing. The conditions for paradigm dialogue were bleak indeed.

Postmodernism and Political Economy

Despite its concern about style, irony, superficiality, and intertextuality, postmodernism was rooted on a surprisingly economistic notion. The epochal transformation of the modern into the postmodern was a function of radical economic and technological change (Baudrillard, 1981; Lyotard, 1984). Where modern societies were industrial and national, postmodern societies became information-based and global. As the information economy and its interactive technologies destroyed the industrial economy and its assembly-line factories, so were the trappings of industrial society destroyed. Industrialism's power relations, categorical hierarchies, and hegemonic processes shattered, leaving the postmodern landscape strewn with bits of the social categories, economic roles, personal identities, cultural definitions, and hegemonic ideologies that once comprised the rigid structure of industrial modernism.

Thus categories like race, class, or gender become options in constructing one's temporary identity. In the same way that Madonna takes on the

appearance of Marilyn Monroe or Marlon Brando, of blessed virgin or material girl, so too a postmodernist population daily constructs itself using
the fragments of the old categories in a mix-and-match process driven by
whim. In this new schizophrenia, we reassemble selves, cultures, texts,
power relations, and societies at will; gone are the social consequences
associated with the old categories. Only the new sensibility and the individual exert power in a postmodern world. With the destruction of
Fordist industrialism, the economic base loses its determinative power,
hence the irrelevance of political economy and all other modernist forms
of knowledge production.

Most U.S. postmodernists base these claims on changes in media artifacts and changes in theoretical discourses about industrialism, postindustrialism, and informatics. With their expertise in texts, their disinterest
in political economic research is understandable. Yet, their claims suggest
a faith in a vulgar economism long renounced by most political economists. The reliance on discourses and artifacts for evidence—without
solid analysis of markets, industries, and transindustrial structures—seriously weakens the postmodernists' position. This is not to suggest that
the postmodernists have been inaccurate in observing cultural shifts.
Rather, they have been quite loose in theorizing the connections between
the cultural and the economic. As Harvey (1990) demonstrates, careful
and detailed analysis can trace connections between shifts in the economy and changes in cultural sensibility. For Harvey, drawing on Bluestone
and Harrison (1982), the key to the postmodernist sensibility lies in the
sociocultural consequences of economic restructuring.

That restructuring began in 1975 through chronic recession, aggressive
exportation of industries, and waves of mergers across and within industries resulting in increased concentration of ownership and operations.
Such conditions increased instability of employment, which was exacer-
bated by deskilling of workers, deregulation of industries, and decertifi-
cation of unions. The expanding service sector failed to absorb all of the
unemployed. Most service sector jobs offered low wages, few benefits,
and job insecurity. Economic restructuring continues at an uneven pace,
affecting some workers, elites, and industries more than others; the pace
is also uncertain as governmental positions shift, affecting policies on
economics, taxes, and globalization.

The uneven and uncertain pace of restructuring creates different working conditions for different strata of the population in different regions of
the nation. For many, the new working conditions have placed considerable pressure on disposable income, forcing an influx of previously protected categories of people into the work place. To the degree that these
changes pervade society, people rethink desires, values, expectations, visions, and possibilities. When communicated and collectivized, this rethinking becomes the basis for emergent sensibilities, subcultures, and
structures of feelings (Thompson, 1963; Williams, 1980, 1982). The result
is not a single emerging sensibility that struggles with a previously domi-

nant one. Rather, multiple ideologies emerge, struggle with each other, and confront extant ideologies that also struggle as the dominant ideologies conflict with alternative or oppositional ideologies rooted in material conditions. This theorization is more complex than that offered by the postmodernists. It builds on materialist cultural theory and accounts for the proliferation of subcultures, both modernist and postmodernist. It also recognizes cultural sensibilities as constructs that are creative, communicated, and grounded.

Discovering how such sensibilities are grounded seems more like a job for ethnographers than political economists. However, much of the postmodernists' evidence depends on readings of media commodities. The elucidation of those grounds—of the economic conditions constraining the manufacture and circulation of media commodities—comprises one specialization of political economists studying communication. Indeed, corporate structure is one economic condition that can have a significant "postmodernist" effect on the construction and circulation of media commodities (Meehan, 1991). Under Reaganism, media corporations increasingly restructured themselves as large, multimedia conglomerates operating on a transindustrial and transnational scope. Cost efficiency dictates that, with so many internal markets to feed, each expensive project must provide products to many of the conglomerate's media subsidiaries. Thus, corporate structure encourages the fragmentation of artifacts and the recycling of those fragments across media, resulting in a commercial version of the postmodernist intertext. A brief sketch of how Warner Communication, Inc.'s (WCI) film "Batman" fed WCI's operations in the music industry illustrates this point.

Costing $30 million, the R-rated "Batman" was generally cited as having symbolically clinched the deal merging Time, Inc. (TI) and WCI. Not only did the film provide onscreen advertising for TI's *Time* magazine, it also generated a considerable amount of product for WCI's subsidiaries in print publishing, videocassette manufacture, video distribution, and licensing. For WCI's subsidiaries in the music industry, "Batman" was a gold mine. WCI's holdings in the music industry include the world's largest operations in song publishing, numerous record labels, video production, and record manufacturing. The company also holds contracts with musical artists across a wide range of genres. All of these resources were brought to bear on "Batman."

Like all blockbuster films, "Batman" had an orchestral score. Danny Elfman's score was broken out of the soundtrack, marketed as a separate album on a Warner label, and pressed by WCI's manufacturing facility; the music was published by a Warner subsidiary. In an unusual move, WCI also hired its contracted artist Prince to write incidental music for the film. Using rap styles, Prince's music provided the background for such scenes as Joker's vandalization of an art museum. Augmented by music inspired by the film, Prince's work was broken out as a crossover album and was marketed to both the white rock market and the black rap market. Like the manufacturing, the music publishing was again handled in house. From the album, WCI broke out Prince's "Batdance" as a single. "Batdance" blended dialogue from the film with Prince's usual concerns about the ambiguity of personal identity, adding rap elements to Prince's rock stylings. "Batdance" was also packaged as a WCI video. Garbed as half-Batman and half-Joker, Prince enacts the song while surrounded by dancers costumed as Batmen, Jokers, and Vicky Vales. This nicely post-modern pastiche, complete with simulacra and schizophrenia, played in heavy rotation on MTV. WCI, which sold MTV to Viacom, retained an in-direct interest through its stock in Viacom. In this fashion, the single arti-fact "Batman" is mined for product capable of feeding WCI's other media operations and interests.

In economic terms, "Batman" was less a film than a product line. In stylistic terms, the product line was a postmodern intertext: A complex web of references, images, and meanings that could be activated at will and be manipulated at whim. Two distinct processes occurred: First, "Bat-man" was designed as product line; and second, elements of that product line were broken out of "Batman" and recycled over WCI's multiple media operations. The release of the film was carefully timed so that it and these recycled fragments generated synergy between the film, its fragments, and the entire product line. Entering that synergy, Batman consumers apply cultural sensibilities created and learned from our so-cioeconomic positionings in order to assemble our own personal Batman. Thus corporate structure not only produces commercial intertextuality but also provides industrial underpinnings for postmodernist style.

Criteria and Methodology

This analysis of WCI's "Batman" suggests how economics influences media and culture. Termed institutional analysis, such work traces indus-trial structures and their effects. Another classic approach, instrumental analysis, traces the personal and business networks within institutions. Both analytic modes rely on the same sources for data: government docu-ments, required corporate disclosures, trade journals, wholesale and re-tail sales lists, statements by employees and journalists, and promotional materials. Such sources are not without problems; analysis is only as reli-able as the data upon which it rests. In this section, we reflect on some problems in data collection and on criteria that undergird our research methods. We focus on methodology as one key to answering the four challenges noted in the beginning of this essay.

Despite complaints that political economists publish few articles with methods sections, we do use specific methods and criteria in our re-search. Unlike positivist paradigms, political economy tends to treat its methods and criteria implicitly; practitioners are expected to follow crite-ria implicit in the paradigm and then to select the method best suited to the problem. In studying media, political economists synthesize the methods of social research (e.g., sociology and history) with the analytic methods of Smith, Ricardo, Marx, Hilferding, etc. As the research focus broadens to include questions of race, gender, and hegemony, the array of methods will undoubtedly expand.

Whatever the method, the process of finding and analyzing data is and should remain as rigorous for political economists as for other media researchers. Research sources and data must be evaluated; the criteria for that assessment must be made explicit. Four criteria for assessing evidence undergird political economic research: authenticity, credibility, representativeness, and meaning (Scott, 1990). While political economists may not always articulate these criteria systematically, they are usually aware of their importance.

The best information comes from direct involvement, or what Scott (1990) calls proximate or direct access. Personal interviews and other primary documentation provide the most authentic and reliable data. But this type of documentation is not always available. More often, various sources mediate or provide indirect access to evidence; such sources differ in credibility and representativeness. For example, trade publications provide extremely useful documentation that cannot be accessed elsewhere. Yet, in terms of credibility, these sources are closely aligned with the particular industries and must be read accordingly. The popular press—too often the only source of "armchair political economists"—is valuable but is unacceptable as the sole source of data. Our own media research has disproved the assumption that the press provides reliable, credible information. Similarly, reminiscences and popularized "insider" stories provide illustrative anecdotes, but such accounts are constrained by the need to entertain while avoiding libel suits. Some even present unsubstantiated dramatic incidents or fabricated dialogue (see Bach, 1985; McClintick, 1982). In contrast, governmental documents are often crafted as collections of data, detailing communication industries and corporate activities. Such "base line" documents are augmented by white papers and hearings, which may try to represent diverse points of view. In both types of documents, however, political agendas can direct the choice and outcome of evidence presented.

Clearly, revealing and understanding the economic dynamics of media is not often as glamorous, dramatic, exciting, or easy as some other forms of analysis. As Guback (1978) notes, texts are infinitely more accessible than data describing the production context. Access to the crucial materials remains one of the most onerous constraints on research; for example, physical proximity to materials is a common problem. More threatening to research is corporate obfuscation, government secrecy, and privatization. If the numerous congressional hearings on corporate disclosure practices found disturbing trends emerging in the 1970s, they would be even more shocked by today's commodification of public information and decreasing corporate accountability. Not only has corporate data become more problematic; increasingly information about "public" institutions is available for pay only (e.g., Disclosure, Inc.'s sale of federal information). Globalization has further dispersed information for many industries, making it more difficult to gather what is available.

In contrast, *infotainment* is easily accessible. In the United States, popular media pay great attention to "show business." Weekly magazines like *Premiere* and *Soap Opera Digest* fill newsstands; television coverage ranges from the daily "Entertainment Tonight" to CNN's hourly "Hollywood Minute." At best, this material offers only superficial coverage of corporate activities; at worst, such infotainment emphasizes voyeurism, consumerism, or fan cultism.

As political economists, we want to *really* know what happened, who was involved, and what interests were served. Since all sources provide information—data strained through a point of view, vested interest, or argumentative stance—we must utilize multiple sources and systematically evaluate data against Scott's (1990) four criteria. As accessibility becomes more problematic, multiple sources and explicit criteria are more crucial to ascertain the reliability of data. But political economists want to know more than who, what, where, and when. We also want to know why. And that requires good, solid analysis to reveal the basic contours of cultural, political, and economic continuity and change.

Change and Continuity

In rethinking political economy, we have examined the essential elements of any paradigm: an assumptive base, research methods, and current challenges. Political economy's grounding in history, moral philosophy, social totality, and praxis orients researchers to the study of social change through economic restructuring. The integration of historical, social, and analytic methods provides tools to uncover and explain structural continuity and structural change. This prepares us to research communism's collapse in the second world, economic stagnation in the first world, and economic bifurcation in the third world. As communications are increasingly central to both change and continuity, communications specialists in political economy have become increasingly central to understanding the social totality.

A strong theoretical challenge to the notion of social totality has been mounted by the postmodernists. Political economists have responded by rethinking the concept and by pursuing research on commercial intertexts. Drawing on labor history and materialist cultural studies, a fuller and more sensitive accounting for apparent changes in sensibility has been advanced. This enriches political economy by fostering interdisciplinary work with fields that trace the history of consciousness and culture, pointing the way to methodological linkages with ethnography. Traditional political economic research into postmodernism opens up the commercial side of intertexts and lays a base for interdisciplinary work with cultural scholars.

These challenges encourage a reassessment of methods and criteria. By specifying the criteria of authenticity, credibility, representativeness, and meaning, we evaluate frameworks traditionally used by political economists and also encourage a greater systematicity among researchers. By using multiple sources, political economists ensure a rich—though potentially contradictory—data base. By evaluating sources and deploying data cautiously, we increase the reliability of our data and the validity of our analysis. This rethinking should strengthen our future research.

Our rethinking of political economy generates a seemingly contradictory injunction: Go forward boldly but, in doing so, proceed with caution. This contradiction is apparent but not real. The best research in political economy generally, and in the political economy of communications specifically, has always involved bold theorizing held in check by careful research and cautious analysis. This combination served our research well in the past; it bodes well for research conducted in an indeterminate future created by the interweaving of change and continuity.

References

Bach, S. (1985). *Final cut: Dreams and disaster in the making of Heaven's Gate*. New York: New American Library.

Baudrillard, J. (1981). *For a critique of the political economy of the sign*. St. Louis, MO: Telos Press.

Bluestone, B., & Harrison, B. (1982). *The deindustrialization of America*. New York: Basic Books.

Golding, P., & Murdock, G. (1991). Culture, communication, and political economy. In J. Curran & M. Gurevitch (Eds.), *Mass media and society* (pp. 15–32). London: Edward Arnold.

Guback, T. (1978, February). Are we looking at the right things in film? Unpublished paper presented at the conference of the Society for Cinema Studies, Philadelphia, PA.

Harvey, D. (1990). *The condition of post-modernity*. Oxford, England: Basil Blackwell.

Lyotard, J. (1984). *The postmodern condition*. Manchester, England: Manchester University Press.

McClintick, D. (1982). *Indecent exposure: A true story of Hollywood and Wall Street*. New York: William Morrow.

Meehan, E. (1991). Holy commodity fetish, Batman!: The economy of a commercial intertext. In W. Uricchio & R. Pearson (Eds.), *The many lives of Batman: Critical approaches to a superhero and his media* (pp. 47–65). New York: Routledge, Chapman, and Hall.

Scott, J. (1990). *A matter of record: Documentary sources and social research*. Cambridge, England: Polity Press.

Smith, A. (1986). *The theory of moral sentiments*. Indianapolis, IA: Liberty Classics. (Original work published 1759)

Smythe, D. W. (1977). Communications: Blindspot of Western Marxism. *Canadian Journal of Political and Social Theory, 1*(3), 1–27.

Thompson, E. P. (1963). *The making of the English working class*. London: Gollancz.

Williams, R. (1980). *Problems in materialism and culture*. London: Verso.

Williams, R. (1982). *The sociology of culture*. New York: Schocken.

Back to the Future: Prospects for Study of Communication as a Social Force

by Dan Schiller, University of California, San Diego

To study communication is no longer to be concerned only with the contributions of a restricted set of media to the socialization of children and youth, or to buying and voting decisions. Nor is it only to engage with the ideological legitimations of the modern state. Now it is also, or can be, to make arguments about the economic logic and cultural forms of social development in general. For a variety of reasons, the study of communication as a social force has converged, directly and at many points, on an encompassing critique of contemporary capitalism.

The analysis of communication as a social force foregrounds the production and distribution of informational and cultural resources. It seeks to set the changing structures of ownership and control of the means of cultural production within larger, historically specific processes of social and cultural change within heterogeneous class society. It attempts, in particular, to comprehend the structural pressures and limits within and against which empirically identifiable social actors shape institutionalized communication processes. Analysis of communication as a social force is in the forefront of efforts both to track the political-economic significance of communication and information and to chart processes of ideological construction.

But does not the collapse of actually existing socialism signal, as has been widely publicized, that the theoretical leverage of approaches emphasizing political economy is dwindling into insignificance? Hardly. It should be recalled that political economy arose before the arrival of any kind of actually existing socialism. Even more important, political economy's emergence as a particularly powerful form of critique is a consequence of its sustained encounter not with socialism but with capitalism. Far from being wedded to any particular national incarnation of the socialist project, political economy has always drawn its inspiration and its force from this continuing theoretical and historical engagement.

Dan Schiller is an associate professor in the Department of Communication at the University of California, San Diego. He wishes to thank Susan G. Davis and Lora E. Taub for their helpful criticisms of earlier drafts of this article.

Capitalism and Communication

This continues to be true today. Capitalism as a political-economic system is in the midst of its most profound and potentially transformative crisis since the 1930s. Involved in this metamorphosis are an unparalleled range of issues: the stagnating economy, heightened economic rivalry, and the fiscal crisis of the state; the location, composition, organization, and wages of work; the viability of the continuing U.S. role as global policeman; continuing struggles for gender and racial equality; and the forms and practices of contemporary cultural production in which these and other issues are figured. Across this great span, analysts must make increasingly significant room for processes of communication. How may we begin to approach this immense and daunting intellectual engagement?

Initially, by clarifying its intellectual historical context. Similar to the critique of advertising's role in subordinating consumption to the requirements of capitalist production, the vital place of communication in the larger and historically dynamic mode of production was first broadly identified within a critique of neocolonialism. "The volume, form, and speed with which current electronic systems transmit intelligence," wrote Herbert Schiller, "have produced a qualitatively new factor in human and group relationships. Telecommunications are today the most dynamic forces affecting not only the ideological but the material bases of society" (1969, p. 33). And could not means of communication be approached as actively developing means of production in their own right? asked Raymond Williams (1980). Thus there began to be envisioned alternatives to the rather mechanical consignment of the media to something called the "superstructure," in accordance with their ascribed character as essentially ideological agencies. Wrote Williams,

> *Cultural work and activity, are not now, in any ordinary sense, a superstructure: not only because of the depth and thoroughness at which any cultural hegemony is lived, but because cultural tradition and practice are seen as much more than superstructural expressions—reflections, mediations, or typifications—of a formed social and economic structure. On the contrary, they are among the basic processes of the formation itself.* (1977, p. 111)

Williams also noted that the economic base, "to which it is habitual to refer variations, is itself a dynamic and internally contradictory process" (1977, p. 82).

These theoretical advances were subsequently idealized and displaced through a complex further devolution of the field toward the realm of culture and ideology. In the emergent intellectual context, communication continued often to be generative—but of "discourse" and "signifying practices," rather than of changing productive processes. A variety of creative efforts nonetheless began to be made to view communication as a growing structural force within the general political economy (Garnham, 1990; Smythe, 1977). These attempts were significantly enriched and compounded through a creative confluence with what had been an entirely separate discussion, originating in economics, regarding the role of information in contemporary society. From a generally conservative position, Daniel Bell, Marc Porat, and other advocates of postindustrial theory proposed that "information" had become the theoretical linchpin of societal evolution. No matter how inadequately conceptualized or empirically dubious—and postindustrial theory is deeply flawed on both counts—it became widely accepted that information lay at the core of the overall social process.

Some critical researchers began to emphasize the increasingly decisive roles played by information in economy and society (Mosco, 1983; Schiller, 1981; Webster & Robins, 1985), while others focused specifically on the still incomplete capitalization of cultural production (Garnham, 1990; Miege, 1989). In short, analysts began to attend to the implications of the fact that capital had not already taken over all cultural production. It became possible to see that in a campaign of historic proportions during the 1970s and 1980s, capital was extruding itself across a completely unprecedented range of cultural production—through privatization, transnationalization, and new information technologies. Communication study was thus extended, complicated, and enriched by a whole gamut of new issues thrown up by the admission of information into the field's effective scope. Such issues have continued to move center-stage, by virtue of their growing strategic policy importance. Sources of structural instability that had long been contained—deepening global stagnation together with mounting intercapitalist rivalry—erupted anew (Bluestone & Harrison, 1988; Kolko, 1988).

The growing centrality of communication to the political economy was occurring within and ramifying back onto massive changes in the social process. Social class relations—at once increasingly transnationalized, and sustaining deepening disparities within the United States itself—comprised the major axis of these changes. How could communication study respond? How could the recognizably generative political-economic potential of information technology be reunited with changing class relations in communication study? Impinging on levels of analysis ranging from phonemic utterances to the structures of transnational cultural production, social class, in particular, was both an objective reality and, perhaps even more important, a "prospect whose 'being' is never permanent or fixed" (Aronowitz, 1992, p. 24). The crucial task became, and remains, to explore the character and repercussions of this protracted, extended, and very complex interlock.

At present, study of communication as a social force, however, is not well suited for this work. This is, ironically, largely the result of this subfield's difficult, politically engaged intellectual history.

From "Mass Persuasion" to Communication Studies

Before the institutionalization of communication research, as late as the end of the 1940s, there developed diverse and often highly critical approaches to the study of communication as a social force. As Jesse Delia observes, numerous scholarly studies were made of "media institutions as parts of a larger social process," and these suggested that agencies of communication were "having broad impact on patterns of everyday life and the creation of a national culture" (1987, p. 33). Relying on the unprecedented documentation produced by a full-blown Federal Communications Commission investigation, for example, Danielian (1939) proffered a graceful but grim portrayal of the economic history and structure of the U.S. telephone industry. In a chapter written for the President's Research Committee on Recent Social Trends, Willey and Rice (1933) noted how growing "concentration" of "agencies of mass impression" (p. 216) was strengthening the ability to control individual behavior. Not "social desirability," the authors noted, but "competitive forces" had resulted in "an all pervasive system of communication from which it is difficult to escape" (Willey & Rice, 1933, p. 217). Pioneering empirical investigators of modern-day community, Robert Lynd and Helen Lynd (1929) charted the explosive growth and disruptive impacts of "inventions remaking leisure," including radio and film. Scholars did not as yet disdain the notion of powerful media.[1]

It was the threat of fascism and, most important, the subsequent realities of a second total war—which brought together many of the leading investigators of communication processes from agencies such as the Office of War Information and the Office of Strategic Services—that focused this variegated, often critical research tradition on the urgent and commanding issue of propaganda. "Nations have seized upon communication as a prime instrument of social control under modern conditions," worried Riegel (1934). "They are assuring themselves of the control of transmission facilities and of news, as well as mobilizing accessory forms of propaganda, with the purposes of forging an obedient and patriotic mentality in the population, and of spreading advantageous propaganda outside of the state as an instrument of national policy" (p. 211). At this time

---

[1] While it may be true, as Chaffee and Hochheimer (1987) have asserted, that the hypodermic needle theory of media impact (against which the limited-effects paradigm purportedly reacted) is itself an invalid reconstruction by subsequent scholars, there existed a widespread belief in powerful media during the 1930s and early 1940s. Part of the confusion may stem from the fact that Chaffee and Hochheimer confine their investigation of the period before 1940 mainly to "empirical investigators of political behavior and mass communication" (p. 95), which artificially (and anachronistically) truncates the range of actually available writing and scholarship.

scholars of diverse stripes continued to bring explicit normative commitments to their work, and without any concession of scholarly rigor.

At first—through the 1930s and into the later 1940s—fears about powerful media in the service of an authoritarian state continued to radiate
expansively outward in scholarship, along spatial and social axes. Propaganda, in the service of national socialism and corporate advertising, was
studied in German and U.S. contexts. Even the term *communication* was
rarely used to describe the emergent field of inquiry until after World War
II (Delia, 1987). Instead, the reality of "mass persuasion" remained prominent within mainstream domestic research. "Never before the present day
has the quick persuasion of masses of people occurred on such a vast
scale," Merton, Fiske, and Curtis observed (1946, p. 1); neither the practitioner nor the academic student of social psychology, they declared flatly,
could "escape the moral issues which permeate propaganda as a means
of social control" (p. 185). Mass persuasion, they concluded (in a study of
Kate Smith's marathon sale of war bonds over CBS radio in 1943), gave
cause for serious concern over the prospects for "democratic values" and
"the dignity of the individual" (Merton et al., 1946, p. 188).

The cold war mobilization of the late 1940s and 1950s drove scholarly
concerns about the political and ideological consequences of mass persuasion to the margins. Mainstream social science, extending its embrace of
American exceptionalism (D. Ross, 1991), likewise replaced *class*—which
had begun to find a new emphasis during the Depression decade—with
stratification, a concept far more congruent with the mechanistic functionalism that had begun to prevail. Depression-era emphases on social
conflict and historical change were abandoned for approaches that
stressed consensus and stability. Within communication study, an individual unit of analysis was codified into orthodoxy, while abstracted patterns
of personal influence supplanted concern with any larger social process.
The field of communication, sustained by a whole host of corporate and
government patrons, was institutionalized as a scholarly discipline just as
analysis of communication as a social force was pushed underground.

In a context that combined McCarthyism, an apparently placid and
complacent domestic working class, and U.S. global hegemony, critical
scholars adopted a tight focus on what C. Wright Mills called "the power
elite" (1957). This emphasis—not surprising given the context—conferred the capacity for meaningful social action on a rather restricted
group of social actors. Its result was a foreshortened conception of communication as a social force, one that stressed the immediate institutional
structure of resource allocation and policy decision—the "ruthless unity"
of the "culture industry" that Adorno and Horkheimer (1972, p. 122) had
perceived[2]—to the general exclusion of social class relations. Apart from

---

[2] I thank Lora E. Taub for reminding me of this reference.

its vitally important emphasis on the very real abilities of the power elite to make undemocratic social decisions concerning resource allocation and policy, critical scholars were forced back on conceptions of social process that were in other respects strikingly similar to the atomized and abstracted portrayals of mainstream researchers. In the context of today's rapidly deepening class division, this emphasis—always more marked in the United States than in Great Britain—has become a major intellectual problem.

It is a problem whose solution has been little assisted by the development of cultural studies over the past generation. The conceptions of cultural practice that came to predominate in this body of work developed at the expense of emphasis on class, which had initially emerged as very powerful—especially in connection with cultural production—in the work of British historians and critics such as Raymond Williams (1961), Edward Thompson (1968), and Christopher Hill (1972).

Social historians, indeed, continue to provide the most substantial aid available in the continuing intellectual engagement with social class. In particular, a variety of empirically grounded studies have analyzed recurrent class conflict over the terms, purposes, and forms of cultural production. Such antagonisms may be found at the specific sites of cultural production, but also far beyond—in law and public policy, technology, and in other realms. Complex relations of social conflict, resistance and, at key points, incorporation, have proved central to the development of the contemporary communications industry. This is so both for the spread of commercial culture—in dime novels (Denning, 1986), movies (E. Ewen, 1980; Rosenzweig, 1983; S. J. Ross, 1991), sports events (Gorn, 1986), theater (Buckley, 1984; Saxton, 1989), parades (Davis, 1986), broadcasting (Cohen, 1990; McChesney, 1992), the newspaper press (D. Schiller, 1981), or style more generally (S. Ewen, 1988)—and for telecommunications (Martin, 1991; Norwood, 1990). This burgeoning literature—still far too little utilized by formal communications scholarship—will be of prime relevance in the larger process of revision we now face.

References

Adorno, T. W., & Horkheimer, M. (1972). *Dialectic of enlightenment.* New York: Seabury.

Aronowitz, S. (1992). *The politics of identity.* New York: Routledge.

Bluestone, B., & Harrison, B. (1988). *The great U-turn.* New York: Basic.

Buckley, P. (1984). *To the opera house: Culture and society in New York City, 1820–1860.* Unpublished doctoral dissertation, State University of New York at Stony Brook.

Chaffee, S. H., & Hochheimer, J. L. (1985). The beginnings of political communication research in the United States: origins of the "limited effects" model. In M. Gurevitch & M. R. Levy (Eds.), *Mass communication review yearbook* (Vol. 5, pp. 75–104). Beverly Hills, CA: Sage.

Cohen, E. (1990). *Making a new deal: Industrial workers in Chicago, 1919–1939*. Cambridge, England: Cambridge University Press.

Danielian, N. B. (1939). *AT&T: Story of industrial conquest*. New York: Vanguard.

Davis, S. G. (1986). *Parades and power: Street theater in nineteenth-century Philadelphia*. Philadelphia, PA: Temple University Press.

Delia, J. G. (1987). Communication research: a history. In C. R. Berger & S. H. Chaffee (Eds.), *Handbook of communication science* (pp. 20–98). Newbury Park, CA: Sage.

Denning, M. (1986). *Mechanic accents*. London: Verso.

Ewen, E. (1980). City lights: Immigrant women and the rise of the movies. *Signs, 5*(3), S45–S65.

Ewen, S. (1988). *All consuming images: The politics of style in contemporary culture*. New York: Basic.

Garnham, N. (1990). *Capitalism and communication*. London: Sage.

Gorn, E. J. (1986). *The manly art: Bare-knuckled prize fighting in America*. Ithaca, NY: Cornell University Press.

Hill, C. (1972). *The world turned upside down*. Harmondsworth, England: Penguin.

Kolko, J. (1988). *Restructuring the world economy*. New York: Pantheon.

Lynd, R. S., & Lynd, H. M. (1929). *Middletown*. New York: Harcourt, Brace & World.

McChesney, R. W. (1992). Labor and the marketplace of ideas: WCFL and the battle for labor radio broadcasting, 1927–1934. *Journalism Monographs, 134*.

Martin, M. (1991). *"Hello, Central?" Gender, technology, and culture in the formation of telephone systems*. Montreal and Kingston: McGill-Queen's University Press.

Merton, R. K., Fiske, M., & Curtis, A. (1946). *Mass persuasion: The social psychology of a war bond drive*. New York: Harper and Brothers.

Miege, B. (1989). *The capitalization of cultural production*. Paris: IG.

Mills, C. W. (1957). *The power elite*. New York: Oxford University Press.

Mosco, V. (1983). *Pushbutton fantasies*. Norwood, NJ: Ablex.

Norwood, S. H. (1990). *Labor's flaming youth: Telephone operators and worker militancy*. Urbana: University of Illinois Press.

Riegel, O. W. (1934). *Mobilizing for chaos: The story of the new propaganda*. New Haven, CT: Yale University Press.

Rosenzweig, R. (1983). *Eight hours for what we will*. Cambridge, England: Cambridge University Press.

Ross, D. (1991). *The origins of American social science*. Cambridge, England: Cambridge University Press.

Ross, S. J. (1991). Struggles for the screen: Workers, radicals, and the political uses of silent film. *American Historical Review, 96*(2), 333–367.

Saxton, A. (1989). *The rise and fall of the white republic*. London: Verso.

Schiller, D. (1981). *Objectivity and the news: The public and the rise of commercial journalism*. Philadelphia: University of Pennsylvania Press.

Schiller, H. I. (1969). *Mass communications and American empire*. New York: Augustus M. Kelley.

Schiller, H. I. (1981). *Who knows: Information in the age of the Fortune 500*. Norwood, NJ: Ablex.

Smythe, D. W. (1977). Communications: Blindspot of western marxism. *Canadian Journal of Political and Social Theory, 1*(3), 1–27.

Thompson, E. P. (1968). *The making of the English working class*. London: Pelican.

Webster, F., & Robins, K. (1985). *Information technology: A Luddite analysis*. Norwood, NJ: Ablex.

Willey, M. M., & Rice, S. A. (1933). The agencies of communication. In President's Research Committee on Recent Social Trends (Ed.), *Recent social trends in the United States* (pp. 167–217). New York: McGraw-Hill.

Williams, R. (1961). *The long revolution*. London: Chatto & Windus.

Williams, R. (1977). *Marxism and literature*. New York: Oxford University Press.

Williams, R. (1980). *Problems in materialism and culture*. London: Verso.

The Past and the Future of Communication Study: Convergence or Divergence?

by Everett M. Rogers, University of New Mexico, and Steven H. Chaffee, Stanford University

Rogers: Ten years back, we began a dialogue on communication as an academic discipline (Rogers & Chaffee, 1983) by referring to Berelson's (1959) "the field is dying" paper. We concluded that Berelson had been wrong, and that the multidisciplinary field of communication was becoming a discipline.

Chaffee: Not exactly. You felt that communication study was a discipline. I argued that the scholars who study communication are too diverse to constitute a single discipline. We lack the intellectual coherence of, say, psychology or literary interpretation—both of which contribute to communication study today, but which make radically different assumptions about what is to be studied, and how.

Rogers: Divergence necessarily comes with growth. In the United States alone, communication programs award 50,000 bachelor's degrees annually (up from 11,000 in 1970), 4,000 master's degrees (up from 1,800 in 1970), and 250 doctoral degrees (up from 145 in 1970). During that period, the number of sociology bachelor's degrees per year dropped from 36,000 in 1970 to 14,000 today (Rogers, in press). There are now about 2,000 schools and departments of communication in the United States, and yet the first so-called department was established at the University of Illinois less than 50 years ago.

Chaffee: But that rapid growth has worked against coherence, which would be the first step toward full recognition in academia. Many of the newer communication programs represent something different from the original conception of the field. And our programs have different names, and often very different departmental traditions and missions, even at a single university. The prospective doctoral student who wants to study

Everett M. Rogers is a professor and the chairman of the Department of Communication and Journalism at the University of New Mexico. Steven H. Chaffee is the Janet M. Peck Professor of International Communication in the Department of Communication at Stanford University.

communication has to choose between titles that carry limiting adjectives—mass communication, speech communication, telecommunication, and so forth. Not many of the 150 or so PhD programs we consider in our field are called just communication. And even those with this singular name do not always resemble one another much.

Rogers: The names don't matter, Steve. All our programs focus on the study of information exchange between humans. It's all communication.

Chaffee: Perhaps, but the names still matter. Many of the greatest research universities, which award most doctoral degrees, have a department of speech communication (or communication arts) *and* a school of journalism (and/or mass communication). These tend to be housed in, respectively, the humanities and the social sciences, which means they take different intellectual perspectives. These units may be located far apart on a large campus, and each may assert that it is the one studying communication. By the norms of a field like botany or French or history, this is bizarre.

The Schramm Legacy

Rogers: It is explained by historical accident. Wilbur Schramm, who we agree was the founder of communication study in the United States, developed his vision for the field when he rubbed elbows with great social scientists like Paul Lazarsfeld and Carl Hovland, while they were planning communication programs in Washington, DC, during World War II. When Schramm returned to the University of Iowa faculty in 1943, he set up the first doctoral program in communication (later called mass communication), as the new director of the existing School of Journalism. He connected the PhD program by committee structure to the existing Department of Speech, as well as to psychology, where he had been a postdoctoral fellow.

Chaffee: Schramm also maintained, through himself personally, ties between his new PhD program and the Department of English, where he had taught literature and writing. Schramm was a man of many parts, so it should not surprise us that the field he envisioned had many parts too. But my point is that Schramm's new paradigm for communication study was added to existing academic traditions of speech and journalism, which never lost their separate intellectual roots. Schramm's later PhDs at Illinois and Stanford mostly came out of speech or journalism—and more often than not they went back to teach in their fields of origin after finishing a doctorate in communication (Rogers & Chaffee, 1992). So we still have two subdisciplines, and tacking the label *communication* onto them doesn't itself add unity of thought or purpose.

Rogers: For that matter, our field's interdisciplinary roots extend to mathematics, political science, electrical engineering, and agricultural extension. Communication study was, as you say, institutionalized in American universities mainly through existing units of speech and journalism. But such divergent beginnings can evolve into a coherent discipline.

Chaffee: Schramm too thought that would eventually happen. In the unfinished book manuscript that he left on his computer disk when he died in 1987 (Schramm, in press), he clearly felt that the field was moving through a transitory phase of separate units—such as speech communication, journalism and mass communication, information science, and telecommunications—into larger schools or colleges that would eventually come to be called communication.

Rogers: We should add that Schramm foresaw integration beginning at the doctoral level. He expected it would occur more slowly, and perhaps only partially, at the undergraduate level—where vocational specialization rather than intellectual integration might be more appropriate.

Chaffee: Wilbur Schramm certainly did as much as one person can to bring this about (Schramm, 1948, 1949, 1954, 1963). He established broad communication doctoral programs at each university where he taught—Iowa, Illinois, and Stanford—and he provided the inspiration for other important early programs, particularly the one at Michigan State. But those PhD programs, and the communication units they are housed in, look quite different from one another today. Do you really think the convergence Schramm predicted is under way now?

Rogers: Yes. I think the pieces of communication study have come together in the past four or five decades, since the first units called communication were founded in U.S. universities. Prior to Schramm's 1940s vision of a unified discipline, the study of communication was far more divergent than is the case today. He saw sociologists studying mass media effects in the field (Lazarsfeld, Berelson & Gaudet, 1944), psychologists running controlled experiments (Hovland, Lumsdaine & Sheffield, 1949), and political scientists content-analyzing propaganda messages (Lasswell, Casey & Smith, 1935). He found related concepts in cybernetics (Wiener, 1950) and information theory (Shannon & Weaver, 1949). By creating books and curricula in communication, the humanist Schramm founded a new social science. It is the first one to be widely accepted in American universities since the five traditional social sciences—psychology, sociology, economics, political science, and anthropology—were established around the turn of the century.

Chaffee: Okay, I'll concede that in the long view communication study is much more unified than it was during World War II. And we are, as you say, a young and growing field. But I still don't see much contemporary movement toward further coherence. A lot of communication faculties would deny that their unit should be considered part of the social sciences at all. They see themselves as a branch of critical studies in literature and philosophy. And at many schools the communication unit is little more than an administrative convenience for a set of vocational training programs that don't square with the mission of a liberal arts college. I know from years of teaching in journalism programs that neither professionals nor our major donors (typically, newspaper publishers) know or care much about communication as an academic discipline. They urge us to train students in technical skills, so they can hire our graduates directly into entry-level jobs.

Specialization and Diversity

Rogers: Perhaps. But alumni donations do support that discipline, even if they don't fully understand its scope. And the discipline is building itself from within. We have the International Communication Association (ICA), a scholarly organization with no limiting adjective on communication. We also have academic journals with unqualified names like the *Journal of Communication, Communication Research, Human Communication Research (HCR),* and *Communication Theory (CT).* None of those existed in 1950.

Chaffee: Granted, but even these new institutions mask the internal heterogeneity of the field, and they are not the greater part of the picture. ICA consists of a dozen or more special-interest groups and divisions. Most of its members come from larger organizations, such as the Speech Communication Association and the Association for Education in Journalism and Mass Communication. These older associations publish journals that do not cite one another much (Reeves & Borgman, 1983; So, 1988). And even our communication journals have developed specific personalities; it is rare to find a mass communication study in *HCR,* for instance, or a theory-testing study in *CT,* despite the generic labels on the covers of the journals.

Rogers: Do you think of that as a problem?

Chaffee: I certainly don't see it as a failing of the journals. Each has found its purpose, and collectively they are enabling us to build an impressive scholarly literature. But we have a problem at the level of the discipline. Divergence between journals, or among divisions within ICA, or between departments or voluntary associations of professors, all illustrate our intellectual separatism (Barnett & Danowski, 1992). The field is balkanized, a kind of academic Yugoslavia.

Rogers: Steve, every field as large as ours is characterized by specialization and diversity. Look at psychology. The American Psychological Association's sections represent even more balkanization than you find in ICA. But most psychologists identify with their discipline, and share a common epistemology and a core paradigm of individualized human behavior. And don't forget that Wilhelm Wundt at Leipzig got psychology started as a discipline in 1889, more than 50 years before communication study became a gleam in Wilbur Schramm's eye. Give us time.

Chaffee: I can't help but wonder if, another 50 years from now, communication study will present any more explicable a picture than it does today. Psychologists, as you say, focus on the individual as their unit of analysis. In communication we have to study more kinds of things than that: relationships, organizations, institutions, even whole cultures. Our competing intellectual traditions, the humanities and the social sciences, pull us in opposite directions as to what is considered acceptable evidence, or what a theory is supposed to accomplish. Meanwhile the communication professions pull against formal academic study of either kind; news reporters and video producers qualify for faculty positions by virtue of their years of experience, which is treated as a third way of knowing about communication. Each camp views the others as inferior, and is willing to say as much.

Rogers: That may be changing too. New communication technologies, one of our most exciting subfields today, almost demands a new, integrative kind of theory.

Chaffee: That's true, and there are good reasons why those should be theories of communication, rather than theories of technology (Chen, Lieberman, & Paisley, 1985). Historically, after all, innovations in communication technology gave rise to this field in the first place. It was the coming of radio (Lazarsfeld & Stanton, 1941) and then television that led people to conceptualize mass media and mass communication back in the late 1930s (Rogers, in press). But often these new technologies have generated new, technically defined departments, or hyphenated units like radio-TV-film departments. I worry that dispersive pressures of new technologies outrun our capacity to develop integrative theory. University budget officers rightly look askance at a field that fails to project an understandable self-image. As last-in, a communication unit may find itself first-out these days.

Rogers: I prefer to place my bet on Schramm's prediction. I think the field will continue to grow and prosper, that university units in our field will increasingly be called just communication, and that they will live up to this label.

Chaffee: Well, we certainly have the people and the energy. Students are, as you say, enrolling in communication programs in huge numbers. Faculty numbers are beginning to catch up with the teaching demand, and these professors are creating exciting new research. But much of our work is ignored outside the field—in the other social sciences, in the communication professions, and in the community at large.

Rogers: I find that communication studies have a natural appeal to the intelligent lay audience, but I agree that more needs to be done to reach those people with the story of what we are doing. Wilbur Schramm, who was the master at writing for a general audience, showed us the way to reach out with our message. He was able to sell a whole new discipline to some great research universities. He also convinced people around the world of the viability of this field, first as an instrument of national development and then as a proper discipline for an up-and-coming university. You will find excellent communication programs, outgrowths of Schramm's vision, in many of the newly industrializing nations today.

Chaffee: Still, too few of our colleagues are out there working to establish communication's place in the standard university curriculum. Communication, which could conceivably become the core of a 21st-century university curriculum, is not being considered in public debates over the academic canon. The most prestigious older private universities of the U.S. ignore communication completely, except for a few instances where they have made a deal with a donor who wants to support a traditional professional school.

Rogers: That means we still have a lot of work to do. The answer to divergence of practice is convergence of theory. For me the central question is, will unifying theories of communication be developed that attract the attention of future scholars, independent of whether channels are the mass media, face-to-face, or interactive technologies?

Chaffee: We shouldn't leave this in the passive voice. Whether or not your rosy prediction comes about in the next few decades lies in our own hands. As the pessimist here, I'd be happy to be proven wrong. If we can make new scholars entering our field aware of its history, and of its promise, you might turn out to be the prophet. Coherence of vision, though, will require that we work hard to devise unifying theories of communication (Berger & Chaffee, 1987; Hawkins, Wiemann, & Pingree, 1988), and that we make clear to our students that they are entering that broad field rather than one of its many competing subdisciplines.

References

Barnett, J., & Danowski, J. (1992). The structure of communication: A network analysis of the International Communication Association. *Human Communication Research, 19*(2), 264–285.

Berelson, B. (1959). The state of communication research. *Public Opinion Quarterly, 23,* 1–6.

Berger, C., & Chaffee, S. (1987). *Handbook of communication science.* Newbury Park, CA: Sage.

Chen, M., Lieberman, D., & Paisley, W. (1985). Microworlds of research. In M. Chen & W. Paisley (Eds.), *Children and microcomputers: Research on the newest medium* (pp. 276–296). Newbury Park, CA: Sage.

Hawkins, R., Wiemann, J., & Pingree, S. (1988). *Advancing communication science: Merging mass and interpersonal processes.* Newbury Park, CA: Sage.

Hovland, C., Lumsdaine, A., & Sheffield, F. (1949). *Experiments on mass communication.* Princeton, NJ: Princeton University Press.

Lasswell, H., Casey, R., & Smith, B. (1935). *Propaganda and promotional activities: An annotated bibliography.* Minneapolis: University of Minnesota Press.

Lazarsfeld, P., & Stanton, F. (1941). *Radio research, 1941.* New York: Duell, Sloan, & Pearce.

Lazarsfeld, P., Berelson, B., & Gaudet, H. (1944). *The people's choice.* New York: Columbia University Press.

Reeves, B., & Borgman, C. (1983). A bibliometric evaluation of core journals in communication research: Networks of communication publications. *Human Communication Research 10,* 119–136.

Rogers, E. (in press). *A history of communication: A bibliographic approach.* New York: Free Press.

Rogers, E., & Chaffee, S. (1983). Communication as an academic disicipline: A dialogue. *Journal of Communication 33*(3), 18–30.

Rogers, E., & Chaffee, S. (1992, August). Communication and journalism from Daddy Bleyer to Wilbur Schramm: A palimpsest. Paper presented at the meeting of the Association for Education in Journalism and Mass Communication, Montreal, Canada.

Schramm, W. (1948). *Communication in modern society.* Urbana: University of Illinois Press.

Schramm, W. (1949). *Mass communications.* Urbana: University of Illinois Press.

Schramm, W. (1954). *Process and effects of mass communication.* Urbana: University of Illinois Press.

Schramm, W. (1963). *The science of human communication.* New York: Basic Books.

Schramm, W. (in press). *The beginnings of communication study: A personal memoir* (S. Chaffee & E. Rogers, Eds.). Newbury Park, CA: Sage.

Shannon, C., & Weaver, W. (1949). *The mathematical theory of communication.* Urbana: University of Illinois Press.

So, C. (1988). Citation patterns of core communication journals: An assessment of the development status of communication. *Human Communication Research, 15,* 236–255.

Wiener, N. (1950). *The human use of human beings: Cybernetics and society.* New York: Houghton-Mifflin.

Genealogical Notes on "The Field"

by John Durham Peters, University of Iowa

The organization of the branches of knowledge in universities is historically variable. Medieval universities taught the seven liberal arts: the trivium (grammar, rhetoric, and dialectic) and the quadrivium (arithmetic,
music, geometry, and astronomy). Through the 19th century, the diverse
inquiries thitherto done mainly by moral philosophers and historians became rationalized into the social sciences as we know them: history, economics, sociology, psychology, political science, and anthropology. This
scheme, so confident of progress and secure in its division of the empire
over "man and world," remains with us today; the idea that communication should be a self-contained and self-sufficient "field" is one vestige of
it. But rather than lamenting that communication studies is not one of the
six social sciences on the 19th-century model, we might more usefully
think of it as a prime example of a newer, nascent way of organizing inquiry. That the field has a curious status is thus less a cause for alarm than
a signal that the ground may be shifting beneath us.

Inquiry is more and more oblivious to the old disciplinary boundaries;
a field is often a synonym for a literature. Areas such as cultural studies,
sports studies, women's studies, race studies, media studies, and film
studies—a list that often appears in critiques of current university
trends—are practiced by scholars having diverse disciplinary identities.
But inquiry is equally crosscutting in seemingly less frivolous fields such
as immunology, public health, social work, pharmacy, or speech and
hearing sciences. The growth areas in the university today are precisely
"areas"; the total system of knowledge today looks less like a branching
tree than a megalopolis or postindustrial belt, with centers of activity
loosely connected by highway, phone, and fax. The 19th-century project
of dividing the cosmos into coherent fields does not claim the credulity it
once did; what commands resources from administrators and funding
agencies are arguments from innovation and importance. Even the old
scheme's ability to provide professional identities for scholars is in trouble: In explaining to strangers that one is a sociologist or economist, one
will inevitably have to resort to description by topic rather than discipline

John Durham Peters is an associate professor in the Department of Communication Studies
at the University of Iowa.

to make clear what one does. One isn't merely a sociologist or economist; one studies stratification or British economic history.

The dynamic of specialization in inquiry has undercut the dream of a total, rational system: We all study *topics* today; "studies" can be attached to almost any area of inquiry (some would prefer "research" as a suffix, but the point holds). Our fields are less and less defined by the professional passport we bear than by the literatures we read, teach, and contribute to. Scholars who study cognition, initial interaction, AIDS prevention, attitude–behavior relationships, resistant readings, the Frankfurt School, and pornography, for example, all meet at the International Communication Association, but the real action is not vertical—along the unifying axis of communication—but horizontal, in specific topics. Participation in these topics is governed less by disciplinary privilege than by the ability to make a contribution and to master the debate and literature of a field. Fields are more and more topic fields rather than discipline fields.

The older vision of disciplines assumed an elite and homogeneous audience for inquiry: peers who had undergone the same training and hence belonged to a common culture of methods, paradigms, questions, and so on. In the new scheme—knowledge ordered by topic—the scholarly audience is less pure. To speak to one's field nowadays is almost always to speak across several fields. The ancients distinguished technical and civic rhetoric: The former was an elite affair, characterized by an in-group and a bounded universe of discourse; the latter prevailed in the marketplace and the law courts, sites whose diverse audiences invited more public means of persuasion. As the differentiation of inquiry and scholarly discourse communities continues, inward specialization may not be the only trend; fields may be compelled to develop quasi-public languages so that scholars having diverse backgrounds will be able to converse in the megalopolises of AIDS research or media studies. Knowledge as a whole may now be a mad swirl of disconnected language games, as Lyotard (1984) argues, but the collapse of the old order does not mean a disappearance of all order. The current arrangement of knowledge fits no grand philosophy of history, but it clearly is shaped by pressures from state, market, society, and professional fashion; this is particularly true for communication studies/research.

The Archipelago of Communication

Communications, or radio research, or mass communication research was conceived as a topic field, not a discipline—a "crossroads" in Schramm's (1963) metaphor. But since that didn't easily fit the older status hierarchy of disciplines, the field has been in a perpetual identity crisis—or rather legitimation crisis—since. And rightly so. Communication, defined with any rigor or imagination, is a topic not amenable to institutional disciplining. It is studied in diaspora; communication studies, in the broad sense, is found in an archipelago throughout the university. This is true not only of the postwar university, but of inquiry in general: What account of human activities could avoid placing linguistic praxis at the center? (cf. the trivium). As James Beniger repeatedly shows in his puckishly titled book-review column in *Communication Research,* "Far Afield," there is fascinating communication theory galore—most of it written by scholars quite oblivious to the existence of a field, per se. The conceit (in either sense) of a discipline field of communication can obstruct vision of the wider expanse of intellectual opportunities.

"I remember one day," said Paul Lazarsfeld, "a friend of mine, in 1937 or so, introduced me to a group of colleagues and said, 'this is a European colleague who is an upmost [sic] authority on communication research,' and saw that no one was particularly impressed, so he wanted to press the point and said 'as a matter of fact, he is the only one who works in this field'" (Morrison, 1978, p. 347). This almost comic account nicely expresses the usual sense of when the field began, who began it, and what it was for. Elihu Katz, most notably, explicitly argues for Lazarsfeld's paternity. With an eye to Lazarsfeld's many recent critics, Katz quips that communication research is "like a (Greek) soap opera" because "the clue to the identity of the true father is revealed by observing whom the offspring are trying to do in" (Katz, 1987, p. S40).

Options in the 1990s—and the 1930s

Arguments about lineage have long been part of the struggle for legitimacy—for kings and scholars alike. Defining paternity is not just a matter of antiquarian rummaging: It locates blame, heirs, and bloodlines. In the past decade or so, discussion on the history and future of mass communication research has produced three major discernable positions. I will identify, for the sake of argument, these three positions with three essays by Stuart Hall (1982), James Carey (1989), and Elihu Katz (1987). Hall's essay is a stocktaking of the sins of American mass communication research, its complicity with the postwar hegemony of the American empire, and a quest for more adequate conceptions of language and society usable for a critical paradigm. Hall wants nothing to do with mass communication or Lazarsfeld; he starts on a different terrain, the genealogy of which runs to the key texts of Western Marxism—Raymond Williams, Althusser, Gramsci, and Marx himself. Carey likewise is more or less willing to concede the effects tradition to Lazarsfeld and his offspring. While finding some value in the tradition, Carey, like Hall, wants to begin in a different place, that of cultural studies. Carey, too, argues genealogically: John Dewey for him would be source, if not father. For both Hall and Carey, the media as such are not a sufficient grounding for a field of study; for Hall, they are one key agent in a larger array of ideological apparatuses, and for Carey (1989, p. 110) they are a "*site* (not a subject or discipline)" for posing the central questions of social theory. By casting a larger net, Hall and Carey seek a nobler lineage and a larger inheritance.

Katz's essay recaptures the breakaway movements represented by Hall and Carey. In a major rethinking of media effects, going far beyond the ABCs of social psychology (attitude, behavior, and cognition), Katz argues that Lazarsfeld's conception of media effects included questions of ideological power and social organization, though he did not carry them out himself. For Katz, Lazarsfeld's legacy is threefold: the study of media effects on (a) information (citizens' agendas), (b) ideology (the distribution of power), and (c) organization (the technological structure of society). In a daring coup, Katz has recreated the diversity of approaches to media studies represented by himself, Hall, and Carey all under the fatherhood of Lazarsfeld. Katz reclaims the center—research on media effects—left as unusable by Carey and Hall.

These names, both the living and the dead, give us a nice map of the current intellectual options in mass communication research (excepting feminism): the empirical study of media effects (Katz–Lazarsfeld), the humanistic study of communication and culture (Carey–Dewey), and the critical study of ideology (Hall–Marx). But once you shift your gaze from the proclamations of these scholars to their research practices, you find that Katz is as interested in meaning as Carey, Carey as interested in power as Hall, and Hall as interested in effects as Katz. Here empirical research can include textual criticism, humanistic studies can include political-economic analysis, and critical approaches can include counting and tabulation.

Are these signs of a new convergence (Curran, 1990)? Perhaps. The 1980s saw the fall of many longstanding walls in intellectual life and geopolitics. But they are also signs of an old convergence—one, unfortunately, that never quite took place. The genealogical stories of Katz, Carey, and Hall come together in one moment: Columbia University in the very late 1930s. Dewey was professor emeritus of philosophy, still active in public and intellectual life; Lazarsfeld was about to be hired by the department of sociology; and the exiled Institute for Social Research (Frankfurt School) was given office space and a home away from home. Key figures of pragmatism, social research, and Western Marxism converged in one place. Their actual interchanges—and the reasons for their rocky character—deserve extensive study. For present purposes, it is enough to note that received labels were wildly scrambled in this moment. Social philosophy, empirical social research, and critical social theory all converged on a common intellectual problem: how to understand new centralized forms of symbolic control over populaces. The key theoretical question for each was how to understand the possibilities for organized social action in that strange collectivity we have learned to call the audience. This was Dewey's concern, as he decried the eclipse of the democratic public, Adorno's as he laid bare the workings of promotional culture, and Lazarsfeld's as he tried to solve the puzzle of how to study a social form that is really visible only via statistical method: the radio audience.

Blurring Critical, Cultural, and Empirical Contrasts

The intellectual convergences and divergences of the late 1930s and early 1940s deserve detailed study. Here I can offer only a few comments in the hope of dissolving some tales of lineage that have created a false sense of both the incommensurability of different research styles and intellectual-political options. First, the critical–empirical divide, sometimes even reified into distinct schools, has become a battle cry, but it is hard to find in the 1930s and 1940s. Granted, much in the ill-starred "collaboration" of Adorno and Lazarsfeld at the Princeton Radio Research Project crystallizes perennial intellectual tensions, but not much can justify the common images of one as a mass society prophet of doom in a black cape and the other as a method-happy sellout in a white lab coat. Only in the 1950s were the sheep separated from the goats: The cold-war filter distorts our picture of prewar political and intellectual alliances.

Quiz: Match author (Lazarsfeld or Adorno) with statement: A. "Mass communications may be included among the most respectable and efficient of social narcotics." B. "Our approach [to present-day television] is practical. The findings should be so close to the material, should rest on such a solid foundation of experience that they can be translated into precise recommendations and be made convincingly clear to large audiences." *Answer*: Statement A is Lazarsfeld and Merton (1948, p. 106); B is Adorno (1954/1991, p. 69). Proof texts rarely prove anything, even in theology, but the point here is that Adorno insisted on the felicitous fusion of social fact and social theory (despite his pronounced disdain for givens), while Lazarsfeld often wondered about the meaning of a capitalist media industry (while rarely venturing very far in the direction of a satisfying answer). Both Lazarsfeld's Bureau of Applied Social Research and Horkheimer's Institute for Social Research were large-scale research enterprises on modern culture and communication. Their rivalry had less to do with the relative weight accorded to theory or data than with the politics and social uses of inquiry.

The other binaries are equally problematic. To see Dewey as a proto-cultural-studies type may miss the way his vision of social science funded programs of administrative research. To see Dewey as opposed to the Frankfurt School may miss the radical edge to his politics and their common philosophical inheritance from German idealism. Dewey's later work, such as *Art as Experience* (1934), and critical theory both cherish a utopian faith in the healing powers of art; both use that faith to berate current mass culture for its dreariness. At times Dewey seems just as enchanted with scientific method as Lazarsfeld and as critical of existing institutions as Adorno, as open to using media for popular education as Lazarsfeld and as glum and historically astute about the origins of mass society as Adorno, and as deep a thinker about the lived experience of communication as anyone.

The convergence of the three traditions in the late 1930s at Columbia was only a microcosm of a much larger and ragged debate in North America and Europe in the years between the wars about what we have come to call—with reluctance, enthusiasm, or habit—mass communication. A diverse company including Dewey, Walter Lippmann, George Herbert Mead, Lewis Mumford, Kenneth Burke, Margaret Mead, Robert Park, Harold Lasswell, Floyd Allport, Robert Lynd, Edward Bernays, Robert Merton, Lazarsfeld, I. A. Richards, F. R. and Q. D. Leavis, Martin Heidegger, Karl Jaspers, Rudolf Arnheim, Georg Lukács, Theodor Adorno, Max Horkheimer, Walter Benjamin, Leo Lowenthal, and Antonio Gramsci, for example, all explored the meaning, in their ways, of new forms of mass culture. Thinkers of this period faced the economic, political, and spiritual fallout of World War I, the rise of mass production, fascist politics, broadcasting, audience measurement, public relations, and survey research, for example. Those listed above raised fundamental issues in communication and mass communication: Why does our intellectual patrimony seem foreign to us?

This returns us to the politics of recognition in academic specialties. The field Lazarsfeld begat had enforced amnesia about the theoretical questions that presided at its birth, only to have them keep popping up in a "return of the repressed" (Hall, 1982). Due to a curious institutional history, communication research inherited only a meager portion of the debate about mass media in the 1920s and 1930s. Genealogical tales by Carey and Hall, among others, are praiseworthy efforts to reclaim the entirety of the debate for students of media and society—not just the small part captured by the term *mass communication*. Hall shows that the taboo idea of powerful media in the effects tradition served as a code word for Marxism; Carey shows that in seeking to refute so-called mass society theory, the effects tradition shunted aside American pragmatism with its worries about the political role of communication in shaping or undermining democracy. The making of the field, then, excluded rival traditions that could have added much to it. In defending the tradition, Katz may be the strongest revisionist of all, in asserting the copresence of all three traditions in the center: He opens the field up to what was lost—but with a certain twist of his own.

A Broader Legacy

The implied debates between Hall, Carey, and Katz and Adorno, Dewey, and Lazarsfeld correspond to larger enduring intellectual and political options. Specifically, each couplet corresponds to a political philosophy and its accompanying mode of social inquiry: liberalism, social democracy, and Marxism—the three main governing philosophies in the 20th century (ignoring fascism here). Katz and Lazarsfeld represent liberal political philosophy and its practice of social science—a practice that is thought not only to study society but to enable civil forms of conversation among citizens. Dewey is a social democrat, and while Carey fits this label less exactly, both use a method of social philosophy for sustained moral reflection on the paths of social reconstruction. Hall, with his predecessors in Western Marxism, subscribes to a Marxist politics and its attendant method of ideology critique. For both social democrats and Marxists, empirical social inquiry is valuable, but only one aspect of the larger activities of philosophy or praxis. In sum, Katz and Lazarsfeld do social science; Carey and Dewey do social philosophy; Hall and Gramsci do social critique. The first pair studies influence, the second theorizes participation, the last unmasks domination. The core object for each is, respectively, the media and minds, democracy and culture, and ideology and power. All three worry about communication, in the broadest sense of that word; theirs is a family quarrel.

Though I here am overly schematic in sketching contrasts—in apparent violation of the point about convergence that I made above—it is to make a point. Liberal, social democratic, and Marxist political philosophies have roots in the 19th century but have not, until recently, been placed side by side as live options in communication research. The future of the field depends in many ways on coming to terms with the past of the field. More specifically, we must see how the institutional field shut down fruitful paths of inquiry into the place of communication in modern life and society. In many ways, the key decision about the future of communication research was already made when Lazarsfeld became its first expert in the 1930s. He made the institution, and the institution wrote the history, suppressing former rivals. The resemblance of our options today to those of the 1930s allows us to recognize Lazarsfeld as one forefather among many and to interpret his role with more nuance, charity, *and* criticism. One ought not attack him just for making a narrow field, but for being a chief agent in giving the state and the market the research tools to operate in the postwar world (see Westbrook, 1983). Here again, the broad social context of communication research has not been appreciated.

Students of communication and society should not restrict the vision of their inheritance to fit a largely specious entity called a field. Hopefully, the range of forefathers—and foremothers—will grow as inquiry is freed to take the best ideas from anywhere, regardless of provenance. Communication pervades the human condition, and mass communication is central to the operating apparatus of the late 20th-century industrial state. We should worry less about fitting a largely irrelevant model of what a discipline is and more about doing intellectually deep work.

References

Adorno, T. W. (1991). How to look at television. In D. Ingram & J. Simon-Ingram (Eds.), *Critical theory: The essential readings.* New York: Paragon House. (Original work published 1954)

Carey, J. W. (1989). Overcoming resistance to cultural studies. In *Communication as culture: Essays on media and society* (pp. 89–110). Boston: Unwin Hyman.

Curran, J. (1990). The new revisionism in mass communication research: A reappraisal. *European Journal of Communication, 5,* 135–164.

Dewey, J. (1934). *Art as experience.* New York: G. P. Putnam's Sons.

Hall, S. (1982). The rediscovery of "ideology": Return of the repressed in media studies. In M. Gurevitch, T. Bennett, J. Curran, & J. Woollacott (Eds.), *Culture, society, and the media* (pp. 56–90). London: Methuen.

Katz, E. (1987). Communications research since Lazarsfeld. *Public Opinion Quarterly, 51,* S25–S45.

Lazarsfeld, P. F., & Merton, R. K. (1948). Mass communication, popular taste, and organized social action. In L. Bryson (Ed.), *The communication of ideas* (pp. 95–118). New York: Cooper Square.

Lyotard, J-F. (1984). *The postmodern condition: A report on knowledge* (G. Bennington & B. Massumi, Trans.). Minneapolis: University of Minnesota Press. (Original work published 1979)

Morrison, D. E. (1978). The beginning of modern mass communication research. *European Journal of Sociology, 19*(2), 347–359.

Schramm, W. (1963). Communication research in the United States. In W. Schramm (Ed.), *The science of human communication: New directions and new findings in communication research* (pp. 1–16). New York: Basic Books.

Westbrook, R. B. (1983). Politics as consumption. In R. W. Fox & T. J. J. Lears (Eds.), *The culture of consumption: Critical essays in American history, 1880–1980* (pp. 143–173). New York: Pantheon.

History, Philosophy, and Public Opinion Research

by Susan Herbst, Northwestern University

In the early days of communication research, when public opinion polling was just beginning to gain respectability, Paul Lazarsfeld (1957) urged his colleagues not to neglect what he called the "classical tradition." For Lazarsfeld, the classical tradition in public opinion research meant several things—a concern with grand theoretical notions, with intellectual history, and with the changing relationship between citizens and their government. Since he wrote, very few scholars in communications have taken his advice. In fact, only rarely do researchers in our field mention ideas about public opinion generated by political philosophers or historians (cf. Noelle-Neumann, 1979, 1984; Price, 1992).

These days, most communication researchers (and most social scientists, for that matter) assume that public opinion is simply the aggregation of individual opinions, as captured by the ubiquitous sample survey. Opinion researchers, so busy measuring public attitudes, seem to have very little patience for grand theory, political philosophy, or normative dialogues. For example, debates about the meaning of "public opinion" in different historical periods—one traditional concern mentioned by Lazarsfeld—is now the domain of social historians, not communication researchers (e.g., Baker, 1987).

There is no "ferment" in the field of public opinion research. On the contrary, there is much consensus. Most researchers employ a small set of social psychological frameworks to understand the elusive "public mood" and are quite confident about their methodological approach—survey research. Despite this complacency, however, public opinion research in communications has become far less interesting than it was in the early years of the field. In those days, there really was a dialogue among historians, political theorists, and social scientists. Now public opinion researchers tend to stay close to their own small subfield, speaking to like-minded colleagues and writing in specialized journals (e.g., *Public Opinion Quarterly*). The goal of this essay is to argue, yet again, for the importance of the classical tradition: for broad, speculative, and historically informed writing and research. It is my belief that a return to classi-

---

Susan Herbst is an assistant professor in the Departments of Communication Studies and Political Science at Northwestern University, Evanston, IL.

cal concerns, which Lazarsfeld described so eloquently, will revitalize the field of public opinion research. The classical tradition also provides badly needed models for the development of grand theory.

A note of warning: This essay concerns mainstream writing about public opinion in communications. There are a few researchers in the field of public opinion who are interested in history and philosophy and who write in the classical tradition advocated by Lazarsfeld. My remarks here are aimed at the more conventional approaches to public opinion found in our journals.

Theory in the Classical Mode

What is the classical tradition? Elaborating somewhat on Lazarsfeld's initial description, I believe that classical public opinion theorizing has three dimensions: It is rooted in political philosophy, underscores historical issues, and, often, includes normative argumentation.

That the most interesting and profound speculation about the public mood and public opinion is based in political philosophy is obvious, and I do not have the space to elaborate on the point here. After Plato, Aristotle, and Machiavelli made their contributions, Mill, Bentham, Marx, and others studied the public mood in the context of constructing grand theory (for a summary of this work, see Herbst, 1993, or Minar, 1960). Classical theory about public opinion focused on political behavior in the broadest sense: Philosophers wondered about the reliability and malleability of public opinion, given fundamental dimensions of human nature.

Beyond the long-standing theoretical disputes with each other, many of which revolved around human nature and leadership, classical theorists were centrally concerned with historical events and their legacies. The most superficial glance at Tarde's *Communication and Social Influence* (1969) or Tocqueville's *Democracy in America* (1969) reveals that these men gave considerable thought to the past. Tarde, for example, developed his ideas about public opinion formation by describing court conversation, salons, and newspapers of 18th-century Paris.

The advantages of the historical approach were obvious to classical theorists. For one, history provided empirical examples on a mass scale. Events like the French and American revolutions enabled rich, textured analyses of the movement of public opinion. Classical theorists thought of these events as their laboratories for understanding political behavior and public opinion development. A preoccupation with the history of states, economies, and ideas also made public opinion theorists more sensitive to the nature of their own surroundings—institutions, popular ideas, and social divisions.

A third component of classical theory is a concern with normative models of social and political behavior. Rousseau, for example, not only described the "general will," he also argued for its importance in political life, detailing how and why it should be heeded by citizens and leaders alike. Similarly, Tönnies (1955) worked to define public opinion, but also worried about how it would fare, as Western societies moved from *gemeinschaft* to *gesellschaft*. One of the latest classical theorists of public opinion, Jürgen Habermas, also adopts a strong normative tone in his most important work on the subject (Habermas, 1989).

These days, public opinion researchers avoid normative arguments, because they view them as unscientific: Such arguments are about preferences and not about discovering the true nature of public opinion formation and change. Yet a concern with the way public opinion *should* look was critical for classical theorists, since it drove them to ask one of the most important philosophical questions: How shall we govern ourselves? The passion of classical theorists was rooted in their normative concerns, and many of them struggled with issues of public opinion because they believed theory could and should change the world. Contemporary researchers have lost this passion, choosing instead to approach public opinion in the most dispassionate ways. Rarely does the typical article or book chapter on public opinion in communications address the "So what?" question: Many articles do conclude with sections labeled "Implications for Democratic Theory" or "Ramifications for American Politics." Yet these brief discussions are usually inserted half-heartedly, and contain few serious attempts to answer the philosophical queries that were once so important to public opinion theorizing.

What Happened to Opinion Research in Communications?

Public opinion research in our field is not about history, building grand theory, or answering weighty philosophical questions. Most articles concern the relationship between media and opinion formation, ways to manipulate survey forms and procedures, or (most often) the nature of public attitudes at one or two brief moments in time. There are many reasons why contemporary communication research about public opinion looks like it does, but I'll concentrate on three: the hegemony of statistics, the absence of genuine interdisciplinary approaches, and our choice of publishing outlets.

With the rationalization of social research in the 20th-century, quantitative techniques for understanding public opinion have gained an enormous amount of respect. Despite the early questions Herbert Blumer (1948) raised about the usefulness of survey research, or Pierre Bourdieu's (1979) sophisticated critiques of opinion polling, communication researchers rarely question their measurement techniques. We recognize all of the problems associated with survey research—its inability to capture anything more than a narrow dimension of public attitudes, or the difficulty in measuring the intensity of opinions—but seldom try to develop new modes of understanding the nature of citizens' opinions and belief systems. Even though public opinion is an inherently *communicative* construct, it is political scientists, sociologists, and anthropologists—more often than communication researchers—who develop innovative ways to understand public attitudes (e.g., Gamson, 1992).

Why are communication researchers so taken with the authority of numbers, when many in other disciplines recognize the limitations of quantitative measures? Perhaps it has something to do with our field's perceived lack of legitimacy: Numbers do tend to make research and theory building seem more "scientific," and maybe some believe that this aura of science will enable our discipline to gain respect.

Another, more serious problem with public opinion writing in our field is that we pay only lip service to the idea of interdisciplinary work. The reason why classical writing on public opinion was so interesting was that theorists seemed to know no disciplinary boundaries. Writers like Tarde and Tönnies borrowed from every field they knew, so their books and essays are difficult to place in any single category: These works are sociological, political, philosophical, social psychological, and ethical, all at the same time. In fact, from the standpoint of late 20th-century social science, these writings seem wild and undisciplined. They are suspended somewhere between the academic fields we have so carefully constructed over the last 50 years. Yet, the wide-ranging nature of classical theorizing is what makes it so fascinating, and so provocative. How could Alexis de Tocqueville have constructed such a rich model of politics had he not taken forays into psychology, aesthetics, and history? In communications, we have the freedom to pursue interdisciplinary approaches to public opinion, but we don't take advantage of that freedom. So often we seem to be imitating social psychologists or sociologists, using their tools, ideas, and units of analysis.

Students who complete their doctorates in communication these days seem only partially educated: They know about "agenda setting" and the "knowledge gap," but often seem not to have read Rousseau, Locke, Bentham, Mill, or Marx for that matter. *These* sorts of knowledge gaps among our students are downright frightening, and as educators we need to take responsibility for such voids. If we don't make political philosophy and history *critical components* of graduate curricula, public opinion research will never assume the character of classical theory.

A final problem in the area of public opinion research and theory concerns the forms we use to present our ideas and findings. Public opinion research has become an "article-driven" field, at least in communications. And communication researchers tend to write for each other. If one monitors the content of history, political science, sociology, and American studies journals, one rarely finds articles by communication researchers. If we are really doing interdisciplinary research, *of interest to scholars in other fields,* why don't we send our work to their journals? My fear is that we are becoming so inbred—with our own journals and books of collected essays—that our work will never be recognized by those in other, related disciplines. Not only do we fail to publish in established journals in other fields, we tend to publish books from a narrow range of specialized publishing houses.

Big Questions, Grand Theory

I think there are some exciting directions for public opinion theory and research, and that communication researchers could *lead* scholarship in this area if we were to pursue these new avenues. Best of all, we might be able to build grand theory about public opinion processes, instead of just "settling" for workable hypotheses.

For one, we need to take history more seriously. Students of public opinion must educate themselves about intellectual history—how philosophers understood public opinion, political behavior, and human nature. But researchers should also explore how the meaning of public opinion changes over time, why it changes, and how such developments are tied to transformations in the social structure. Tönnies tried to trace the evolving definition of public opinion given the rise of industrialization, the decline of communities, and changes in social norms. We need to make the same connections between institutional development and conceptions of public opinion. Furthermore, there is much work to be done on the progression of public opinion expression and measurement techniques. Some of us have started to do this work, but there is room for much more historical research along these lines.

We need to start building theory at the level of states, instead of concentrating so intently on individuals. Dicey (1920), writing in the early years of the 20th century, developed a wonderfully resonant comparative theory of public opinion, arguing that when governments adopt a policy, the citizenry grows more favorable toward that policy. A few political scientists have conducted research on the topic (e.g., Zaller, 1992), but this is fertile ground for more extensive theory building about public opinion, media, conversation, and state legitimacy. Along these lines, we need to link our research about public opinion more closely to democratic theory. It is very rare that public opinion researchers directly address classical *or* recent arguments about democracy (e.g., Barber, 1984). In fact, I rarely see references to this body of work in the typical articles about public opinion published in our journals. We need to initiate a more direct dialogue with democratic theorists if we are to place our work on public opinion in its proper context.

Finally, we should take a cue from Tarde (1969), Habermas (1989), Gamson (1992), and others who place *conversation* at the center of public opinion processes. Conversation is difficult to study: It is dynamic and complicated. Observing political discussions is also incredibly labor-intensive. Yet if we are really interested in public opinion formation—how people arrive at opinions through their contact with media and with friends—we must take conversation more seriously. Survey research is a poor tool for understanding the dynamic process of opinion *formation,* and while it is useful for some research endeavors, it won't help us understand how and why public opinion crystallizes.

My intention here has been to remind public opinion researchers in communication about the classical tradition. Classical theory should not be treated as a dinosaur but as a source of inspiration. Asking the large, normative questions about public opinion processes, trying to build grand theory, and taking history seriously, are not outdated approaches to knowledge: These strategies will enrich our field and command the attention of scholars in others.

References

Baker, K. (1987). Politics and public opinion under the Old Regime. In J. Censer & J. Popkin (Eds.), *Press and politics in pre-revolutionary France* (pp. 204–246). Berkeley: University of California Press.

Barber, B. (1984). *Strong democracy: Participatory politics for a new age.* Berkeley: University of California Press.

Blumer, H. (1948). Public opinion and public opinion polling. *American Sociological Review, 13,* 242–249.

Bourdieu, P. (1979). Public opinion does not exist. In A. Mattelart & S. Siegelaub (Eds.), *Communication and class struggle* (pp. 124–130). New York: International General.

Dicey, A. (1920). *The relations between law and public opinion in England during the 19th century.* London: Macmillan.

Gamson, W. (1992). *Talking politics.* New York: Cambridge University Press.

Habermas, J. (1989). *The structural transformation of the public sphere: An inquiry into a category of bourgeois society.* Cambridge, MA: MIT Press.

Herbst, S. (1993). *Numbered voices: How opinion polling has shaped American politics.* Chicago: University of Chicago Press.

Lazarsfeld, P. (1957). Public opinion research and the classical tradition. *Public Opinion Quarterly, 21,* 39–53.

Minar, D. (1960). Public opinion in the perspective of political theory. *Western Political Quarterly, 13,* 31–44.

Noelle-Neumann, E. (1979). Public opinion and the classical tradition: A re-evaluation. *Public Opinion Quarterly, 43,* 143–156.

Noelle-Neumann, E. (1984). *The spiral of silence: Public opinion—our social skin.* Chicago: University of Chicago Press.

Price, V. (1992). *Public opinion.* Newbury Park, CA: Sage.

Tarde, G. (1969). *On communication and social influence* (T. Clark, Ed.). Chicago: University of Chicago Press.

Tocqueville, A. (1969). *Democracy in America* (J. P. Mayer, Ed.). New York: Anchor.

Tönnies, F. (1955). *Community and association* (C. Loomis, Trans.). London: Routledge & Kegan Paul.

Zaller, J. (1992). *The nature and origins of mass opinion.* New York: Cambridge University Press.

Communication in Crisis: Theory, Curricula, and Power

by Pamela J. Shoemaker, Ohio State University

The traditional growth and diffusion of communication-related studies across U.S. universities has interacted with recent countrywide university budget crises to call into serious question the legitimacy and survival of communication units within the university. As communicators, our first impulse is to "communicate" better with others in the university about what we are doing, but our response to these challenges must include more than better communication about and justification of our scholarly mission. We must take a critical look at the nature of our scholarship and take steps to improve it. We must critically examine our curricula and ensure that, in the terms of Ernest Boyer's (1990) book *Scholarship Reconsidered: Priorities of the Professoriate,* there is scholarship in our teaching as well as in our research. We must build strong alliances with other academic units in our universities and with communication practitioners in the field to ensure that we will have academic and practitioner friends if we are threatened.

Crisis of Legitimacy

The discipline of communication includes the study of its practitioners, process, and products, as well as its effects on people and society. Such scholarship takes place in university departments, schools or colleges under various names, including communication, journalism, communication arts, and mass communication *plus* in sociology, psychology, political science, economics, history, law, and others.[1] Communication courses are virtually everywhere in the modern university, including in departments of agriculture, natural resources, nursing, business, human ecology, and engineering. One could say that communication is both everywhere and therefore nowhere, a status that has potentially disastrous

---

[1] I use the term *communication* in this article to represent the breadth of communication-related academic units.

Pamela J. Shoemaker is a professor and director of the School of Journalism, Ohio State University, Columbus.

consequences for communication departments as they battle for their slice of an ever-smaller university budget pie. We have an urgent need to articulate the scholarly focus of our discipline to others within the university, rather than allowing them to think that we exist merely to create effective communicators. While this is a noble and necessary role, I have found out first-hand that this does not much impress our friends in disciplines such as sociology or psychology. Over the last two years, I have found the need to argue that journalism is central to the mission of my college[2] and university and to justify the legitimacy of my discipline—something I naively thought would never be necessary.

Our familiarity with and love for our discipline can make us blind to the realities of university politics, blind to what the realities of university budget crises can mean for our discipline. Our departments may be decimated, crushed, or eliminated. It has happened already and it will happen again. Communication departments that were smart and fortunate enough to have large endowments are relatively cushioned from the budget blow, but even the most fortunate of us apparently must be content with little or no growth.

Even more serious than the loss of revenue, however, is the loss of campus status and legitimacy faced by those in universities where differential cuts are made, in lieu of an across-the-board reduction. When administrators are forced to make differential cuts in the budgets of academic units and when they identify communication units as less central, less important, or as being of lower quality than units such as sociology, psychology, and political science, the discipline of communication is in trouble. Once made, these negative evaluations tend to stick like bad cologne and have lasting effects on all sorts of administrative decisions, including promotion and tenure, approval of proposed curricula, and inclusion into the scholarly life of the university.

We must act now to improve our departments, which will ultimately mean improving the discipline. We must be more theoretical in our research and more scholarly in our curricula. We must build a power base of support within and outside of the university, by making as many connections as possible between our communication departments and other academic units.

Crisis of Theory

Research in communication topics is not a modern invention (e.g., rhetorical studies), but as a practical matter we can date communication as a separate university discipline back to the early 1950s, when the first PhDs

---

[2] At Ohio State University, the School of Journalism is within the College of Social and Behavioral Sciences.

in communication were graduated and set out to do research and get tenure in universities across the country. Being only 4 decades old as a domain of scholarly research at the doctoral level puts us at an immediate theoretical disadvantage relative to our competitors in producing communication research—departments such as sociology, psychology, and political science. Our body of research is integrally tied to these other departments, however, because our first theoretical work is derived from theories in these disciplines—whether in research done by PhDs in communication departments or by those who are interested in communication topics but have completed sociology and other more traditional doctoral programs.

Researchers in communication departments really face two problems. First, we must find ways to build more theory ourselves rather than relying only on theories borrowed from other disciplines. For example, in mass communication research we are stuck on a theoretical plateau that hinders the growth of communication theory (Shoemaker & Reese, 1991). This is at least partially because exposure to specific mass media content has been largely ignored (with the notable exception of experiments where content is the manipulated variable and hence may deliberately differ from real-world media content). Exposure of real people to real media content does not play much of a part in mass communication research, where the emphasis has been on measuring general exposure to the media rather than to specific types of content. Home-grown theories such as agenda setting (McCombs & Shaw, 1972), cultivation analysis (Gerbner, Gross, Morgan, & Signorielli, 1986), and uses and gratifications (Blumler & Katz, 1974) are rare examples in a discipline that generates so many books and journal articles.

But even among those theories that we can identify, where can we go to *read* the theories of our field? Although many of us studied and now teach theory-building techniques (e.g., as outlined in Hage, 1972), our theories are generally not very well organized or specified in much detail. As a graduate student, I vividly remember reading a study that tested a uses and gratifications hypothesis and innocently asking where I could go to read this theory. Having just studied Hage's (1972) structure for elaborating theories, I expected to find a formalized uses and gratifications theory. Where is the formal theory of uses and gratifications laid out or *systematically* specified? What are its assumptions and hypotheses? What are its concepts, and how are these theoretically and operationally defined? Uses and gratifications is not unique in lacking this level of specification; the fact is that our theories reside in dozens and sometimes hundreds of individual journal articles and book chapters, and synthesizing these is difficult and often not done. Our field lacks a systematic theory-building effort, and therefore we are at an intellectual disadvantage when compared to other disciplines.

The need for theory building that considers key communication variables (such as exposure to specific content) rather than borrowing theories from sociology, psychology, and the like cannot be overemphasized. We generate many quantitative studies that test hypotheses, and some of them are derived from theory. But what contribution do these studies make to building theory? Authors frequently fall short of tying their results back to the theory; they fail to make judgments about the extent to which the theory is supported or should be modified. We talk about inductive and deductive processes in our research and theory classes. The reality, however, is that very little of either takes place if we define these as theory-building processes with important feedback mechanisms from theory to data and the reverse. Theories, when used, are often mere justifications for testing hypotheses that have no clear and direct impact on the theories. For example, papers read today at meetings of the Communication Theory and Methodology Division of the Association for Education in Journalism and Mass Communication (AEJMC), an early interest group that focused on mass communication research, often explicitly deal with neither theory nor methodology.[3] Rather, many of these papers present one isolated hypothesis test after another and contribute little to theory building or to the development of research methods. The same could be said for the bulk of journal articles published on communication topics.

The second problem we face is that the theoretically derived and excellent research that we *have* produced is often invisible to our colleagues in other departments who do communication research. In general, we are more likely to read and cite journals in sociology, psychology, and political science than scholars in those fields are to cite communication journals. We must investigate and strengthen the intellectual connections with these other disciplines on campuses, in our journals, and among our associations. We must create ties that bind communication scholarship no matter where its academic home, and we must ensure that our colleagues in other disciplines are aware of our own research.

We must build linkages between communication and other intellectual disciplines. These linkages can be as simple as holding joint colloquia, working together on graduate students' projects, and reading each others' work. They can be as intellectual as finding ways in which to work on common research projects and forming transdepartmental partnerships to seek outside funding.

Communication studies lend themselves naturally to interdisciplinary research, and we must try to overcome all of the difficulties and lack of incentives that make good interdisciplinary research unlikely. These include poor communication of interests across academic units, lack of rewards for working with someone in another department, the issue of "credit" or which department gets how much grant money, and the fact that interdisciplinary journals are often new, unknown to many faculty,

---

[3] As a member and former head of this division, I bear as much responsibility for this as anyone.

and therefore difficult for faculty committees to evaluate in tenure and promotion cases.

If we are able to overcome the obstacles of interdisciplinary research, we will build relationships with our colleagues in other disciplines, and we will be able to better inform them about the good work being published in communication journals. This will facilitate the growth of theory in all of these disciplines and will enrich our understanding of communication in every way. If we are acknowledged as intellectual partners of other disciplines within the university, the centrality of our mission to that of the college or university is less likely to be questioned.

Crisis of Curricula

Although our doctoral programs are only about 4 decades old, our curricula may be even older and in many cases were defined prior to the establishment of communication departments as doctorate-granting units. The very organization of our departments is an outgrowth of the industrial revolution and may not be useful in the information age.[4] For example, the traditional division of journalism departments into industrial sequences (e.g., broadcasting or print media) and the development of curricula designed specifically to train students in each sequence to fill particular job slots in the working world is not unlike the process of training workers to fill slots on an assembly line. The task is assumed to be routine and repetitive, for example, a worker trained as a reporter on one newspaper "assembly line" could easily move to another organization with minimal retraining.

This method of educating students worked quite well for a long time, but as the communication industry has changed and continues to change at a rapid pace, we can no longer anticipate exactly what the job "slot" will be 5 or 10 years from now (Blanchard & Christ, 1993). A person trained only to fit a slot may be unable to adapt to new job conditions and may drop out of the field. Flexibility must be a valued characteristic of communication workers, and generating flexibility requires a different sort of education than that needed to train somebody to fill a slot. The need for increased critical thinking skills cannot be underestimated (Shoemaker, 1993). It is the ability to analyze, synthesize, and evaluate information that will allow communicators to train themselves to take on future jobs. Courses in critical thinking should be part of communication curricula and should come early in the major program, so that students can be better prepared to meet the intellectual demands of their upper-division courses. Critical thinking skills will also help our graduates as they seek their place in the work force.

---

[4] I am indebted to Wayne Danielson, University of Texas at Austin, for these ideas, although any errors in application are my own.

We know that new graduates from one sequence get jobs in others, and that the second or third job may be even farther afield from the original, narrow university training. The walls of the sequences need to come down—or at least be lowered—in recognition of what is happening in the field. We need a large common core of classes that address what *every* student in our departments needs to know, and then we must build the sequence skills courses on this broad base. And, since originality may be hampered by thinking primarily in terms of the existing curriculum, we should "throw out" all of our existing courses, a process that could be called "zero-based" curriculum development. We should consider the concepts and skills that all students in our departments need to know. Once we have defined the knowledge base we want our students to acquire, we can rebuild the curriculum, sometimes reinstituting or revising existing courses and at other times inventing new ones to accomplish our new goals.

Educating students for *slots* is unacceptable in a world in which our graduates experience so much change in their *careers*. We must finally put into place what the "Oregon Report" (Dennis, 1984) told us to do years ago: We must give our students a general communication education with a large conceptually based core of courses. There will still be a place for classes that give students technical skills for entry-level jobs, but these must be subordinate to classes that teach critical thinking, law, history, mass media and society, international communication, and so on. These will prepare our students for new careers no matter where fortune should take them. Our graduates must leave campus with the ability to train themselves when changes in the field require it, to work effectively with others to solve problems, and to think critically about the world around them.

Crisis of Power

Our current weaknesses in theory and curricula have made us relatively weak when it comes to defending ourselves against those who would eliminate or decimate communication departments. We are living in a lost world—a world in which everyone assumes that professorial work is valuable, that autonomy is understood as necessary to the production of intellectual products, and that intellectual products are themselves judged to be valuable to society. Questions about faculty workloads are one symptom of a larger suspicion among the general population and communication practitioners that we lead an undeservedly privileged life and that, as one newspaper editor recently told me: "You guys are living on my tax dollars, and I want to make sure I get my money's worth." The desire for more accountability manifests itself in public and legislative calls for outcomes assessment (i.e., tests of students' proficiency in the academic discipline), faculty workload policies, and post-tenure reviews. The "ivory tower" is being shaken and its mission questioned.

The very validity of our work is called into doubt by both the general public and by communication practitioners. They ask: Are you doing what you're supposed to be doing? And what exactly is that? Many practitioners perceive little relevance between what we teach in the classroom and what they need from communication graduates in their first jobs. This indicates serious trouble for communication education: If our "customers" (those who hire our graduates) have lost faith in us, who will come to our defense if we are threatened?

We need strong connections between communication departments and communication practitioners. They can not only help us with advice on curriculum issues, funding, and internships and placement, but they can also act as our advocates when budget cuts threaten or our existence is questioned.

Our power as an intellectual discipline is only as strong as our theory, our curricula, and our connections with intellectual and practitioner partners. To remain as vital university disciplines, to grow as we know we can, we must reach out both to our university colleagues and to our practitioner friends. And the nature of this outreach must be to ask not only how we can be strengthened, but also what we can do for others. This must be a time of external surveillance and action—external both in terms of other departments within the university and to communication practitioners. We must ask how our research and instructional activities can complement what is being done elsewhere in the university and perhaps open our doors to more nonmajors. We must also ask how our research and instructional activities can benefit practitioners.

If our day of ultimate reckoning comes, we must have working beside us our university colleagues and communication practitioners. These linkages are the power that can help resolve crises in our favor.

References

Blanchard, R. O., & Christ, W. G. (1993). *Media education and the liberal arts: A blueprint for the new professionalism*. Hillsdale, NJ: Lawrence Erlbaum Associates.

Blumler, J. G., & Katz, E. (Eds.). (1974). *The uses of mass communications*. Beverly Hills, CA: Sage.

Boyer, E. L. (1990). *Scholarship reconsidered: Priorities of the professoriate*. Princeton, NJ: Carnegie Foundation for the Advancement of Teaching.

Dennis, E. E. (1984). *Planning for curricular change in journalism education*. Eugene: School of Journalism, University of Oregon.

Gerbner, G., Gross, L., Morgan, M., & Signorielli, N. (1986). Living with television: The dynamics of the cultivation process. In J. Bryant & D. Zillmann (Eds.), *Perspectives on media effects* (pp. 17–40). Hillsdale, NJ: Lawrence Erlbaum Associates.

Hage, J. (1972). *Techniques and problems of theory construction in sociology*. New York: John Wiley & Sons.

McCombs, M. E., & Shaw, D. (1972). The agenda-setting function of mass media. *Public Opinion Quarterly, 36,* 176–187.

Shoemaker, P. J. (1993). Critical thinking for mass communication students. *Critical Studies in Mass Communication, 10,* 98–111.

Shoemaker, P. J., & Reese, S. D. (1991). *Mediating the message: Theories of influences on mass media content.* New York: Longman.

The Curriculum Is the Future

by Lana F. Rakow, University of Wisconsin–Parkside

It has been more than 10 years since the field of communication has been put under its own lens for study. These 10 years have been marked by calls for change in research priorities, political alliances, and methodology. To what effect? Has the field of communication studies changed substantially as the result of the past decade of "ferment"? And how would we know if it had?

To support the position that the field has undergone a substantial change, we certainly can point to the burst of new scholarship appearing in old journals, new journals, and books. We can point to edited volumes that argue about our methodological and political differences. We can attend panels at the conferences of our professional associations that feature critical and interpretive approaches to the study of communication, where previously quantitative and scientific approaches ruled the day. We can note that feminist and ethnic studies scholarship has won a place in some quarters of the discipline. Does all this amount to a revolution in the field?

One way to find out is to look at the curricula of our undergraduate and graduate programs. The way we conceptualize and divide up our field of study, the priorities we give to certain approaches and methodologies, the inclusion of some topics and exclusion of others, the theories and skills we deem necessary for students to learn—all of these and more result in, and in turn are produced by, the curriculum of a program. If we are looking for a significant change in the way the field has been mapped and in the assumptions upon which it rests, the curriculum should tell us about that change. Has there been a wholesale transformation of communication curricula? In a word, no. Are there some signs of "ferment in the curriculum"? Yes, but actual change is yet modest at best (see Dickson & Sellmeyer, 1992; Wartella, 1991). To test my conclusion, take a look at the

Lana F. Rakow is associate vice chancellor for undergraduate studies and associate professor of communication at the University of Wisconsin–Parkside. Prior to taking her current position, she was chair of the Communication Department at the same university. The author would like to acknowledge the faculty and staff of the UW–Parkside Communication Department who produced the department's current curriculum, its mission, and its expectations in the form of competencies. They are Wendy Leeds-Hurwitz, Judy Logsdon, Tomas Lopez-Pumarejo, and Monika Strom. This article represents the opinion of its author, however, and should not be taken to represent the opinions of the Communication Department.

department you are sitting in right now. Is it doing business much as usual—that is, much as it was doing "pre-ferment"? Browse through the position advertisements in the newsletters of the International Communication Association, Speech Communication Association, and Association for Education in Journalism and Mass Communication (few though they may be in these economic times). Are departments still looking to fill vacancies in narrow professional tracks? The next time you conduct a search to fill a position in your department, consider the academic preparation of the applicants fresh out of graduate schools. How many have had some course work in critical and interpretative approaches to communication? How many have been in graduate programs that did not have a feminist scholar or ethnic studies scholar on the faculty?

No, our curricula—even more so at the undergraduate level—have remained remarkably unchanged, suggesting that no shift in power away from the traditionally dominant approaches to the field has occurred or that any significant accommodation has occurred. Yet if we want to see a change occur in the field—and I, for one, believe a change is imperative—curricular change is where we should begin. The curriculum is not simply a thermometer that measures the state of the field. It is a thermostat that regulates its future direction. If we want to change the future, we must change the curriculum.

What is the current state of the communication curriculum? By and large, it is (a) gridlocked by real and imaginary obligations to students and employers, (b) fragmented and overly specialized, (c) partial and incomplete, and (d) passive in response to the pressing need for global change. In contrast, a curriculum of the future, a curriculum *for* the future, would be (a) independent, (b) integrated, (c) inclusive, and (d) visionary.

The Independent Curriculum

The first step in making curricular change is to recognize its political nature. Programs in professional journalism, public relations, advertising, and broadcast production provide the most obvious illustration of how political and economic ties constrain and shape curricula. In these programs, students are trained to take up their occupational places in large, powerful organizations that exist primarily to make a profit (despite how good individuals and professions may feel about the value of what they do). To serve businesses and media industries is profoundly political. Their interests should not be served above others. A university should serve the interests of the public, first and foremost, rather than those parts of society with the most money and influence. There are those, of course, who claim that serving the needs of media industries in fact serves the interests of the public. However, the conflict of interest inherent in designing the communication curriculum to meet the needs of industry is brilliantly displayed by the revelation that newspaper editors put mass communication courses about history and theory at the bottom of a list of courses they believe journalism graduates should take (Henley, 1992). Should we have supposed that those in media industries would want their employees to have a critical perspective on their employing institutions? We must come to the realization that if communication graduates cannot engage in a critical and informed analysis of the mass media, we cannot expect that others will. If media industries shape our curricula, it is unlikely our graduates will be doing their job.

Those who support the ties of professional programs to media industries make the argument that these professional programs do serve the public's interests in that these industries provide *employment* for communication graduates. The success of our students is itself a public service, it could be said. Furthermore, freedom of speech, represented by a vigorous media system fueled by communication graduates, is a noble, public good.

The employment argument is an increasingly weak reason for specialized professional programs. Blanchard and Christ describe the "corn-hog cycle" (a concept used by economists) of communication programs (1993, p. 39). In essence, this phrase refers to a cycle of supply and demand of graduates from certain majors, where popularity of the major waxes and wanes because of cycles of over- and undersupply of graduates. The current waning interest of students in print journalism in favor of advertising and public relations exemplifies the cycle. Eventually, the authors argue, demand for advertising and public relations will decline with the oversupply of graduates, and programs will face too many and highly paid faculty in a dwindling curricular area.

An important counter to the employment argument to justify the curriculum is the fact that technologies, media industries, and careers are changing so rapidly and in such unpredictable directions that preparing students for today's media occupations is shortsighted and ill-advised. Preparing students for "successful" careers is just as troubling. In making an argument that the world is actually worse off as a result of our educational system rather than better, Orr makes a compelling statement about the potential value of education:

> *The plain fact is that the planet does not need more "successful" people. But it does desperately need more peacemakers, healers, restorers, storytellers, and lovers of every shape and form. It needs people who live well in their places. It needs people of moral courage willing to join the fight to make the world habitable and humane. And these needs have little to do with success as our culture has defined it.* (1992, p. 54)

The argument that support of the media is of importance to the public good because of the fundamental role in a democracy of freedom of speech and the public's right to be informed also can be refuted. The freedom of speech principle that undergirds most communication programs may not serve the public interest. It should be exchanged in favor of a more equitable principle: *the right to communicate.* Unlike the principle of freedom of speech, which has been interpreted to mean freedom from *government* restrictions on speech, the principle of the right to communicate is based on the premise that everyone has the right to free and equal *voice,* that is, to the presence and weight of their participation in public discussion. The freedom of speech principle has protected the voices of those with power and resources to speak, even at the expense of the speech of those without power and resources. The right to communicate would ensure that even those without power and resources have the opportunity and the mechanism to participate in public discourse.

A communication program needs a curriculum that is independent of the narrow political and economic interests of industry and the fickleness of the latest employment demands of students. It should exist to serve a larger social good than the individual career success of students and the hiring wishes of media industries. Blanchard and Christ (1993) argue that our current professional programs are really occupational programs, having abdicated a fundamental component of professionalism—dedication to public service. Their point should be well taken.

The Integrated Curriculum

The curriculum of and for the future should be holistic, that is, it will need to integrate what are now fragmented subfields of speech, interpersonal, organizational, and mass communication, all of which are further fragmented into specializations. Becker (1992) reminds us that the goal of our field is—or should be—to understand the whole. While American universities have inherited the European model of compartmentalization of knowledge, communication studies should serve as a model in breaking down disciplinary walls.

One reason for integrating the study of communication in and through the curriculum is that the divisions are becoming less and less sustainable. The development of new technologies that straddle the realms of interpersonal communication and mass communication provide an obvious illustration. As a consequence of our artificial boundaries, little attention has been given to these technologies even as they are embedding themselves further into our social, political, and economic lives. In short, the categories we have created for the study of human communication have emphasized some things at the expense of others, have made our understanding of the world incomplete. How can we hope to understand people's relationships with each other if we do not look at the means (technologies) of communication that hold people together? How can we look at the impact of certain media texts if we do not consider the role of stories and rituals in people's lives?

Another reason to build an integrated curriculum is that we owe it to our students and to ourselves to make sense of our field of study. We must find the commonalities, the patterns, the maps, if we expect our students to leave us with anything more than a collection of disparate courses. Smith chastises us by saying that "students ought not to be asked to organize and integrate what the faculty will not" (1983, p. 15). As we hear more and more about the need for undergraduate curricular reform, at the level of general education and of the major, we are hearing about the need to integrate, to require that students leave us having integrated and made sense of their education. If we want to move beyond our narrow career specializations and if we hope to be successful in our responses to calls for greater accountability, we must be able to articulate what the study of communication is and who and what it is good for. As universities are being urged or required to institute vigorous assessment programs to measure and improve student learning, we will be forced to specify what our students should be learning and how their course work is interrelated in bringing about the common educational objectives of our programs. We are being faced with a wonderful opportunity to make sense of what we are up to as a field in order to rebuild our curricula.

What should our students be learning? We should, for starters, want them to understand how humans make meaningful worlds to live in, how our identities are created through our memberships in cultural categories, how opportunities and problems are produced through our definitions and interpretations, how conflicts are produced through our conflicting definitions and interpretations, how humans interact with each other and the role of technology in changing those patterns of interaction. They need to study knowledge and information, what it is, who is believed to have it, and the means by which people come by it. Ultimately, in all of the subfields of our discipline we are concerned with ontology (the study of how we are) and epistemology (the study of how we know) or, even more simply, with being and knowing.

The Inclusive Curriculum

Despite calls from feminist scholars and race and ethnic studies scholars over the past 10 years, attention to gender and race remains superficial and tangential to most core curricula. It is true that more and more programs are offering a course that deals with gender and one that deals with race or minority communication issues. Most teachers and scholars, however, seem to regard race and gender as subspecialty areas of research that should be, at the most, tolerated or accommodated by the field and the curriculum. They are unaware that the end result that must follow an understanding of the role of race and gender in human communication is a transformation of the field and the curriculum. Race and gender will need to be put at the *center* of the curriculum if we are ever to achieve a curriculum that is not partial and incomplete, one that acknowledges and seeks to understand the broadest range of human communication experience.

A curriculum that seeks to understand the broadest range of human communication experience will need more than a simple addition of a section of speeches by famous American women in a public address course or a section on African American newspapers in a media history course. Giving race and gender their full consideration in the curriculum will require us to consider the standpoint of all the material we currently teach. Whose experiences are enshrined in the canons of theory and research now being taught? Who wrote the theories, who did the research, who was the subject of the research? Whose speech, whose rhetorical strategies, whose relationships, whose experiences with technology, whose media, whose definitions of news are your students learning about?

There are those who argue that the curriculum cannot hope to include the experiences of everyone; it must always be partial and incomplete because a finite number of courses cannot accommodate the infinite variety and complexity of human experience. We must make difficult choices about who and what to include. There simply may not be enough room in the curriculum to include more about the experiences of women and people of color.

If this is the case, then I propose we *do* make hard choices. Here's one. Instead of designing the communication curriculum around the theories, research, and experiences of those who are white and those who are men, let us then design our programs around the theories, research, experiences, and communication needs of the majority of the world's human inhabitants—those who are women and those who are people of color. What would remain of what is now in our curricula? What would replace all the material about whites and men? Few teachers and scholars, I am afraid, could supply the answer, illustrating the degree to which both our curricula and our own knowledge of communication are narrow and inadequate. Can we truly claim to understand human communication when we have examined the experiences of only a small portion of the population? Is it true that studying the experiences of white men somehow illuminates the human condition more than studying African American women?

The Visionary Curriculum

It has become a cliché in education to advocate changes in the curriculum in the name of "preparing students for the 21st century." Even the study of race and gender has been advocated by university administrators, business, and civic leaders under the rubric of preparing students to be part of and to manage the diverse work force of the future. Projects to provide students with access to computers are sold to legislators and benefactors on the basis of the argument that students need to adapt to the latest technological advances of society. The taken-for-granted assumption inherent in these arguments is that somehow the future has already been determined, that it is waiting for us, preformed. All that is necessary is for us, as educators, to look ahead and to prepare students to jump on the moving sidewalk that will take them there. In this regard, communication faculty are no different than faculty in other disciplines, except that predictions about changes in communication needs and technologies are so central to descriptions of the future that the role of communication education has been taken as central to the new world waiting for us tomorrow.

It is time for universities to assert a different relationship to the future. We should not abdicate our responsibility to critique the present and *imagine* the future. If we are not fulfilling this function, who is? Business and military leaders and politicians, all of whom have a vested interest in the future turning out in particular ways. Their ability to define the future and to get us to act as if it is inevitable is the exercise of power. Universities must become those places where what is taken for granted is examined and critiqued, where questions are raised about what should be retained from the past and what is in need of change, where the future can be imagined as a consequence of paths taken and not taken. The study of communication, as ontology and epistemology, should be central to this enterprise of inquiry and imagination. We will have to stop thinking of how to prepare our students for the next century; rather, we will need to prepare them to create it.

Is a communication curriculum that is independent, integrated, inclusive, and visionary practical, even possible? Have I simply described a utopian dream that could never be achieved in the waking reality of university budgets and politics, student vocational interests, and entrenched senior faculty?

These impediments to change will make curriculum transformation difficult, but not impossible. The Communication Department at the University of Wisconsin–Parkside is one program that is self-consciously designed to avoid the common pitfalls of communication departments. Students study communication theory across all subfields of the discipline. They are required to develop an understanding of the role of cultural differences of race and gender in social life. They consider the need for social change as much as they learn about the way things are. They can apply what they have learned in any setting, private or public, personal or professional. To ensure that our program has coherence and that students in fact leave us with what we want them to learn, we have articulated our goals for student learning. Our mission and the 12 competencies we have agreed upon (see Appendix) provide the framework for a curriculum that is independent, integrated, inclusive, and visionary. Students are assessed on their level of competency in each area in our Senior

Seminar course, where they are required to integrate and make sense of their entire program of study in the major.

Surely other programs are working toward a curriculum that is independent, integrated, inclusive, and visionary. When these characteristics can accurately be applied to the majority of communication programs in the U.S., we will know that the ferment in the field at last has produced a lasting and revolutionary change.

Appendix

Mission Statement: University of Wisconsin–Parkside Communication Department

The goal of the Communication Department is to aid students in developing *cultural competence*. Cultural competence is the ability to understand the taken-for-granted communication patterns that create and sustain human relations in their cultural contexts and the ability to apply this understanding to creatively negotiating and/or changing the worlds in which our graduates live and work. The Department fosters cultural competence through experiential learning opportunities within the context of theoretical course work. Students are expected to be able to demonstrate two levels of cultural competence, professional and theoretical, by the end of their program of study.

Professional Competencies:

1. Creating and critically evaluating information, observing patterns and synthesizing points of view, and developing and supporting original positions;

2. Attending to detail and negotiating situational contexts, and reflexively observing one's own place within a context;

3. Identifying and critically analyzing rules and patterns at work in cultural texts ranging from social interaction to media products;

4. Working constructively in small and large groups;

5. Defining, choosing, and implementing solutions to problems;

6. Creating oral, written, visual, and technological messages for a variety of audiences and purposes.

Theoretical Competencies:

7. Understanding communication as the symbolic production of reality;

8. Understanding the self as a social product, sustained through language and interaction, and acting back upon its cultural environment;

9. Understanding the role of communication in creating, sustaining, and changing (negatively and positively) cultural differences (of race, ethnicity, nationality, gender, and social class);

10. Understanding individuals as members of groups, organizations, and communities, and understanding ways groups, organizations, and communities communicate with each other;

11. Understanding the political and rhetorical role of myth, narrative, and ritual in creating meaningful lived worlds, shared within particular communities;

12. Understanding the connections between technology and patterns of human association and consciousness.

References

Becker, S.L. (1992). Celebrating spirit, commitment, and excellence in communication. *Southern Communication Journal, 57,* 318–322.

Blanchard, R. O., & Christ, W. G. (1993). *Media education and the liberal arts.* Hillsdale, NJ: Lawrence Erlbaum Associates.

Dickson, T. V., & Sellmeyer, R. B. (1992). Responses to proposals for curricular change. *Journalism Educator, 47*(3), 27–36.

Henley, D. C. (1992). What do students need more of—theory or skills? *Journalism Education, 128*(16), 12.

Orr, D. (1992, Winter). What is education for? *In Context, 27,* 52–55.

Smith, J. Z. (1983, July/August). Questioning the great, unexplained aspect of undergraduate education. *Change,* pp. 12–15.

Wartella, E. (1991, August). *The integration of journalism and speech in communication: Transcending the professional/non-professional divide.* Paper presented at the meeting of the Association of Schools of Journalism and Mass Communication, Boston, MA.

Fragmentation, the Field, and the Future

by David L. Swanson, University of Illinois at Urbana–Champaign

No one can foretell the future, as I was reminded recently by an economist, of all people. Taking the reminder to heart, this short essay concentrates instead on the past and especially on some of the changes that have taken place since "Ferment in the Field" was published in 1983. The goal is to identify what these changes may imply for the prospects of the academic field of communication in the years just ahead.

One problem faced by essayists who ponder the field's future is disagreement about what constitutes "the field." As some critics noted at the time (Cronkhite, 1984; Wiemann, Hawkins, & Pingree, 1988), "Ferment in the Field" sidestepped this question, with most contributors concerning themselves only with one subfield—mass communication. In fact, it is not uncommon for authors to predict the coalescence of a unified or at least coherent field of communication simply by excluding subfields that don't fit the new unity. For example, rhetorical studies, our oldest subfield, seems to be exiled to irrelevance by Wiemann et al.'s (1988) conception of our emerging disciplinary identity as a social science and by Berger and Chaffee's (1988) prediction that the various subfields will be held together by a common interest in cognition.

In fact, disciplinary identity always has been problematic for communication, and speculation about the field's future must recognize the importance of long-standing identity issues. Communication benefited from the rapid growth and reorganization that occurred in American universities following World War II, when many of our departments were established. But the field emerged from that era with an identity that in some ways struggled against the then-developing, more specialized conceptions of how academic disciplines do their work. As the sorting of disciplines into humanistic and social scientific modes of operation was being completed, communication described itself as a humanity but soon began to develop a social scientific thrust. As arts and sciences colleges were being subdivided into more narrowly drawn confederations based on more differentiated models of disciplinary practices, communication committed itself simultaneously to conventional modes of humanistic

David Swanson is a professor and associate head and the director of graduate study in the Department of Speech Communication, University of Illinois at Urbana–Champaign.

and social scientific scholarship (as in rhetorical studies and the newly developing communication research), to performance (in theatre and interpretation), to professional media training and policy concerns, and to biologically and behaviorally focused research and training in speech pathology and audiology. Because its far-flung alliances cut across the emerging boundaries that organized academic work, the identity of the field of communication was indistinct and confusing. In the intervening years, programs in areas such as theatre and speech and hearing science have gone their own way as autonomous departments on many campuses, but uncertainty remains about what the field of communication is and therefore by what standards its accomplishments and value should be judged.

In many respects, the years since "Ferment in the Field" have brought good news for the broad field of communication, however defined. Questions about communication have been foregrounded by pressing social concerns and political, institutional, and technological changes: development of new technologies of communication and information; privatization and globalization of mass media; growing internationalization of commerce; democratization in the former Soviet Union and elsewhere; efforts to develop more successful multicultural societies and institutions; desire to redefine traditional social roles, expectations, and forms of interaction; and so on. Such developments have given new importance and relevance to several domains of communication research, and the last decade saw a great outpouring of work on both new and traditional topics filling an expanding array of journals and book series.

Rising interest in questions about communication has occurred across a range of academic disciplines and in many countries of the world. As a result, there has been a steady increase in scholarship about communication conducted in many different disciplines and in a growing number of nations. The popularity of communication as a research topic has created new opportunities for those working in the field of communication to expand conversation and collaboration with interested colleagues across disciplinary and national boundaries. Indeed, communication research has become more international in scope as the number of international exchanges, conferences, publications, and collaborative projects grows each year. Within the American academy, links between researchers in communication programs and scholars in other disciplines also have expanded on a scale not seen before.

Heightened interest in the study of communication, particularly its practical applications, also led to large increases in undergraduate enrollments in communication curricula throughout the 1970s and 1980s. The number of bachelor's degrees granted annually in the U.S. in all the various communication subfields increased by 385 percent from 1970–71 to 1989–90 (that is, from 10,324 to 50,063 graduates per year), more than three times the rate of increase in total baccalaureate degrees awarded in all disciplines (National Center for Education Statistics, 1992). Although increases in master's and doctoral degrees awarded during this period have not kept pace (130 percent and 81 percent, respectively), the dramatic growth in undergraduate enrollments has given communication programs new visibility and prominence.

The present popularity of communication research and instruction may offer opportunities finally to consolidate our field's position within the academy. These opportunities are timely, as American universities have entered what is likely to be a prolonged period of intense, ongoing evaluation and restructuring to deal with the increasingly keen competition for scarce resources. It is important to take full advantage of present opportunities because resource priorities established during the era of reassessment and reallocation now underway might well influence the size, scope, and structure of our field for years to come. Toward that end, the following account describes some developments that have taken place in our field since "Ferment" and suggests their implications on the question of how to capitalize on our present circumstance.

Accelerating Fragmentation

One striking feature of the last few years has been rapid growth of interest in problem-centered and context-based communication study, as reflected in the expansion and proliferation of applied, context-defined subfields of communication research and instruction. In the time since "Ferment" was published, five new divisions and interest groups have been created within both the International Communication Association (ICA) and the Speech Communication Association (SCA), but not the same five in each association. The latest additions continue the process—which has been underway for some years—of institutionalizing applied subfields such as political, intercultural, instructional, health, and organizational communication. Other applied areas, such as family and marital communication, are awaiting their turn. With no commensurate increase in the size of departments, new subfields have developed through reallocation of resources from what used to be regarded as foundational, core areas such as interpersonal communication and communication theory.

In one sense, the proliferation of subfields may be seen as resulting from normal disciplinary development toward increased specialization. Indeed, the emergence of specialized research communities devoted to pursuing subjects in great depth might be taken as a mark of disciplinary maturation. However, many communication subfields seem to be evolving along a trajectory that, over time, leads away from identification with any parent discipline. The current trend appears to be for subfields to define themselves as interdisciplinary undertakings. They gauge their success in part by their ability to attract researchers from a range of disciplines, and they endeavor to legitimate themselves by developing distinctive theories, methods, or syntheses of multiple disciplinary perspectives that will differentiate them from parent disciplines and from other subfields. This pattern of development can be seen in Nimmo and Swanson's (1990) description of the evolution of the subfield of political communication, which has emerged from a coalition of researchers formerly working in discipline-based specialties in communication (e.g., political rhetoric, media effects), political science (e.g., politics and media), and other fields. The same aspirations for autonomy and independence are evident in calls for distinctive and specific theory and research in the interdisciplinary subfields of marital communication (e.g., Fitzpatrick, 1987), health communication (e.g., Pettegrew & Logan, 1987), and intercultural communication (e.g., Casmir & Asuncion-Lande, 1989), to name but a few examples.

As interdisciplinary subfields grow and develop their own organizations and publication outlets, participants may come to orient to the subfield more than to their parent discipline as the primary site of scholarly work and interaction. The subfield may become inward-looking and self-absorbed as it searches for its own center. As a result, there tends to be perhaps less intellectual exchange than we might expect between the subfields and the core domains of the disciplines out of which they arose, as Berger (1991) has noted. The core domains begin to decompose into narrowly defined subfields that struggle against disciplinary traditions and limitations in order to build more focused, discipline-spanning research communities. Subfields developing in this way transfer the field's intellectual capital from the center to the periphery. This is why, on balance, the proliferation of interdisciplinary subfields has been a centrifugal force, straining communication's already problematic disciplinary identity.

Centrifugal forces fragmenting the field have been energized in a somewhat different way by the recent coalescence of new subfields devoted to culturalist approaches (e.g., critical and cultural studies, feminist scholarship). None of these approaches has any special interest in communication; all have generated work across the humanities and social sciences. But the field of communication has been especially hospitable to culturalist approaches and has moved perhaps farther and more rapidly than some other fields in legitimating and supporting this work. At one level, culturalist approaches might be thought to exert a centripetal influence on the field of communication through their contention that every subfield should be redefined to focus on questions of power and representation. That is, the thematic preoccupations of culturalist approaches might be viewed as a meeting ground for uniting disparate, context-based communication subfields. In fact, culturalist approaches turn out to have further fragmented the field by their essential posture of critique. In rejecting as wrongheaded or trivial the viewpoints, methods, and questions of other subfields, and in insisting that all subfields be redefined in terms of culturalist themes (e.g., Bowen & Wyatt, 1992), these approaches have distanced themselves from the context-based subfields and simply added more centrifugal voices to the cacophony in the field of communication.

The foregoing description is not intended to object to recent developments in the field or to suggest that fragmentation is a new phenomenon. The field of communication always has been a loose confederation of independent enterprises pulling in different and sometimes opposing directions, and fragmentation has long been a concern. Some readers will recall that the theme of the 1977 annual SCA meeting concerned the search for "a center which holds." The theme responded to "uneasiness" that, having grown to 9 divisions, 5 sections, and 4 subsidiary groups, the association (and the field) lacked "a commonly agreed-upon central focus or perhaps even central foci" (Blankenship, 1978). Today, that disciplinary organization has become a conglomerate of 13 divisions, 5 sections, and 25 assorted caucuses, commissions, committees, and councils.

The fragmentation of communication studies is not a bad thing insofar as it reflects responses to pressing issues and developments, as well as the high level of specialization that is needed to foster research and teaching of quality. Moreover, the particular kind of fragmentation we have experienced has led to the development and rapid expansion of beneficial ties with other disciplines. These connections are occurring within particular interdisciplinary subfields, where growing numbers of scholars from a range of disciplines have established formal and informal relationships with colleagues in communication to explore common interests and pursue collaborative projects. The late Wilbur Schramm probably was right when he observed in "Ferment" that whether or not communication "comes to be recognized generally as a discipline . . . is not a matter of first importance. More important is whether it continues to provide a center of scholarly excitement" (1983, p. 14).

What does seem to be new in the present circumstance is a greater degree of fragmentation than the field has seen before, and a different, more substantive kind of fragmentation created by the transfer of resources from disciplinary specialties and core areas to interdisciplinary subfields and culturalist research communities. These new elements of fragmentation have important implications for the future of the field.

Departmental Specialization

Those who predict the future in American higher education seem to agree that the present scarcity of resources will continue and will force reevaluation of current commitments to comprehensiveness in institutions and programs. Some states already are moving to distinguish more sharply among the missions of their colleges and universities, and many comprehensive universities are devising plans to narrow their scope by reallocating resources to fewer programs that can be sustained at a high level of quality.

Within communication programs, processes of differentiation and specialization have been underway for some time. Virtually none of the units has had the resources both to maintain programs of depth and quality in all the major, traditional domains of communication study and to create faculty groups devoted to each new subfield as it has come on line. Instead, units have had to choose which disciplinary areas will be supported. As a result, students' choices of graduate schools today are shaped by which units offer programs in their desired area of study to a much greater extent than was the case as recently as a decade ago.

It seems likely that, because of economic constraints and continuing subfield proliferation, units will have to become even more selective in deciding what areas to support in the years ahead. Their choices will reflect their existing strengths and local circumstances and opportunities. Different units will make different choices, and communication programs will resemble each other less and less. Because of the natural affinities of certain groups of subfields, a few basic models and patterns of co-occurrence no doubt will emerge, such as cultural studies, feminist scholarship, philosophy of communication, and popular culture; mass and political communication; and organizational communication, communication technology, and information systems (see Barnett & Danowski, 1992). But, increasingly, the rule across units devoted to communication study is likely to be difference, not similarity.

As communication programs become more specialized and more sharply differentiated, nagging questions about the field's identity, assumptions, and structures may become key issues to be confronted. When programs become less comprehensive, students have fewer opportunities for exposure to the full range of concerns represented in the field at large, and graduates of differently focused programs are likely to hold different conceptions of what the field is. Inevitably, "the field" loses meaning and salience as an object of orientation; instead, particular subfields and clusters of related subfields become the primary structures of identification and reference. In this context, traditional "umbrella" disciplinary associations and scholarly journals that try to represent all subfields may face an uncertain future, unable to compete with more narrowly drawn organizations and publications based in particular subfields or clusters of related subfields.

The consequences of departmental specialization and differentiation seem to lead in the direction of breaking the field apart, to be replaced by free-standing subfields configured in various ways in different departments. Set against the powerful forces pushing the field apart, the old nostrums that have been offered to define a discipline that holds us all together—for example, "we are bound together by an interest in and concern for 'persons communicating,'" and "[we are] concerned with research *and* teaching *and* practice" (Blankenship, 1978, pp. 20–21)—no longer impress.

The Golden Future Time?

In the foregoing analysis, we recognize communication as an interdisciplinary field composed of a number of diverse subfields that are connected politically by the organization of communication programs and historically, but not intellectually. At a global level, these subfields do share an interest in "communication," each in its own terms and for its own purposes, but it would be disingenuous to portray that shared interest as having substantive content and providing a common disciplinary perspective. The field's development in recent years has tended in the opposite direction, to ever-greater differentiation and wider dispersion rather than to convergence.

This does not mean that the subfields have not influenced each other. Cohabitation within the same departments and colleges does have its effects. For example, due largely to the influence of culturalist approaches, researchers in other subfields have come to better appreciate the political dimensions of communication forms and institutions and have been sensitized to the importance of attending to the broad social and political context in which their empirical questions are posed. And, there have been attempts to create connections between subfields, such as efforts to link mass communication and interpersonal communication (e.g., Hawkins, Wiemann, & Pingree, 1988; Rubin & Rubin, 1985) and to build bridges between rhetoric and critical theory (e.g., McKerrow, 1989) and between cultural studies and mass media social science (e.g., Hay, Wartella, & Grossberg, in press). In the main, however, these efforts have come to little, which is perhaps an indication of the major differences in orientations and interests that separate the subfields.

So understood, how may our diverse field position itself to succeed and thrive in the increasingly heated competition for resources within the academy? The future of communication in the more intensely quality-conscious, cost-conscious, and "accountable" academic environment can be good, so long as we recognize what is required to succeed and exploit our strengths and legitimate claims to support.

Communication is not likely to succeed by claiming to be what it is not, that is, by insisting on its centrality as a discipline. It is not a discipline, at least in any traditional sense, and it will be helpful to discard the contrary view once and for all. That view is wrong, has led to much energy being wasted over the decades in a vain search for the "foundations" or "essential content" of "our discipline," and gets in the way of understanding clearly the nature of our field. It may also be harmful insofar as it encourages us to be inward-looking and stay at home intellectually.

Although not a discipline, communication nevertheless is quite central, both in terms of demand for instruction and in its growing intellectual exchanges and partnerships with disciplines across the humanities and social sciences. Communication is an interdisciplinary field where intellectual

work proceeds free of the hegemony of an imposed (or opposed) disciplinary perspective that, in some other fields, rules certain ideas or methods
out of bounds or at least suspect. This is precisely why interdisciplinary
subfields developing out of communication are proving so attractive to so
many researchers from other disciplines. The space for innovative, interdisciplinary work provided within our subfields is a major strength of the
field that, if recognized and nurtured, draws us toward the center of intellectual life in the academy from our traditional perch on the periphery.

The reasons for our existence as a field do not depend on clinging to
the fiction that we are a discipline. It does not follow from recognizing
our status as an interdisciplinary field that, therefore, our various subfields should break apart and seek independent status. Something like the
present organizational structure of the field, shaped and elaborated by
whatever new subfields come along, ought to be retained. The politics of
academic life are such that it is better to be larger than to be smaller.
Small, unidimensional, multidisciplinary programs have no future. Although the field's work will be differentially represented and configured
in departments that will be somewhat more narrow and specialized, it
will continue to be the case that units will be more secure to the extent
that their programs and missions are multifaceted. The traditional structure of alliances that makes up the field has not hampered, and indeed
has encouraged and supported, the development of subfields of all sorts,
and it needs to be retained in some fashion.

Perhaps the most important factor in our future success as an avowedly
interdisciplinary enterprise will be the quality of our work, as is proper.
Even high demand for instruction is unlikely to save programs of doubtful
quality from reduction to service roles, where their mission will become
delivering undergraduate instruction on a large scale at bargain basement
prices. The turn to a more interdisciplinary orientation means that our
work is addressed to wider audiences in more cosmopolitan forums. We
will continue to succeed in these forums and to expand our interactions
across disciplinary boundaries to the extent that our work aspires to the
most demanding standards of quality. Appropriately, we have been very
concerned with inclusiveness, representativeness, and broad participation in organizing our conferences and doing our other work. These important goals can and should be pursued without developing local, more
forgiving standards of quality that conflict with those of the larger academic community.

The future of the field as an academic enterprise also is linked directly
to the quality of the education we provide. Graduate and undergraduate
education pose particular problems in a field that is expanding rapidly, is
balkanized, and lacks a disciplinary center that would define an intellectual tradition to relate its diverse undertakings. As programs become less
comprehensive, the tension between breadth and specialization in students' programs of study will become a difficult issue to resolve. An
equally thorny issue will concern the appropriate relationship between

the field's graduate teaching and research directions on the one hand, and its undergraduate curricula and support base in basic skills instruction on the other hand.

On the foregoing reading, the field of communication enjoys access to the ingredients of success in the academic environment of the 1990s. The field commands a large base of instructional demand, intellectual connections with an expanding range of disciplines, subfields structured to support innovative and exciting work, and a subject matter that has come to be recognized as being of enormous social, political, economic, and cultural importance to the worlds of today and tomorrow. The task of holding these various elements together will become increasingly difficult, but the field is well positioned to thrive if it recognizes and capitalizes on its strengths, and if its commitment to quality in its intellectual work and its instructional mission is uncompromising. The field's future may not quite be the "Golden Future Time," but it promises to reward those who are equal to its challenges.

References

Barnett, G. A., & Danowski, J. A. (1992). The structure of communication. *Human Communication Research, 19,* 264–285.

Berger, C. R. (1991). Communication theories and other curios. *Communication Monographs, 58,* 101–113.

Berger, C. R., & Chaffee, S. H. (1988). On bridging the communication gap. *Human Communication Research, 15,* 311–318.

Blankenship, J. (1978, February). Presidential message. *Spectra* (newsletter of the Speech Communication Association), pp. 1, 20–22.

Bowen, S. P., & Wyatt, N. (1992). *Transforming visions: Feminist critiques in communication studies.* Cresskill, NJ: Hampton Press.

Casmir, F. L., & Asuncion-Lande, N. C. (1989). Intercultural communication revisited: Conceptualization, paradigm building, and methodological approaches. In J. A. Anderson (Ed.), *Communication yearbook 12* (pp. 278–309). Newbury Park, CA: Sage.

Cronkhite, G. (1984). [Review of *Ferment in the field: Communications scholars address critical issues and research tasks of the discipline,* by G. Gerbner & M. Siefert, (eds.)]. *Quarterly Journal of Speech, 70,* 468–473.

Fitzpatrick, M. A. (1987). Marital interaction. In C. R. Berger & S. H. Chaffee (Eds.), *Handbook of communication science* (pp. 564–618). Newbury Park, CA: Sage.

Hawkins, R. P., Wiemann, J. M., & Pingree, S. (1988). *Advancing communication science: Merging mass and interpersonal processes.* Newbury Park, CA: Sage.

Hay, J. A., Wartella, E., & Grossberg, L. (Eds.). (in press). *Towards a comprehensive theory of the audience.* Boulder, CO: Westview.

McKerrow, R. E. (1989). Critical rhetoric: Theory and praxis. *Communication Monographs, 56,* 91–111.

National Center for Education Statistics. (1992). *Digest of education statistics 1992* (NCES Publication No. 92-097). Washington, DC: U.S. Department of Education, Office of Educational Research and Improvement.

Nimmo, D., & Swanson, D. L. (1990). The field of political communication: Beyond the voter persuasion paradigm. In D. L. Swanson & D. Nimmo (Eds.), *New directions in political communication* (pp. 7–47). Newbury Park, CA: Sage.

Pettegrew, L. S., & Logan, R. (1987). The health care context. In C. R. Berger & S. H. Chaffee (Eds.), *Handbook of communication science* (pp. 675–710). Newbury Park, CA: Sage.

Rubin, A. M., & Rubin, R. B. (1985). Interface of personal and mediated communication: A research agenda. *Critical Studies in Mass Communication, 2,* 36–53.

Schramm, W. (1983). The unique perspective of communication: A retrospective view. *Journal of Communication, 33*(3), 6–17.

Wiemann, J. M., Hawkins, R. P., & Pingree, S. (1988). Fragmentation in the field—and the movement toward integration in communication science. *Human Communication Research, 15,* 304–310.

The Purebred and the Platypus: Disciplinarity and Site in Mass Communication Research

by Anandam P. Kavoori, University of Georgia, and Michael Gurevitch, University of Maryland

Aboard *The Beagle,* Charles Darwin mulled over the history of the finches, their breeding habits, their constitutiveness as a species. His ruminations, enshrined as the theory of evolution, also begat an idea about what constitutes natural order: taxonomic discreteness, coherence, and hierarchy. That legacy and much of the history of science has traditionally determined our understanding of the social order—ideas about race, religion, culture, and even ways of thinking (i.e., paradigms).

Today, of course we know that zoological classifications are, like other taxonomies, constructs—culturally relevant labels and designations (e.g., mass communication research) designed to impose intellectual order on a chaotic universe. That knowledge is a social product and is amenable to change is a truism. What is less often acknowledged is the contingent, necessarily political nature of institutional knowledge, which has come to be characterized by that most crucial of terms—*discipline.*

In this essay we offer some thoughts about some of the problems facing mass communication as a cultural practice by mapping briefly its historical constitutiveness (and the problems therein) and then discuss the dimensions of mass communication research as site. While not advancing a formal thesis, we do offer a diagnosis of how to view the avowed "fragmentation" of the field. Thus, the aim of these reflections is to provide a starting point for a discussion of the now and future state of the field of mass communication.

The Purebreds

The ongoing ferment over the past, the present, and the future of mass communication research is tied to an ongoing urge for the imposition of

Anandam P. Kavoori is an assistant professor in the College of Journalism and Mass Communication at the University of Georgia. Michael Gurevitch is a professor at the College of Journalism, University of Maryland, and an associate editor of the *Journal of Communication.*

order—a tendency that we have labeled *disciplinarity*. By disciplinarity we mean an essentialist tendency in the production of academic knowledge that produces a set of theoretical and methodological axioms, and then formalizes them as dogma.

The history of mass communication reads nearly perfectly as a negotiation with disciplinarity. Sanctified in the "Ferment in the Field" issue of this journal, the dominant metaphor for discussion of mass communication research has been that of conflict, being played out ceaselessly in scholarly conferences, college assemblies, classroom discussions, and appointment committee meetings. At issue here is that while differences between the social scientific and the culturalist/critical traditions have drawn upon different intellectual traditions, it is the conscious ideologization of those traditions that has generated the problem of an unbridgeable divide. In both critical and social scientific discourses there have been conscious attempts to create a set of presupposed truths about social reality. Both approaches have attempted to create "true fictions"—fictions of essentialized discourses, of being purebreds. They have attempted to do so by employing traditional strategies of staking out territories. These include, as Grossberg, Nelson, and Treichler (1992) put it, "developing a unique set of methodological practices, and carrying forward a founding tradition and lexicon" (p. 2).

In the social scientific tradition, this was achieved through the active incorporation of mass communication research as an empirical, applied, and scientific discipline. By inserting itself into the dominant academic discourse of postwar America, mass communication research insinuated itself powerfully into the academic domain as a field in its own right. In drawing on pragmatism as its philosophical context, it added to the celebration of instrumental values and the practicality of human action that reflected the dominant values of liberal pluralism (Hardt, 1992, p. 5).

Central to this stage in the historical striving of mass communication research toward the status of a legitimate discipline were the institutionalization of applied research and of survey research as the method of choice (Delia, 1987). In many ways this was a distorted picture, for it ignored studies by Lazarsfeld, Berelson, and Gaudet (1944), Merton and Kendall (1946), and others who deployed in their work open-ended interviews and other qualitative methods. Equally ignored in this characterization was the interdisciplinary background of the "founding fathers" of communication research, who came from diverse established disciplines. There was a tension, as Delia points out, between the researchers' background and their "formation of an autonomous area of study" (1987, p. 72).

But regardless of the multidisciplinary reality of mass communication research at this early stage, an ideology of theoretical and methodological coherence was adopted. The tendency toward disciplinarity grew out of this early institutionalization of mass communication research as an autonomous and coherent field, thus obscuring its more eclectic origins. In its desire to be a purebred, social scientific research disavowed all of its original mongrelization.

A similar, albeit more restricted movement toward disciplinarity took place in the critical tradition, especially as it came in contact with the mainstream American tradition. Born of multiple, often embattled traditions (such as Marxism, structuralism, semiotics, psycholinguistics) critical media work, in its American avatar, drew almost iconically on the Birmingham School, and evolved a lingua franca: ideology, hegemony, alienation, resistance, empowerment. It drew an essentialized picture of itself as it rigorously critiqued the ahistorical, unreflective nature of American mass communication research, and presented itself as a continuous alternative to the dominant tradition, which it represented as inert and unchanging. It criticized the inclination of the mainstream tradition for being oblivious to the commercial and political interests that generated and shaped its basic research models (Delia, 1987). In practice, however, both the critical and the social scientific schools were closely tied to institutional agendas: funding agencies on the one hand and academic institutions with a manifest agenda (e.g., the New School for Social Research) on the other.

Overall, while there was a divergence of intellectual traditions within the critical school, there was also a great deal of symbiosis. For instance, Curran (1990) points out that despite substantial differences between the political economy and culturalist strands of critical scholarship, they both worked with a "neo-Marxist model of society; both perceived a connection between economic interests and ideological representations and both portrayed the media as serving dominant rather than universal societal interests" (p. 139). This need for commonality, and for closure, became especially evident in the interaction between the mainstream and the critical approaches. In the course of this interaction, critical scholarship increasingly assumed a tendency toward disciplinarity, while often claiming *interdisciplinarity* as its emblem. The purebreds, whether or not they existed in reality, certainly existed as fictions, as ideologies reflecting essentialized notions of intellectual knowledge.

The Platypus

As U.S. mass communication research developed, some of the tensions inherent in its (avowed) initial constitution began to become evident. The purebred began to show considerable impurities. With the onset of a range of intellectual influences from Europe, it rapidly began to turn into a platypus. Today, it is full grown.

The platypus, of course, is a zoological embarrassment; half bird and half mammal, it defies classification. The ceaseless bemoaning of the fragmentation of the field is symptomatic of its practioners' discomfort with its current constitution. The marginalization of the platypus is evident in how the "mother" disciplines treat it—by largely ignoring it ("nobody knows who we are," goes the lament)—and in the self-image of communication scholars ("we don't know who we are").

At issue is whether the current bemoaning over the field's fragmentation indeed reflects the predicament of communication research today or whether in the lament of the platypus lies the continuing search for the purebred. Clearly, as mass communication research develops in the future it will continue to search for some shared understanding as to its identity. The question is whether the nature of this debate should be framed in terms of a "crisis" of disciplinarity. That need prompted the "Ferment in the Field" issue, as well as this issue. Such deliberate soul-searching every 10 years or so is exciting intellectual stocktaking, but it also fuels the institutional and ideological quest for disciplinarity. Once the discussion is framed around issues of boundary maintenance and institutional definition, the issue of constitutiveness, of disciplinarity, is never far behind.

In fact, our present-day platypus is at risk of becoming an even stranger creature by calls to order, such as this issue. As Hardt puts it, "notions of compromise or friendly accommodation in the spirit of common interests [have] clouded the potential for the emergence of real differences and radical changes" (1992, p. 21). It seems to us that while such real differences and radical changes are in evidence in the range of work that is being done in the field, what is equally evident are the commonalities in a number of research areas that reflect a *composite* of theoretical, methodological, and paradigmatic continuities (rather than the radicalization of traditions that seems to reflect a continuing negotiation with disciplinarity). We discuss some examples of these research areas below.

What is needed in evaluating these composites is a new language that allows us to see how mass communication is being constituted as a *site*. The notion of site is not new, but it is one that needs to be problematized without erring into the academic equivalent of "anything goes," or reiterating arguments about media as cultural forum, and academe as a marketplace of ideas.

A first step is to point out that neither the purebred nor the platypus are adequate *descriptive* categories. They are prescriptive, ideological variants asserting disciplinarity in the first instance and bemoaning the lack of it in the second. A more useful way to characterize mass communication research is as a site, where site is not equated with "domain" but rather is seen as a *configuration of research influences.*

The operational term is *configuration;* it alludes to a range of intellectual traditions rather than a single-focus vision driven by disciplinarity. Moreover, that configuration changes when we look at different research areas or sites. The multiplicity of interconnections and influences in mass communication research can then be adequately represented by a range of research sites, all of which are differently configured. Equating the field with a single, specific site is still a negotiation with disciplinarity. We propose that a plurality of sites, themselves differently configured, can be seen as constituting mass communication research.

To operationalize site(s), one could identify a specific research area (such as reception analysis or media effects) or an entire research tradition (such as social scientific or critical) and consider the terms in which they are configured. That configuration could be studied by looking at two attributes of a site: its *boundedness* and its *contestedness*. Simply put, we need to identify what constitutes the range of influences (boundedness) and the range of disarray (contestedness) within the research area being described. Looking at social scientific research, for example, one could point out that it is bounded by its behaviorist/functionalist framework, but that within that framework there is a surprising diversity of theoretical and methodological approaches. Similarly, critical research is bounded by a Marxist/ neo-Marxist framework, within which there is an enormous variety of emphases, ranging from debates about the dynamics of base superstructure to issues of ideological contestation and the nature of hegemonic coding. The general point here is that these oppositional pulls are not independent of each other but are, in fact, corollaries; it is their duality that can be used to understand how sites are constituted. They also provide for a more *complex* reading of the field's history rather than one framed exclusively in terms of the two major paradigms.

Configuring Reception Analysis as Site

The ethnographic tradition of reception analysis is a research area whose configuration provides a contemporary example of the complexity of communication research sites. Viewed in terms of disciplinarity, reception analysis appears as an exclusively culturalist and postmodernist enterprise. However, its boundedness reveals a more complex configuration.

Usually identified as a move in interpretive research from a literary analysis of texts to the ethnographic analysis of audiences, reception analysis is equally a move toward an empirical, behaviorist orientation for the study of mass communication. Reception analysts attempt to learn how people deal with different media contents. This brings reception analysis close to the study of *effects* and *uses and gratifications*—both studied media consumption in terms of the definitions provided by the consumers. In a sense, then, both empirical and behavioral concerns are implicit in the literature of reception analysis. The methodologies still differ in their relative emphases on qualitative versus quantitative analysis, but there is increased traffic of mutual borrowing and cross-fertilization between the approaches.

This implicit return to behaviorist concerns in reception analysis is also evident in a range of other issues. We discuss some of these, drawing in part on the works of Curran (1990), Evans (1990), and others.

First, underlying much of the work in reception analysis is a postmodern prescriptive that emphasizes the open and unfettered quality of meanings. In many reception studies the media are freed from their role as a class-based institution, shifting it to one that exists primarily in reader–text interactions, characterized by disaggregation of ideological meanings by virtue of their variability of use. What emerges is an examination of the media that does not focus on links of the text to conditions of power, but rather one that denies those links. Unwittingly, underlying the theoretical sophistication of postmodernism is a view not that different from the view of the American liberal tradition that both the media and the audience are autonomous. The popular notion of *semiotic democracy* implies that people draw from a range of subcultures to mediate media texts, thus creating *interpretive communities*. These communities are not defined exclusively in terms of class, ethnicity, or gender but as a mix of these and other categories. They vary from interpretation to interpretation and from individual to individual. The result is a view similar to that of the autonomous individual, deploying individual free will and choice (characteristic of the behavioral approaches) to interpret media texts.

Second, there are continuities with behaviorist approaches in the way in which interpretive and reception approaches look at media institutions. Schlesinger (1989), for example, has argued that while the media may act as primary definers of meaning, they present conflicting frames of reference to the audience because they, in turn, are offered a range of meanings by powerful institutions and interests. This view is similar to the conventional pluralist portrayals of the media as a forum for public debate. Ironically, pluralist research has itself moved away from that position, repudiating the idea of the media as an autonomous fourth estate.

Third, notions of audiences' *pleasure* and *play*—of central importance to many reception theorists—are not exclusive to culturalist analysis. Curran (1990) points to the similarities between Ang's (1985) study of soap opera fans (*Watching Dallas*) and Herzog's (1944) classic uses and gratifications study of the audience of radio serials. Both studies tried to find what made the shows pleasurable for the viewers. In fact, Curran argues that Herzog's study was more oriented to the socially situated ways in which the audience constructed pleasure because she chose to conduct interviews rather than look at letters as Ang did. Similarly, he argues that Radway's (1987) study of readers of romance novels is as much about socialization (which would put it in the pluralist/behaviorist camp) as it is in the radical camp of cultural studies (by adopting a feminist theoretical orientation).

Fourth, there are also some philosophical similarities between the reception-type interpretive approaches and the uses and gratifications approach. Evans (1990) has argued that despite the distinction made between the hermeneutic and positivistic roots of these two approaches, they share a common perception of the individual as a rational creative being. By rational he means that both approaches view the receiver of the media message as actively decoding and manipulating the text. This view of an active, if not hyperactive viewer, is also linked to the romanticization of the power of the viewer over the text—a tendency that is tied to the pluralist celebration of individualism.

These different dimensions of the boundedness of reception analysis suggest how we can configure it as a site. A configuration of the field's contestedness would include the range of culturalist traditions—poststructuralist, feminist, subaltern, ethnic studies with their often fierce debates about representations, identity, and ideology—and the social scientific tradition with its debates on questions of effects, motivations, and cognitions. Thus, examining the boundedness and contestedness of reception analysis reveals how it is configured as a site. Such an exercise offers a more complex and composite account of reception analysis than its rendering in terms of disciplinarity.

Future Directions

The aim of such a rendering of site, then, is to begin to formulate a language with which to speak about how mass communication research is set up. In allowing for complexity of development one begins to see how different areas in mass communication research are configured. Accounts similar to the one above can be provided for other areas of research. For example, in a recent essay McLeod, Kosicki, and Pan (1991) attempted an articulation of the site (though they did not use that term) of media effects research and configured it in terms of both the critical/cultural tradition and the dominant behaviorist tradition.

Of all the recent developments in mass communication research, cultural studies reflects the idea of site most comprehensively. Despite its recent success and its attendant perils of institutionalization, the field of cultural studies still charts an idiosyncratic, antidisciplinary path. It configures a research tradition that—while bounded by a commitment to culture, theory, and politics—experiences highly contested debates in the theoretical and substantive issues it tackles. This is of special significance for the future of the field, since it can be argued that cultural studies is more than a source of important influence in mass communication research; it is the dominant constitutive element in the configuration of the various sites of research. It may be that, as Hardt puts it, the influence of cultural studies may lead to the absorption of mass communication research

> *by a series of cultural and political and economic interests in the relationship between human subjects and society, whose ideas of Cultural Studies embrace questions of communication and who rely on communication research purely for topic references or methodological expertise. There is also the possibility, however, that the domi-*

> *nant perspective of communication research will adapt to the evolv-*
> *ing conditions of multidisciplinary analysis of the social environ-*
> *ment by yielding to alternative definitions or visions of studying so-*
> *ciety with the result of changing the field forever. In either case, the*
> *boundaries of communication research are shifting to accommo-*
> *date the diverse sites of communication in the study of culture and*
> *society.* (1992, p. 8)

This, we would argue, is already taking place in the current configura-tions (in our terms) of the many sites of mass communication research.

Finally, the usefulness of the concept of site becomes clearer when we look at how mass communication research as a cultural practice is situat-ed today. Just as the notion of disciplinarity reflected issues of American (cultural) pragmatism and European (cultural) criticism, the notion of site(s) mirrors the complexity of cultural constitution, fashioned by both cultural localism and forces of transnationalism. In their larger context, the site(s) of mass communication research need to be understood in terms of how communication itself draws from some preeminent condi-tions of cultural constitution. These include its internationalism and cul-tural interpenetration and the generative role of mass communication in creating these cultures.

Are we seeing the end of "paradigm battles"? Is the purebred spent? What will it be replaced by? As we have argued earlier, concerns with the future of the field—hovering between fragmentation and cohesion—re-flect the continued longing for the disciplinarity of the purebred. Viewed through the kaleidoscope of site(s), the future of mass communication re-search is seen instead as a series of divergences and convergences, of shifting coalitions of theoretical and methodological perspectives that are shaped by historical influences and changing traditions. That kaleido-scope will inevitably offer us ever-changing pictures.

References

Ang, I. (1985). *Watching "Dallas."* London: Methuen.

Curran, J. (1990). The new revisionism in mass communication research: A reappraisal. *European Journal of Communication, 5,* 135–164.

Delia, J. G. (1987). Communication research: A history. In C. R. Berger & S. H. Chaffee, *Handbook of communication science* (pp. 20–99). Beverly Hills, CA: Sage.

Evans, W. A. (1990). The interpretive turn in media research: Innovation, iteration, or illu-sion. *Critical Studies in Mass Communication 7,* 147–168.

Grossberg, L., Nelson, C., & Treichler, P. (1992). Introduction. In L. Grossberg, C. Nelson, and P. Treichler (Eds.), *Cultural studies.* New York: Routledge.

Hardt, H. (1992). *Critical communication studies: Communication history and theory in America.* New York: Routledge.

Herzog, H. (1944). What do we really know about daytime serial listeners? In P. Lazarsfeld & F. Stanton (Eds.), *Radio research 1942–1943*. New York: Duell, Sloan and Pierce.

Lazarsfeld, P., Berelson, B., & Gaudet, H. (1944). *The people's choice*. New York: Columbia University Press.

McLeod, J. M., Kosicki, G. M., & Pan, Z. (1991). On understanding and misunderstanding media effects. In J. Curran & M. Gurevitch (Eds.), *Mass media and society* (pp. 235–259). London: Edward Arnold.

Merton, R. K., & Kendall, P. L. (1946). The focused interview. *American Journal of Sociology, 51,* 541–557.

Radway, J. (1987). *Reading the romance*. London: Verso.

Schlesinger, P. (1989) Rethinking the sociology of journalism. In M. Ferguson (Ed.), *Public communication* (pp. 61–83). London: Sage.

Communication Research: New Challenges of the Latin American School

by José Marques de Melo, University of São Paulo

The first research carried out by Latin American scholars on communication processes in Latin American appeared in the 1940s. Early studies focused on journalism and advertising, the hegemonic categories in the development of the emerging cultural industry. These studies coincided with the formation of the first schools of journalism, created in Argentina in 1934, Brazil in 1935, and Cuba in 1942. In fact, the studies were written mainly by faculty members of these schools.

These early studies followed the disciplines of history and law but were influenced by the social sciences. Examples include the trilogy on journalism by Cuban scholar Octavio de la Suarée (1944, 1946, 1948); works on advertising and propaganda by Brazilian scholars Ernani Macedo de Carvalho (1940) and Ary Kerner (1943); the work of Argentinean Carlos Juan Zavala Rodriguez (1947); and studies of the press in Mexico (Miquel y Verges, 1941), Brazil (Rizzini, 1946), and Argentina (Ballester, 1947).

The following decade witnessed an increased diversity of research on the continent's growing communication industries. In Brazil, Salvyano Cavalcant de Paiva (1953) and Alex Vianny (1959) worked on cinema, while Saint-Clair Lopes (1957) researched radio broadcasting. In Argentina, Mouchet and Radelli (1957) studied the artistic copyrights of media. Other scholars, included Brazilian Genival Rabelo (1956), Chilean Alfonso Silva Delano (1960), Mexican Salvador Borrego (1951), Equadorian Gustavo Adolfo Otero (1953), and Venezuelan Julio Febres Cordero (1959). In addition, Brazilian Luiz Beltrão (1960) examined the treatment of current events and printed communication.

In 1959, the establishment of the International Center of Higher Journalism Studies for Latin America (CIESPAL), located in Ecuador under the sponsorship of UNESCO, gave communication research a new face, the face of empirical sociology. This meant the Latin American scholars would assimilate the concepts and methodologies originating from mass communication research at the Chicago School and the *sciences de l'in-*

José Marques de Melo is a professor in the Communication and Arts School at the University of São Paulo.

formation from the Paris School. Research following these two schools prospered in Latin American through the late 1960s. Studies of morphology and press content (J. Fernandez, 1967; Marques de Melo, 1972a) were inspired by Schramm, Kayser, and Deutschmann; studies of audiences and social effects (Da Via, 1977; Samaniego, 1968; Santoro, 1969) by Danielson, MacLean, Dumazedier, and Rogers; and sociographical diagnoses (J. Fernandez, 1965; Mujica, 1967; Verga, 1965) by the studies of Nixon, Leaute, and Rovigatti.

Concurrently, or shortly after the development of this functionalist school, new epistemological trends developed. Originating mainly in Europe, they included Frankfurt School critical theory (Cohn, 1973; Miceli, 1972; Pasquali, 1963, 1967; Sodré, 1972), French structural linguistics (Verón, 1969), Piercean semiotics (Blanco, 1980; Pignatari, 1968, 1974), different Marxist interpretations (Dorfman & Mattelart, 1971; Esteinou Madrid, 1980; Genro Filho, 1987; Mattelart, 1971; Niezen Matos, 1980; Silva, 1970; Taufic, 1973), and studies following the work of Paulo Freire (Freire, 1967, 1970, 1971).

Latin American scholars, however, began to note the insufficiency of foreign axioms when applied to their region. The "entanglement" of these axioms' premises in the social systems in which they were generated resulted in their theoretical as well as methodological inadequacy (Beltrán, 1979). Some Latin American researchers introduced contextual variables, either to allow for the correction of deviations in informative structures (Beltrán, 1972; Beltrão, 1967) or to explain the cultural domination processes (Marques de Melo, 1972b). Others suggested macro alternatives to generate new communications systems within the framework of existing societies (Aguirre e Bisbal, 1980; Capriles, 1976; F. Fernández, 1982).

Little by little researchers attempted to overcome foreign dependency. Yet, immersed in a culture marked by miscegenation, they did not hesitate to practice methodological syncretism (Lins de Silva, 1982; Marques de Melo, 1981; Martin Barbero, 1978), combining procedures from the Chicago and Frankfurt schools, and other traditions popular in Paris, Moscow, and Rome. At the same time, researchers tried to create mechanisms for self-sufficiency, forging an academic community beyond national boundaries and regional differences (Aguirre e Bisbal, 1981), in spite of traumatizing political instability and the lack of economic resources. The trademark of Latin American research became its policy of "engagement" in the public interest (Marques de Melo, 1983).

The creation in 1978 of the Latin American Association of Communication Research (ALAIC), represents a step toward the consolidation of the Latin American communication research school. Although appearing at a critical moment when the historical contradictions anticipating the "lost decade" became more pronounced, ALAIC managed to take part in international forums and give support to democratic communication policies. Concerned with the "memory of the field," ALAIC promoted documentation centers on communication research in Argentina, Bolivia, Brazil,

Chile, Colombia, Mexico, Panama, and Peru (Almegor, Arauz, Golcher, & Tuñon, 1992; Anzola & Cooper, 1985; Beltrán, 1990; Fuentes, 1988; Marques de Melo, 1984; Munizaga & Rivera, 1983; Peirano & Kudo, 1982). It encouraged the diffusion of Latin American research on communication in universities and graduate programs, neutralizing the influence of European and North American theories and methodologies.

International Projection

The singularity characterizing Latin American communication research aroused the interest of international analysts. Their remarks and criticisms supported the efforts of an intellectual community often working under adverse conditions. Miquel de Moragas was among the first to note that an autochthonous communication science was emerging in the region:

> *Almost all Latin American countries currently have specialized centers to form communicators and carry out communication research. The advice which came from UNESCO in the fifties [sic] with regard to the importance of such centers had widespread and rapid acceptance in Latin America, which in this area occupies the front line of the developing countries. Although it has been said that this proliferation of schools was the consequence of a domination strategy, the truth is that things are never so simple. The critical reflection on mass communication that evolved in these schools and centers through several different stages has managed to put itself, in many aspects, at the top of current world communication research. In Latin America, because of the dynamism of social changes and the communication transformations, the political consequences of communication research have appeared more clearly than in any other world context.* (Moragas, 1981, p. 199)

Emile McAnany also accentuated the innovative character of Latin American research and suggested that it be carefully considered by the North American community:

> *Latin American scholars have produced recurrent self-examination centering on their dependence on foreign sources for their ideas and research models. The ideal has been to create a new social science, a new economic approach, and a new communication science that would be appropriate to the Latin American context and its historic necessities. . . . Latin Americans know their own needs and are attempting to build a coherent scientific communication model for the purpose of knowledge building as well as policy change. Given the long history of social frustration and political repression in Latin*

America, most communication researchers are not naive about the possibilities of social change. Yet they refuse to disengage the attempt to build a more relevant and valid social science for their own environments and historical realities from the effort to create in their societies. (1986, pp. 40, 41)

Everett Rogers made an epistemological evaluation of the region's research trends, identifying a pragmatically hybrid model located in the confluence of the North American empirical studies and European critical reflection:

These are generally not very "happy" societies: health and nutrition are relatively poor, especially in rural areas and in urban slums; poverty is a very serious social problem; and the mass media are heavily oriented to urban, educated elites. There is a high degree of economic penetration by foreign-owned multinational corporations (especially those headquartered in the USA), including American films, television programs, and magazines. Under these conditions, it is not surprising for a communication scholar to question whether the mass media are very functional for his society. At least to some degree, a critical stance is only natural for many Latin American scholars as they look at their society. But many Latin American communication scholars also realize that they are seldom able to convince government officials or politicians to change a communication policy unless they can present empirical evidence about the suggested policy change.

In Latin America, the positivistic potential of quantitative communications research in ameliorating social problems make a synthesis of certain elements of the critical school and the empirical school seem to be a logical and promising direction for the future. If a synthesis of the empirical and critical approaches is to be forged, it may be most likely to occur in Latin America. (Rogers, 1985, p. 230)

Philip Schlesinger dedicated an issue of *Media, Culture, & Society* to Latin American communication research, selecting articles by the continent's most renowned scientists. Elsewhere, Schlesinger observed:

No doubt, the main worry which unifies much of what has been written, notwithstanding the theoretical and methodological orientation is, precisely, the effort to develop a correct Latin American approach to communication problems and the culture of that continent. As in any other field of research, the research of culture and media in Latin America has had its own and characteristic development stages, and has been the reason for more large movements, either socio-political, economical or intellectual, which are behind the emergence of the new issues. At the heart of the recent Latin American research history, a battle against intellectual dependency has been fought. (Schlesinger, 1989, p. 55)

Robert White is another North American scientist concerned with the Latin American communication research trends. White's analysis of recent trends indicates the degree of (pluralist) cohesion that exists within this academic community:

One of the most remarkable features of the Latin American communication research—contrasting somewhat with what happens in Europe and in other parts of the world—is the notable intercommunication between researchers, the cooperative investigation projects and the connection among several organizations, institutes, publications and schools. Latin Americans tend to consider their task as a continental investigation enterprise, which gives place to a relatively high degree of mutual knowledge of what researchers are doing. Obviously very often there are discussions and bitter disagreement, however many Latin Americans refer [affectionately] to their continental connection as the "mafia" . . . . (White, 1989, p. 44)

New Challenges

Today, Latin American communication researchers face new challenges: the free trade agreements and the privatization of traditional public services, the fragile reconstruction of democratic institutions, growing poverty, and the new economics roles of communication and culture. In the past, different sociopolitical contexts marked different periods of theoretical development and research. This phase of Latin American history, too, will leave its mark on the development of communication research.

Latin America's recent history of military dictatorships, foreign debt, rapid inflation, overurbanization, and unemployment represents a very different context for mass communication and hence for communication scholarship. Critical approaches are particularly likely to flourish under such conditions, in contrast to the more industry-oriented and neutrally scientific approaches that have been established in the United States. Latin American scholars will continue to respond to the demands of underdevelopment with an evolving mode of research unique to the region, a distinct brand of communication scholarship unlike the North American and European roots from which it originally sprang.

Research on communication phenomena of Latin America preserves its critical attitude in the formulation of theoretical hypotheses and in the delimitation of analytical premises. However, it intensifies the use of empirical procedures (quantitative and qualitative) to describe and diagnose situations. This position corresponds to a consensual feeling that scientific research represents a vital instrument in building prosperous and pluralist democratic societies, a utopia that mobilizes Latin American communication scientists standing at the threshold of the 21st century.

References

Aguirre J. M. e Bisbal, M. (1980). *El nuevo cine venezolano* [The new Venezuelan film]. Caracas: Editorial Ateneo.

Aguirre J. M. e Bisbal, M. (1981). Tendencias de los estudios Latinoamericanos en el analisis de los medios masivos [Trends in Latin American studies of the mass media]. In *La ideologia como mensaje y masaje* [Ideology as message and massage], pp. 11–59. Caracas: Monte Avila.

Almegor, M., Araúz, J., Golcher, I., & Tunón, M. (1992). *La investigación en comunicación social en Panamá* [Research in social communication in Panama]. Panama: Instituto Nacional de Cultura.

Anzola, P., & Cooper, P. (1985). *La investigación en comunicación social en Colombia* [Research in social communication in Colombia]. Lima: DESCO/ACICS.

Ballester, E. C. (1947). *Derecho de prensa* [Press freedom]. Buenos Aires: El Ateneo Editorial.

Beltrán, L. R. (1972). *Communication in Latin America: Persuasion for status quo or for national development?* Unpublished doctoral dissertation, Michigan State University, East Lansing, MI.

Beltrán, L. R. (1979). Premisas, objetos y métodos foráneos en la investigación sobre comunicación en América Latina [Foreign premises, objects and methods in research on communication in Latin America]. In M. de Moragas (Ed.), *Sociologia de la comunicación de masas* [Sociology of mass communication]. Barcelona: Gustavo Gili.

Beltrán, L. R. (1990). *Bibliografia de estudios sobre comunicación en Bolivia* [Bibliography of communication studies in Bolivia]. La Paz: PROINSA.

Beltrão, L. (1960). *Inicação à Filosofia do Jornalismo* [Initiation to the philosophy of journalism]. Rio de Janeiro: Agir.

Beltrão, L. (1967). *Folkcomunicação, um estudo dos agentes e dos meios populares de informação de fatos e expressão de ideias* [Folkcommunication: A study of the agents and the popular media of information and of the expression of ideas]. Unpublished doctoral thesis, Universidade de São Paulo, São Paulo Brazil.

Blanco, Desiderio y Bueno, R. (1980). *Metodologia del análisis semiótico* [Methodology of semiotic analysis]. Lima: Universidad de Lima.

Borrego, S. (1951). *Periodismo transcendente* [Transcendental journalism]. Mexico: Editorial La Esfera.

Capriles, O. (1976). *El estado y los medios de comunicación en Venezuela* [The state and the mass media in Venezuela]. Caracas: ININCO.

Carvalho, H. M. (1940). *Publicidade e propaganda* [Advertising and propaganda]. São Paulo: Companhia Editora Nacional.

Cohn, G. (1973). *Sociologia da communicação* [Sociology of communication]. São Paulo: Pioneira.

Da Viá, S. C. (1977). *Televisão e consciencia de classe* [Television and class consciousness]. Petropólis, Brazil: Vozes.

Delano, A. S. (1960). *Los abusos de la publicidad* [Abuses of advertising]. Santiago: Editorial Juridica de Chile.

Dorfman, A., & Mattelart, A. (1971). *Para leer el Pato Donald* [How to read Donald Duck]. Santiago: Universidad Católica de Chile.

Esteinou Madrid, J. (1980). *Aparatos de comunicación de masas, estado y puntas de hegemonia* [Mass media, the state and hegemony]. México: TICOM.

Febres Cordero, J. (1959). *Tres siglos de imprenta y cultura Venezolanas* [Three centuries of Venezuelan printing and culture]. Caracas: Imprenta Nacional.

Fernández, F. (1982). *Los medios de difusión masiva en México* [Mass media in Mexico]. México: Juan Pablos Editor.

Fernandez, J. (Ed.). (1965). *Enseñanza de periodismo y medios de información colectiva* [Teaching journalism and the mass media]. Quito: CIESPAL.

Fernandez, J. (1967). *Dos semanas en la prensa de América Latina* [Two weeks of the press of Latin America]. Quito: CIESPAL.

Freire, P. (1967). *Educação como prática da liberdade* [Education as a practice for freedom]. Rio de Janeiro: Paz e Terra.

Freire, P. (1970). *Pedagogia do oprimido* [Pedagogy of the oppressed]. Rio de Janeiro: Paz e Terra.

Freire, P. (1971). *Extensao ou comunicação?* [Extension or communication?]. Rio de Janeiro: Paz e Terra.

Fuentes, R. (1988). *La investigación de comunicación en México* [Research in communication in Mexico]. México: Ediciones de Comunicación.

Genro Filho, A. (1987). *O segredo da pirámide—Para uma teoria Marxista do jornalismo* [The secret of the pyramid—a Marxist theory of journalism]. Pôrto Alegre, Brazil: Tehe Editora.

Kerner, A. (1943). *Nos bastidores da publicidade* [The advertisers]. Rio de Janeiro: Gráfica Olimpica.

Lins da Silva, C. E. (Ed.). (1982). *Comunicação, hegemonia e contra-informação* [Communication, hegemony and counter-information]. São Paulo: Cortez/INTERCOM.

Lopes, S. (1957). *Fundamentos juridicos—Sociais da radiodifusão* [Legal bases of broadcasting]. Rio de Janeiro: Editora Nacional de Direito.

Marques de Melo, J. (1972a). *Estudos de jornalismo comparado* [Studies of comparative journalism]. São Paulo: Pioneira.

Marques de Melo, J. (1972b). *Fatores socio-culturais que retardaram a implantação da imprensa no Brasil* [Socio-cultural factors that delayed the introduction of the printing press in Brazil]. Unpublished doctoral thesis, Universidade de São Paulo, São Paulo, Brazil.

Marques de Melo, J. (Ed.). (1981). *Populismo e comunicação* [Populism and communication]. São Paulo: Cortez/INTERCOM.

Marques de Melo, J. (Ed.). (1983). *Teoria e pesquisa em comunicação–Panorama Latino-americano* [Theory and research in communication: A Latin American panorama]. São Paulo: Cortez/INTERCOM.

Marques de Melo, J. (Ed.). (1984). *Inventario da pesquisa em comunicação no Brasil* [Inventory of communication research in Brazil]. São Paulo: INTERCOM.

Martin Barbero, J. (1978). *Comunicación masiva: Discurso y poder* [Mass media, discourse and power]. Quito: CIESPAL.

Mattelart, A. (1971). *Comunicación masiva y revolución socialista* [Mass communication and the socialist revolution]. México: Diogenes.

McAnany, E. (1986). Seminal ideas in Latin American critical communication research: An agenda for the north. In R. Atwood & E. McAnany (Eds.), *Communication and Latin American society*. Madison: University of Wisconsin Press.

Miceli, S. (1972). *A noite de madrinha* [The night of the godmother]. São Paulo: Perspectiva.

Miquel y Verges, J. M. (1941). *La independencia Mexicana y la prensa insurgente* [Mexican independence and the rebel press]. CITY: Colegio de Mexico.

Moragas, M. (1981). *Teorias de la comunicación—Investigaciones sobre medios en América y Europa* [Communication theories—Media research in America and Europe]. Barcelona: Gustavo Gili.

Mouchet, C., & Radelli, S. (1957). *Los derchos del escritor y del artista* [Copyright]. Buenos Aires: Editorial Sudamericana.

Mujica, H. (1967). *El imperio de la noticia* [The empire of the news]. Caracas: Universidad Central de Venezuela.

Munizaga, G., & Rivera, A. (1983). *La investigación en comunicación social en Chile* [Research in social communication in Chile]. Lima: DESCO/CENECA.

Niezen Matos, G. (1980). *Bases para una teoria Marxista de la comunicación* [Bases for a Marxist theory of communication]. Lima: Centro de Investigación de la Comunicación.

Otero, G. A. (1953). *La cultura y el periodismo en America* [Culture and journalism in America]. Quito: Casa Editora Liebmann.

Paiva, S. C. (1953). *O gangster no cinema* [The gangster in film]. Rio de Janeiro: Editorial Andes.

Pasquali, A. (1963). *Comunicación y cultura de masas* [Communication and mass culture]. Caracas: Universidad Central de Venezuela.

Pasquali, A. (1967). *El aparato singular* [The singular apparatus]. Caracas: Universidad Central de Venezuela.

Peirano, L., & Kudo, X. (1982). *La investigación en comunicación social en el Peru* [Research in social communication in Peru]. Lima: DESCO.

Pignatari, D. (1968). *Informação, linguagem, comunicação* [Information, language and communication]. São Paulo: Perspectiva.

Pignatari, D.(1974). *Semiótica e literatura* [Semiotics and literature]. São Paulo: Perspectiva.

Rabelo, G. (1956). *Os tempos heróicos da propaganda* [The heydays of advertising]. Rio de Janeiro: Empresa Jornalistica.

Rivera, J. (1986). *La investigación en comunicación en Argentina* [Communication research in Argentina]. Lima: DESCO/ASAICC.

Rizzini, C. (1946). *O livro, o jornal e a tipografia no Brasil* [Books, newspapers and printing in Brazil]. Rio de Janeiro: Kosmos.

Rogers, E. (1985). The empirical and critical schools of communication research. In E. Rogers & F. Balle (Eds.), *The media revolution in America and Western Europe* (pp. 229–230). Norwood, NJ: Ablex.

Samaniego, R. (1968). *Manual de investigación por encuestas en la comunicación* [Research manual for interviews in communication]. Quito: CIESPAL.

Santoro, E. (1969). *La televisión Venezolana y la formación de estereotipos en el niño* [Venezuelan television and the formation of stereotypes in the child]. Caracas: Univerdidad Central de Venezuela.

Schlesinger, P. (1989). Aportaciones de la investigación Latinoamericana—Una perspectiva Británica [Contributions of Latin American research: A British perspective]. *Telos, 19,* 55.

Silva, L. (1970). *La plusvalia ideológica* [Ideological surplus value]. Caracas: Universidad Central de Venezuela.

Sodré, M. (1972). *A comunicação do grotesco* [Communication of the grotesque]. Petropólis, Brazil: Vozes.

Suarée, O. (1944). *Manual de psicologia aplicada al periodismo* [Manual of psychology applied to journalism]. La Habana, Cuba: Cultural S.A.

Suarée, O. (1946). *Moralética del periodismo* [Moral ethics of journalism]. La Habana, Cuba: Cultural S.A.

Suarée, O. (1948). *Socioperiodismo* [Socio-journalism]. La Habana, Cuba: Cultural S.A.

Taufic, C. (1973). *Periodismo y lucha de clases* [Journalism and class struggle]. Santiago de Chile: Quimantu.

Verga, A.(1965). *El periodismo por dentro* [Inside journalism]. Buenos Aires: Ediciones Libera.

Verón, E. (1969). *Lenguaje y comunicación social* [Language and social communication]. Buenos Aires: Ediciones Neuva Vision.

Vianny, A. (1959). *Introdução ao cinema Brasileiro* [Introduction to Brazilian film]. Roi de Janeiro: Ministério da Educação e Cultura.

White, R. (1989). La teoria de la comunicación en América Latina [Communication theory in Latin America]. *Telos, 19,* 43–54.

Zavala Rodriguez, C. J. (1947). *Publicidad comercial* [Commercial advertising]. Buenos Aires: Editorial Depalma.

Index

A

ABC, 202
Abrahams, R., 324
Accuracy in Media (AIM), 66
Adler, M. J., 123
Adorno, T. W., 112, 239, 240, 363, 378, 379
Affective processes, 119
Agenda setting
    Batesonian field guide in, 135–40
    and Durkheimian theory, 166–70
    and hierarchy of institutional values,
        158–64
    and implications of public relations,
        172–79
    and interpersonal relationships, 127–33
    and research on communication and cul-
        ture, 149–56
    and multiple–process theories of commu-
        nication, 118–24
    and third stage of information society,
        141–47
Akuto, H., 312
Alexander, J. C., 121, 122
Allen, P. M., 147
Allport, F., 379
Almegor, M., 426
Altheide, D. L., 67n
Althusser, L., 232, 258, 376
Altman, I., 182
American model of communication, 11, 26–28
American pragmatism, 40
American Telephone & Telegraph Co., 218
Anderson, B. R., 231n
Anderson, J. R., 286, 287
Anderson, W. T., 48
Ang, I., 256, 258, 420
Antonelli, C., 142, 145, 146
Anzola, P., 426
Applewhite, H., 306
Arauz, J., 426
Arbogast, R. A., 231
Archer, M. S., 146
Arnheim, R., 379
Arnold, C. C., 168
Aronowitz, S., 361
Articulation, 332
Association for Education in Journalism and
    Mass Communication (AEJMC), 370, 391
Asuncion-Lande, N. C., 408

Atkin, C. K., 173
Audience research
    active audience theory in, 255–60, 298
    and design of communication interfaces,
        301–9
    developments in, 14–22, 247–53
    and framing, 293–300
    historical reception studies in, 262–68
    information processing and media effects
        in, 284–90
    Japanese perspective of, 311–19
    and journalism, 322–27
    in mass communication research, 415–22
    and perspectives on communication,
        100–6
    realism and romance in, 278–82
Authoritarian culture, 178
Automatic effects theory, 271–72
Avery, R. K., 183

B

Babe, R. E., 143
Babrow, A. S., 119, 120
Bach, S., 355
Baecker, R., 306
Bagdikian, B., 188, 222
Baker, K., 382
Bakhtin, M., 85
Ballester, E. C., 424
Baran, S., 188
Barber, B., 386
Barbero, M., 425
Barkin, S., 325
Barnett, G. A., 161, 410
Barnett, J., 370
Barnett, S., 243
Barnouw, E., 32, 127, 218
Barsamian, D., 343–44
Barthes, R., 325
Basil, M., 288
Bateson, G., 135–36
Baudrillard, J., 145, 350, 351
Bauman, R., 324
Baxter, L., 84
Becher, J., 93
Becker, H., 279
Becker, S. L., 182, 183, 185, 399
Belenky, M. F., 46

Bellah, R., 326
Beltràn, L. R., 424, 425, 426
Benedikt, M., 301
Benford, R. D., 294
Beniger, J. R., 54, 100, 376
Benjamin, W., 379
Bennett, W. L., 191, 192, 325
Berard, D. H., 20
Berardo, F. M., 20
Berelson, B. R., 44, 101, 102, 367, 369, 416
Berger, C. R., 35–36, 40, 83, 84, 86, 122, 127,
   131, 158, 162, 164, 182, 372, 405, 408
Berger, J., 83
Berger, P. L., 45, 152
Berkeley, G., 91
Bernays, E., 379
Berry, C., 287
Bertaux, D., 263
Besen, S. M., 198
Best, S., 58
Bhabha, H., 325
Bibliometric studies, 26, 27, 83
Billig, M., 250
Biocca, F., 301, 304–6, 308
Biological cognitivists, 46
Bird, S. E., 325
Bisbal, A., 425
Blanchard, M. A., 142
Blanchard, R. O., 392, 398, 399
Blankenship, J., 409, 410
Bluestone, B., 352, 361
Blumer, H., 173, 384
Blumler, J. G., 20, 44, 103, 108–9, 111–12,
   182, 190, 192, 211, 248, 390
Bochner, A. P., 84, 87–88, 130, 160
Bogart, L., 188
Boorstin, D. J., 45–46, 94, 95
Bopp, M. J., 124
Borchers, H., 247, 248, 253
Borgman, C. L., 83
Borrego, S., 424
Boserup, E., 226, 229
Bourdieu, P., 20, 384
Bowen, S. P., 408
Bower, G., 288
Bowers, J. W., 168
Boyer, E., 388
Braithwaite, R., 94
Braman, S., 143, 145
Bridges, J., 203
Brody, R. A., 194, 296–97
Broom, G. M., 173
Brosius, H. B., 18, 287
Brouwer, M., 314
Brubaker, J. S., 216

Bruner, J., 54, 58
Brunsdon, C., 257
Buckley, P., 364
Budd, B., 257, 258
Budd, M., 298
Bullinger, H., 306
Bulmer, M., 164
Bureau of Applied Social Research, 378
Burgoon, M., 161–62
Burke, K., 379
Burleson, B. R., 182
Burrell, G., 14–15, 16, 18
Bush, G., 201
Busterna, J. C., 200
Buxton, W., 306

C

Cacioppo, J. T., 119
Cadish, K., 20
Campbell, R., 325
Cantril, H., 263–64
Capital Cities/ABC, 201
Capitalism
   and communication, 360–61
   critiquing, 344–45
Capriles, O., 425
Carbaugh, D., 250
Card, S., 306
Carey, J. W., 146, 152, 168, 210, 220, 240, 241,
   243, 289, 323, 333, 376–77, 380
Carpenter, E., 66n
Carragee, K. M., 325
Carter, R. F., 60
Carvalho, H. M., 424
Casey, R., 369
Casmir, F. L., 408
Castro, F. G., 20
Cavalcant de Paiva, S., 424
CBS, 202
Chaffee, S. H., 35, 83, 109, 122, 127, 173, 182,
   286, 288, 362n, 367–72, 405
Chaiken, S., 119
Chandler, A. D., Jr., 142
Chen, M., 371
Cherry, C., 306
Chicago School of sociology, 163–64, 263
Children's Television Act (1990), 239
Children, understanding of television, 46–47
Chomsky, N., 343
Christ, W. G., 392, 398, 399
Christians, C., 222
Clapper, J., 288
Clark, A., 87

Clifford, B. R., 287
Clinchy, B. M., 46
Cobb, R. W., 173
Cognition
    and autonomy, 49
    contributions to theory, 83–89
    and dissonance, 35, 315
    information as central to, 26–33
    and message–driven explanations, 42–50
    science in, 86
Cohen, B. C., 192
Cohen, E., 364
Cohen, J., 66
Cohn, G., 425
Columbia University, 377
Commission on Freedom of the Press, 217
Communication
    and capitalism, 360–61
    as cross–disciplinary field of study, 96
    disciplinary status for, 100
    fragmentation in, 405–13
    hierarchy of institutional values in, 158–64
    journals specializing in, 127
    message–driven conceptions of, 44
    multiple–process theories of, 118–24
    personal and social relationships in, 128–31
    political economy of, 347–57
    as social force, 359–64
    subdisciplines in, 83
    theoretical synthesis of, 30–31
Communication curriculum
    crisis in, 388–94
    future for, 396–403, 405–13
    legitimacy of, 388–89
    power in, 393–94
    theory in, 389–92
Communication design, as theory and research, 306–9
Communication education, characteristics of, 218–19
Communication policy, reconnecting communication studies with, 207–13
Communication research
    agenda for, 152–53
    Batesonian field guide in, 135–40
    building disciplinary status for, 91–99
    and centrality of information, 26–33
    coherence in, 84–89
    community focus of, 153–56
    consequences of expertise in, 80–82
    convergence and divergence in, 367–72
    and consequences of vocabularies, 75–82
    crisis in, 340–45, 388–94
    cultural studies in, 106, 331–39
    in design of communication interfaces and systems, 301–9
    dimensions of, 14–16, 53–54, 55, 63–74, 77–80
    Durkheim's contributions to, 167–70
    and feminism, 226–35
    and future of curriculum, 396–403
    genealogical notes on, 374–80
    and harmonization of systems, 141–47
    historical reception studies in, 262–68
    implications of public relations in, 172–79
    in interpersonal relationships, 127–33
    interpretive approach to, 44–45
    Japanese perspective of, 311–19
    and journalism, 322–27
    Latin American school on, 424–29
    and media economics, 198–205
    message–driven, 42–50
    multiple–process, 118–24
    need for commitment, 48–50
    paradigms in, 14–23, 188–96
    perspectives of, 100–6
    problem–centered, 88
    and public opinion research, 382–87
    and the "real world," 182–86
    reasons for number of theories in, 34–40
    reconnecting with policy, 207–13
    reflexive explanation of, 42
    social models of, 219–22, 359–64
    status of, 76–77, 208–9
    subdisciplines in, 83
    and telecommunications, 215–23
Communication Research, 370, 376
Comstock, G., 211
Condit, C., 256, 325
Connectedness, 159, 161–63
Constraint recognition, 174
Constructivists, camps of, 45–46
Content analysis, 43
    ferment and harmony in, 63–74
    and framing, 293–300
    and information processing and media effects, 284–90
    and legitimacy gap, 108–16
Control, 28
Converse, P., 289
Cook, T. E., 192
Cooke, J., 326
Cooper, P., 426
Cordero, J. F., 424
Corner, J., 250, 256, 258, 259, 266
Coulter, J., 86, 87
Coupland, N., 130
Craig, R. T., 35, 40, 60, 182
Creedon, P. J., 177

Critical pluralism, theory of, 111
Critical Studies in Mass Communication
     (CSMC), 183–84
Critical theory
     and communication as social force,
          359–64
     and cultural studies, 331–38
     at crossroads, 340–45
     and political economy, 347–57
Cronen, V., 128
Cronkhite, G., 405
Cross–level theory and research, 120–23
Culbertson, H. M., 173
Cultural studies
     agenda for, 149–56
     coherence/cohesion in, 83–89
     communication as social force, 359–64
     at the crossroads, 340–45
     information as central to, 26–33
     link of communication research to, 106
     perspectives of, 100–6
     place of, in communication, 331–39
     and political economy, 347–57
     and silence, 238–45
Curran, J., 16, 22, 109, 219, 248, 256–57, 258,
     264, 377, 417, 419
Curtis, A., 363
Cuzzort, R. P., 167
Cybernetics, 28, 369
Cyberspace, 301, 303–5
Czitrom, D. J., 218

D

Dance, F. E. X., 85–86
Danielian, N. B., 362
Danowski, J. A., 161, 370, 410
Dardenne, R. W., 325
Darwin, C., 415
da Silva, L., 425
Da Via, S. C., 425
Davis, D. K., 152, 153, 154, 287
Davis, S. G., 364
Dayan, D., 102, 324
de Certeau, M., 258
Decoding, 43, 335
Deference–emotion system, 120
DeFleur, M., 266
Delaney, B., 304
Delano, A. S., 424
de la Suarée, O., 424
Delia, J. G., 83, 84, 86, 164, 182, 362, 363, 416,
     417
Dennett, D., 87
Denning, M., 364

Dennis, E. E., 110, 393
Derrida, J., 85, 256
Dervin, B., 16*n*, 34, 60, 83, 84, 190
Descartes, R., 93, 95
de Sola Pool, I., 227
Development Alternatives with Women for a
     New Era (DAWN), 229–30
Development communication, and feminism,
     226–35
Development support communication (DSC),
     231–32
Dewey, J., 173, 376, 377, 378–79, 380
Dialectical perspectives, 123–24
Dicey, A., 386
Dickson, T. V., 396
Dillard, J. P., 119
Dindia, K., 128
Disciplinary status
     Batesonian field guide for, 135–40
     building, 91–99
     and consequences of vocabularies, 75–82
     convergence and divergence in, 367–72
     in crisis, 388–94
     and Durkheimian theory, 166–70
     ferment and harmony in, 63–73
     fragmentation of, 405–13
     future of, 42–50
     genealogical notes on, 374–80
     hierarchy of institutional values in, 158–64
     legitimacy gap in, 108–16
     in mass communication research, 415–22
     and number of communication theories,
          34–40
     policy research paradigm for, 188–96
     and telecommunications study, 215–23
     theory in, 14–23, 26–33, 83–89
Discursive colonization, 230
Distributive ethics, 50
Dobash, R., 266
Dominant ideology thesis, 256
Donohew, L., 44, 119
Dorfman, A., 425
Douglas, M., 18
Dozier, D. M., 174–75, 177
Drake, W. J., 211
Drew, D. G., 287
Dual–process models of persuasion, 119
Durkheimian theory, 166–70

E

Eadie, W. F., 183, 184, 185
Eagly, A. H., 119
Eason, D., 326
Edelman, M. J., 45, 294, 325

Edelstein, A. S., 311, 313, 314
Ehling, W. P., 175
Ehrenreich, B., 343
Eisenberg, E. M., 84, 87–88, 160
Elaboration likelihood model (ELM), 119
Elder, C. D., 173
Elliot, P., 325
Ellis, C., 84, 130
Ellul, J., 222, 308
Empiricism, 29–30
Encoding, 43
Encoding/decoding model, 335
Entman, R. M., 193, 200, 257, 258, 294, 295, 297, 298, 299
Environments, media as, 69–71
Epstein, B., 343
Epstein, E. S., 152, 154
ESPN, 201
Ethnocentric universalism, 230
Ethnography, and active audience theory, 255–60
Ettema, J. S., 324, 325
Eulau, H., 121
Europe, and the legitimacy gap, 108–16
Evans, W. A., 255, 419, 420
Ewen, E., 364
Ewen, S., 364
Expertise, consequences of, 80–82

F

Fairness and Accuracy in Reporting (FAIR), 66
Falsifiability, 35
Farace, V., 128
Featherstone, M., 21
Federal Communications Commission (FCC), prime–time access rule, 203
Fejes, F., 247
Feminism, challenge of, for development communication, 226–35
Feminists for Free Expression (FFE), 66
Fenton, N., 266
"Ferment in the Field" issue, 14, 15, 34, 75, 100, 108, 127, 158, 159, 161, 182, 183, 198, 207, 220, 284, 340, 396, 405, 416
Fernández, F., 425
Fernandez, J., 425
Festinger, L. A., 35, 315
Filho, G., 425
Fine, E., 324
First Amendment theory, 103
Fischer, M., 259
Fish, R., 306
Fisher, W. R., 325

Fishman, M., 278, 323
Fiske, J., 255, 257, 258, 260, 298
Fiske, M., 363
Fiske, S. T., 294, 295
Fitzpatrick, M. A., 128, 129–30, 408
Flores, F., 301
Foerster, H. von, 46
Fogel, A., 250
Forester, T., 227
Formal models, 17–18, 19
Foucault, M., 46, 232
Fourth estate, power of, 322–23
Fowler, R., 323
Fox Network, 202
Framing, concept of, 293–300
Frankfurt School, 377, 378, 425
Freeman, F. E., 173
Freire, P., 226, 228, 233, 425
Frith, S., 343
Frow, J., 332
Frye, M., 229
Fuentes, R., 426
Fujitake, A., 311
Fukuyama, F., 344
Fulk, J., 143
Functionalist sociology, 14
Furuhata, K., 312

G

Gadamer, H. G., 99
Galileo, 94
Gallagher, M., 231
Gamson, W. A., 278, 279, 281–82, 294, 297, 385, 386
Gannett Group, 201
Gans, H. J., 192, 211, 323
Gardner, H., 290
Garfinkel, H., 86
Garnham, N., 212, 222, 361
Gaudet, H., 102, 369, 416
Gay, G., 306
Geertz, C., 36, 38, 85, 86, 143, 241, 243, 263
Geiger, S., 285, 286, 287
Gender differences, in public relations, 177
General Agreements on Tariffs and Trade (GATT), 145
General Electric, 218
Geras, N., 342
Gerbner, G., 32, 127, 207, 390
Giddens, A., 50, 54, 57, 152, 259, 290
Giesen, B., 121
Giles, H., 130
Gitlin, T., 46, 111, 159, 193, 278, 323

Glasersfeld, E. von, 46
Glasgow University Media Group, 323
Glasser, T. L., 325
Goffman, E., 86, 152, 294
Golcher, I., 426
Goldberger, N. R., 46
Golding, P., 258, 350
Gomery, D., 199, 200
Gonzalez, H., 232
Gorn, E. J., 364
Graber, D. A., 294, 295, 323
Grammar analysis, 53–61
Gramsci, A., 232, 350, 376, 379, 380
Grandi, R., 213
Gray, A., 259
Great Britain
    Broadcasting Research Unit in, 239, 242
    communication scholarship in, 238–45
    cultural studies in, 331
    mass observation studies in, 264
    radio in, 265
Greca, A. J., 20
Green, M., 267–68
Greene, J. O., 87
Greenwood, J. D., 131, 132
Grimes, T., 287
Groeben, N., 16–17
Gross, J. L., 18
Gross, L., 32, 127, 390
Grossberg, L., 16*n*, 34, 83, 84, 85, 190, 331,
    332, 338, 411, 416
Gross–Ngate, M., 306
Grown, C., 229, 230
Grunig, J., 174, 175, 176, 177, 178
Grunig, L., 174–75, 177, 178
Guback, T., 355
Gunter, B., 244, 287
Gurevitch, M., 44, 103, 109, 111, 190, 192,
    219, 280*n*, 325

H

Habermas, J., 54, 151, 154, 222, 232, 308, 384,
    386
Hage, J., 390
Haight, T. R., 208
Halbwachs, M., 325
Hall, J., 94
Hall, S., 56, 85, 219, 232, 243, 255, 257, 259,
    318, 331, 334, 376–77, 380
Hallin, D. C., 189, 192, 193
Hamelink, C. J., 208
Hampton, H., 68
Hancock, P. A., 306

Haraway, D., 232
Hardt, H., 323, 418
Harre, R., 131
Harrison, B., 352, 361
Harvey, D., 352
Hata, E., 311
Hawkins, R. P., 83, 84, 160, 372, 405, 411
Hay, J. A., 411
Headrick, D. R., 142
Heeter, C., 306
Hegemony, 279, 350, 363
Heidegger, M., 222, 379
Henley, D. C., 398
Hepworth, M., 144
Herbst, S., 383
Herman, E. S., 344
Hermeneutics, 131
Hersh, H., 306
Herzog, H., 420
Hesse, M., 39
Higgins, E. T., 119
Hill, C., 364
Hinkle, R. C., 169
Hirsch, E., 247, 250
Hirsch, P., 258, 323
Hirst, W., 286
Historical reception studies, 262–68
Hobson, D., 219
Hochheimer, J. L., 362*n*
Hoggart, R., 243, 334
Hoijer, B., 21
Holman, R. H., 20
Homma, N., 313
Hon, L. C., 177
Horkheimer, M., 363, 378, 379
Horowitz, I. L., 238, 240, 241
Horwitz, R., 222
Hovland, C., 368, 369
Huber, P., 145
Hughes, H. S., 241
Human Communication Research (HCR), 370
Human–computer interaction, human factors
    in, 305–6
Humanistic inquiry, into journalism, 323–27
Humanities, boundary between social sci-
    ences and, 36–38
Hunt, T., 176
Husni, S. A., 202
Hymes, D., 326

I

Ikeda, K., 311
Ikuta, M., 312

Information, as central to culture, cognition, and social behavior, 26–33
Information behavior model, 312–15, 318–19
Information processing, and media effects, 284–90
Information–seeking paradigm, 44–45
Information society, third stage of, 141–47
Information theory, 369
Infotainment, 356
Innis, C., 289
Innis, H. A., 212, 222
Institute for Social Research (ISR), 377, 378, 425
Institutional change, 155–56
Institutional values, hierarchy of, in communication discipline, 158–64
Institutions, effects of mass media on, 275
International Center of Higher Journalism Studies for Latin America (CIESPAL), 424
International Communication Association (ICA), 223, 267, 370, 375, 407
International Encyclopedia of Communications, 32, 167
Interpersonal communication
    and extracted information, in Japanese conversations, 312
    versus mass communication, 83
    new world of, 127–33
    research in, 127–28
    social approaches to, 85–86
    theory in, 83–89
Interpretive community, journalism as, 326–27
Interpretive sociology, 14–23
Intimacy, ideology of, 129–30
Irvine, R., 66
Iser, W., 250
Isocrates, 93
Ito, Y., 144, 311, 312, 313, 314, 315*n,* 316–17
Iwabuchi, Y., 311
Iyengar, S., 102, 192, 294, 298

J

Jackson, S., 128
Jacobs, S., 86, 128
Jacoby, R., 238, 244
Jaggar, A. M., 230, 233
Jamieson, K. H., 186
Jankowski, N. W., 21, 263
Jantsch, E., 147
Japanese perspective, on political communication research, 311–19
Jaquette, J., 229, 234*n*
Jasinski, J., 154
Jaspers, K., 379

Jennings, B., 290
Jensen, J. W., 219
Jensen, K. B., 16, 21, 262, 263, 266
Jhally, S., 266
Johansson, T., 20, 21
Johnson, M., 45, 64
Joho Kohdo model, 312–15, 318–19
Joreskog, K. G., 18
Journalism
    alternative ways for examining, 327
    communication explanation of, 322–27
    framing in, 293–300
    humanistic inquiry into, 323–27
    instruction in, 217
    objectivity in, 298–99
    public research policy in, 188–96
Journal of Applied Communication Research, 183, 184, 185
Journal of Communication, 370
Jowett, G., 218
Jussawalla, M., 227

K

Kahneman, D., 294, 295, 296, 298
Kalmar, D., 286
Kammen, M., 325
Kato, H., 312
Katz, E., 15, 20, 44, 102, 112, 120, 160, 248, 250, 253, 256, 278, 314, 324, 376, 377, 380, 390
Kawatake, K., 312
Kay, A., 304
Kellner, D., 58
Kendall, P. L., 416
Kepplinger, H. M., 18, 311, 313, 314
Kerner, A., 424
Kim, K., 312
Kimball, B. A., 93
Kimura, Y., 312
Kinder, D. R., 102, 192
King, E. W., 167
Kishwar, M., 231
Kitamura, H., 312
Klapper, J. T., 44, 101
Knapp, M., 127
Knight Ridder Group, 201
Kohei, S., 311, 315*n*
Kohlstedt, S. G., 216
Kolko, J., 361
Kosicki, G. M., 294, 421–22
Kramarae, C., 130
Krattermaker, T. G., 198
Kraut, R. E., 306

Kreutzner, G., 247, 248, 253
Krippendorff, K., 49, 50
Kudo, X., 426
Kuhn, T. S., 35, 305
Kurz, K., 288

L

Lacan, J., 258
Lacy, S., 200
Lakoff, G., 45, 64
Lambert, B. L., 87
Lang, A., 286, 288
Lang, G. E., 15, 213
Lang, K., 15, 213
Languages, media as, 66–68
Lannamann, J. W., 84, 85
Lasswell, H. D., 43, 103, 207, 369, 379
Lather, P., 56, 233
Latin American Association of Communication Research (ALAIC), 424–25
Latin American school, of communication research, 424–29
Laurel, B., 306
Lazarsfeld, P. F., 101, 102, 112, 239–40, 314, 368, 371, 376, 377, 378–79, 379, 380, 382, 383, 416
Leavis, F. R., 379
Leavis, Q. D., 379
Leeds–Hurwitz, W., 84, 85, 130
Legitimacy gap, as problem in mass media research, 108–16
LeGrow, C., 303
Lent, J. A., 227
Lerner, D., 227
Levy, M. R., 301, 306, 323
Lewin, K., 39
Lewis, J., 266
Lieberman, D., 371
Liebes, T., 120, 250, 253, 256
Limited effects theory, 271–77
Lindlof, T., 263, 326
Lippmann, W., 103, 379
Liska, J., 130
LISREL analyses, 18, 20, 28
Literary genres, 37
Litman, B., 203
Livingstone, S. M., 247, 250, 252, 253
Locke, J., 93, 95
Logan, R., 408
Long, E., 278, 282
Loos, V. E., 152, 154
Loov, T., 21
Lopes, S.–C., 424

Lowe, A., 219
Lowenthal, L., 379
Lowery, S., 266
Lubiano, W., 343
Lucaites, J. L., 325
Luckmann, T., 45, 152
Lüdtke, H., 21
Luká, G., 379
Lule, J., 323, 325
Lull, J., 120, 263
Lumsdaine, A., 369
Lunt, P. K., 252
Luthra, R., 233n
Lynd, H., 362
Lynd, R., 362, 379
Lyotard, J.–F., 351, 375

M

MacAloon, J., 324
MacIntyre, A., 98
MacPherson, C. B., 345
Mader, D. C., 160, 163
Mader, T. F., 160, 163
Madsen, R., 326
Maibach, E., 288
Makita, T., 313–14
Malinowski, B., 168
Mancini, P., 115
Mandel, M., 306
Mander, M. S., 325
Manoff, R. K., 232, 325
Mansell, R. E., 207, 209
March, K., 235
Marcus, G., 259
Marques de Melo, J., 425, 426
Marquit, E., 123, 124
Martin, M., 364
Martin–Barbero, J., 256
Marvin, C., 218
Marx, K., 15, 207, 232, 350, 355, 376
Marxist scholarship, 347
Mass communication
    effects of, 101, 271–77
    versus interpersonal communication, 83
    weakness of field, 110
Mass communication research
    administrative tradition, 247, 249
    critical tradition, 247–48, 249
    disciplinarity and site in, 415–22
Mass Communication Review Yearbook, 111
Massey, D., 259
Mass persuasion, 362–64
Materialist scholarship, 347

Matos, N., 425
Mattelart, A., 213, 425
Maturana, H. R., 46
McAnany, E., 426–27
McCarthyism, 27, 363
McChesney, R. W., 364
McClintick, D., 355
McCombs, M. E., 45, 102, 390
McKerrow, R. E., 411
McLean, P. E., 233
McLeod, J. M., 20, 121, 122, 173, 421–22
McLuhan, M., 281, 289
McPhee, W. N., 101
McQuail, D., 113, 199, 203, 248
McRobbie, A., 278
Mead, G. H., 379
Mead, M., 46, 379
Media
    active audience theory for, 255–60
    analysis of, 69, 70–71
    and audience research, 247–53
    disciplinary and site in research, 415–22
    implications of public relations for, 172–79
    and legitimacy gap, 108–16
    metaphors of, 64–71
    old versus new, 7
    perspectives on, 100–6
    policy research paradigm for news, 188–96
    and scholarship as silence, 238–45
    and telecommunications study, 215–23
    theory in, 83–89
Mediacentrism, 111
Media economics, centrality of, 198–205
Media effects
    and information processing, 284–90
    Japanese perspective on, 311–19
    limited factors on, 272–76
    minimal effects model, 101–2
    strong effects model, 271
Media research, 108–16
    assessing state of, 7
    basis for, 219
    changing paradigmatic guard, 9–11
    definition of, 7
    disciplinarity and site in, 415–22
    legitimacy gap between Europe and United States, 108–16
    reconnecting with communications policy, 207–13
    re–imaging, 71–72
    strength of field, 8–9
    vocabulary in, 63–74
Meehan, E., 130, 353
Melkote, S. R., 231–32
Melody, W. H., 207, 209

Menneer, P., 244
Merton, R. K., 35, 363, 378, 379, 416
Message–driven explanations, 42, 43, 44, 45–46, 47, 48
Message–processing modes, 118–24
Metaphors, of media, 64–71
Metzger, A. R., Jr., 198
Meyer, K., 306
Meyrowitz, J., 222
Miceli, S., 425
Midooka, K., 312
Miege, B., 361
Miegel, F., 20, 21
Miliband, R., 342
Miller, G. R., 127, 128, 182, 209
Millison, D., 303
Mills, C. W., 238, 240, 259, 363
Minami, H., 312
Minar, D., 383
Miquel y Verges, J. M., 424
Mitchell, A., 20
Münch, R., 121
Modernism, agenda for, 149–52
Modigliani, A., 278
Mohanty, C. T., 229, 230–31
Monopoly, 201
Moores, S., 265
Moral Majority, 66
Moran, T. P., 306
Morgan, G., 14–15, 16, 18
Morgan, M., 390
Morgas, Miquel de, 426
Morley, D., 255, 258, 259, 263
Moro, A., 324
Morris, M., 332
Morrison, D. E., 242, 243, 376
Mosco, V., 211, 361
Mouchet, C., 424
Mujica, H., 425
Multiple–process theory, 118–24
Mumford, L., 301, 302, 305, 379
Munizaga, G., 426
Murdock, G., 257, 258, 350

N

Naisbitt, J., 227
Nakano, O., 312, 316
Narrative, journalism as, 324–25
Nass, C. I., 121
Nationwide Audience project, 256
Natural language philosophy, 47
NBC, 202
Nelson, C., 85, 331, 416

Nelson, D., 306
Network economics, 145
Network firms, 145
Neuman, W. R., 104, 266, 267, 287
Newcomb, H., 258
Newcomb, M. D., 20
Newell, A., 306
Newhagen, J., 287, 288, 289
New Information Order, 207
Newly industrialized countries (NICs), 144, 348
News
    framing in political, 297
    limits on media effects, 274–75
    policy paradigm for, 188–96
    realism and romance in, 278–82
Newton, I., 93
New World Information Order, 144, 223
Nicholaïdis, K., 211
Nilan, M. S., 306
Nimmo, D., 408
Nishihira, S., 311, 315*n*
Noam, E. M., 207, 211
Noble, D., 222
Noelle–Neumann, E., 18, 45, 318, 382
Nord, D. P., 323
Nordenstreng, K., 318
Normative democratic theory, and public
    opinion, 299–300
Normative dimension of European Research,
    113–16
Norris, C., 256
Norwood, S. H., 364
Nyerere, J., 228

O

Objective journalism, 142
Objective linear models of information pro-
    cessing and communication, 27
O'Brien, D., 323
Oda, M., 311
Ohler, P., 22
Okamoto, S., 311
O'Keefe, B. J., 16*n*, 34, 83, 84, 86, 87, 190
O'Keefe, D. J., 119
Oligopoly, media, 201–2
Oliver, R. T., 184
Ong, W., 222
Opaschowsky, H. W., 21
Oregon Report, 393
Organizational communication research,
    34–40
Orr, D., 398

Otero, G. A., 424
Owen, B. M., 201, 202

P

Pacanowski, M. E., 44
Pacey, A., 222
Page, B. I., 193, 297, 299
Paisley, W., 121, 371
Paletz, D., 191
Palmgreen, P., 20
Pan, Z., 121, 122, 294, 421–22
Panitch, L., 342
Park, R., 164, 323, 379
Parks, M., 129
Parsonian social theory, 35
Pasquali, A., 425
Pauly, J. P., 326
Peacock, A., 242
Pearce, B., 128
Pearson, R., 173
Peirano, L., 426
Performance
    analysis of, 198–205
    journalism as, 324
Peters, J. D., 95
Petersmann, E., 145
Peterson, T., 217
Pettegrew, L. S., 408
Petty, R. E., 119
Picard, R. G., 200
Pingree, S., 83, 84, 160, 372, 405, 411
Plato, 93
Pluralism, 29–30, 159–60
    reactions to, 161–62
    and Durkheimian theory, 166–70
    and hierarchy of institutional values, 158–64
Point–to–point telecommunications, 208
Political advertising, 274
Political communication
    Japanese perspective on, 311–19
    and legitimacy gap, 108–16
Political economy
    and capitalism, 360–61
    change and continuity, 356–57
    criteria and methodology, 354–56
    definition of, 349–51
    and postmodernism, 351–54
    and use of power in communication indus-
        tries, 31, 226, 234–35
Popper, K. R., 35
Popular culture, 30
Postman, N., 245

Postmodernism
    and harmonization of systems, 141–47
    and political economy, 351–54
Pratt, L., 193
Preferred reading, 255
Press agentry model of public relations, 176
Price, V., 120, 121, 173, 382
Princeton Radio Research Project, 378
Problem–centered research, 83–89
Public information model of public relations,
    176
Public opinion, and normative democratic
    theory, 299–300
Public Opinion Quarterly, 382
Public opinion research
    and framing, 293–300
    history, philosophy and, 382–87
Public policy
    and centrality of media economics,
        198–205
    and feminism, 226–35
    paradigm for, 188–96
    and the real world, 182–86
    reconnecting communications study with,
        207–13
    and telecommunications study, 215–23
Public relations
    effects of, 173, 178–79
    gender differences in, 177
    implications of, 172–79
    macro level of, 177–78
    meso level of, 175–77
    micro level of, 172–75
    models of, 176–77
    segmentation of publics, 173–74
    strategic management of, 174
Puckett, T., 152
Purcell, W. H., 158
Purcell, W. M., 36n
Putnam, H., 131
Putnam, L. L., 44
Pye, L., 227

Q

Qualitative reception analysis, 21, 22, 23

R

Rabelo, G., 424
Radcliffe–Brown, A. R., 168
Radelli, S., 424

Radical constructivism, 46
Radical humanism, 14
Radical individualism, 150
Radical structuralism, 14
Radiotelevisione Italiana (RAI), 112–13
Radway, J., 247, 256, 259, 282, 326, 420
Ragin, C. C., 19
Rakow, L. F., 158, 177
Rawlins, W. K., 123, 124
Rayner, S., 18
RCA/NBC, 218
Reardon, K. K., 83, 84, 158, 162
Reception analysis, 21–23, 262–68
Recipient participation, 226–35
Redding, W. C., 182, 3636n
Redundancy, 43
Reeves, B., 83, 121, 285, 287, 288, 289
Reeves, J. L., 325
Reflexive theory, 49–50
Reimer, B., 21
Repper, F., 174, 175
Research Foundation of the International As-
    sociation of Business Communicators, 175
"Rhetorical turn" theory, 37
Ricardo, D., 349, 355
Rice, R. E., 83, 173, 306
Rice, S. A., 362
Richards, I. A., 256, 379
Richards, J., 264
Richardson, K., 266
Riegel, O. W., 362
Riker, W. H., 297, 299
Ritchie, L. D., 121
Rivera, A., 426
Rizzini, C., 424
Roach, C., 223
Robbins, S. P., 178
Robins, K., 361
Robinson, J. P., 153, 287, 323
Rogers, E. M., 83, 84, 109, 158, 160, 162, 164,
    182, 367–72, 425, 427
Rogers, L. E., 128
Rojecki, A., 294, 299
Rokeach, M., 158
Root, R. W., 306
Rorty, R., 93, 97
Rosenfield, L. W., 160, 163
Rosengren, K. E., 14, 16, 17, 18, 20, 21, 23,
    248
Rosenzweig, R., 364
Roshco, B., 323
Ross, D., 199, 363
Ross, S. J., 364
Rossi, A. M., 128
Rothenbuhler, E. W., 168

Rowland, W. D., Jr., 219
Rubenstein, R., 306
Rubin, A. M., 411
Rubin, R. B., 411
Rucinski, D., 153
Rudy, W., 216
Rychlak, J. F., 123

S

Salience, 295
Salmon, C. T., 173
Samaniego, R., 425
Santoro, E., 425
Sappington, A. A., 131
Saussure, F. de, 262, 309
Savage, J., 343
Saxton, A., 364
Scannell, P., 265
Schechner, R., 324
Scheff, T. J., 119, 120, 122
Scherer, F. M., 199
Schieve, W. C., 147
Schiffman, L. G., 20
Schiller, D., 222, 361, 364
Schiller, H. I., 198, 207, 226
Schlesinger, P., 266, 323, 420, 427–28
Schleuder, J., 286, 288
Schramm, W., 32, 54, 108, 127, 212, 217, 227,
    289–90, 302–3, 368–70, 371, 372, 375, 409,
    425
Schroder, K. C., 247
Schudson, M., 167, 258, 267, 280, 323, 325,
    326
Schutz, A., 151
Scott, J., 355, 356
Seamann, W. R., 256, 258, 343
Seiter, E., 247, 248, 253
Sellmeyer, R. B., 396
Sen, G., 229, 230
Sennett, R., 154
SeyMour–Ure, C., 265
Shakespeare, W., 93–94
Shannon, C. E., 27, 43, 369
Shanor, D. R., 193
Shaw, D. L., 45, 102, 390
Sheffield, F., 369
Sheridan, D., 264
Shiga, T., 311
Shimizu, I., 311
Shneiderman, B., 306
Shoemaker, P. J., 392
Shorter, E., 128
Siebert, F. S., 217

Sigal, L., 192
Sigman, S., 86
Signorielli, N., 390
Silverstone, R., 247, 250, 258, 263
Simmel, G., 20
Simonds, W., 278, 282
Simons, H. W., 37
Singer, J. L., 285
Situational theory, 172–79
Skinner, Q., 38n
Slavendy, G., 306
Smelser, N. J., 121
Smith, A., 223, 349, 350, 355
Smith, B., 369
Smith, J. Z., 400
Smythe, D. W., 160, 218, 222, 226, 350, 361
Snavely, L., 128
Snow, D. A., 294
So, C. Y. K., 83
Social communication, 219–22
    expansion of environment, 303–4
Social constructivism, theory in, 34–40, 45–46,
    83–89
Social relevance, and hierarchy of institution-
    al values, 158–64
Social research, agenda for, 149–56
Social sciences
    boundary between humanities and, 36–38
    and verbing communication, 53–61
Social sciences journals, organization of, 60
Social theory, Durkheimian, 166–70
Sociological research, 14–23
Sodré, M., 425
Sorbom, D., 18
Sorrentino, R. M., 119
Specialism, risk of, 110–13
Speech Communication Association (SCA),
    183, 370, 407
Speech programs, emphasis in, 217
Spigel, L., 264
Spivak, G., 232
Springston, J. K., 177
Sriramesh, K., 178
Stanton, F., 371
Staudt, K., 234n
Steeves, H. L., 231
Stein, G., 95
Steiner, L., 326
Steinfield, C., 143
Steinman, C., 257, 258, 298
Stich, S., 87
Strickwerda, M., 286
Suchman, L. A., 87
Sugiyama, M., 312
Sullivan, W., 326

Sumner, J., 286
Sunnafrank, M. J., 182
Sussman, G., 227
Swanson, D. L., 20, 21, 408
Swidler, A., 278, 279, 326
Sypher, H. E., 119

T

Taqqu, R., 235
Tarde, G., 101, 103, 383, 385, 386
Tarule, J. M., 46
Taub, L. E., 363*n*
Taufic, C., 425
Taylor, F., 306
Taylor, S. E., 294, 295
Taylor, T. J., 87
Technocentric approach, 306
Telecommunications
    and capitalism, 360
    communication research and teaching
        model for, 220–22
    emergence of, as academic field, 217
    implications of communication research
        for, 215–23
TeleCommunications, Inc., 201
Tetlock, P. E., 296–97
Thayer, L., 39, 182
Theory. *See also* Communication research
    definition of, 35–36
    relationship to practice, 38–40
Thompson, D., 163
Thompson, E. P., 239, 243, 352, 364
Thompson, P., 263
Time, Inc., 353
Times Mirror, 201
Time Warner, 201
Tipton, L., 44
Tipton, S., 326
Tönnies, F., 103, 384, 385
Tocqueville, A., 383
Today's Speech, 184–85
Toga, T., 311
Tokinoya, H., 311
Toth, E. L., 177
Towler, B., 244
Transmission–persuasion tradition, 227–28
Tripolar *kuuki* model, 315–19
Truman, H., 102
Tsujimura, A., 312, 316
Tuchman, G., 46, 156, 192, 278, 294, 299, 323,
    325
Tunstall, J., 182, 198, 211, 222–23

Tuñon, M., 426
Turner, V., 324, 325
Tversky, A., 294, 295, 296, 298
Two–step flow model, 44
Two–way asymmetrical model of public rela-
    tions, 176
Two–way symmetrical model of public rela-
    tions, 176

U

Uemura, S., 313–14
Uses and gratifications approach, 18, 19, 20,
    22, 44, 282*n,* 419

V

Valdivia, A. N., 229
van Dijk, T. A., 250, 323
Van Dinh, T., 160
Varela, F. J., 46
Varis, T., 318
Vassilou, Y., 306
Veblen, T., 20
Veltri, J. J., 20
Verbality, 94
Verbing communication, 53–61
Verga, A., 425
Verón, E., 425
Vianny, A., 424
Vickers, J., 234*n*
Virtual reality, 301, 307
Vocabularies
    consequences of, 75–82
    of media studies, 63–73
Vogel, H. L., 201
Vorderer, P., 16–17

W

Wagner, R., 47
Wagner–Pacifici, R., 324
Wallis, R., 188
Warner Communication, Inc. (WCI), 353–54
Wartella, E., 16*n, 34,* 83, 84, 182, 190, 239,
    241, 396, 411
Warth, E.–M., 247, 248, 253
Watzlawick, P., 46
Weaver, C., 266
Weaver, W., 43, 369
Weber, M., 20
Webster, F., 361

Weeks, G. R., 124
Welles, Orson, 263
Wenner, L. A., 20
Westbrook, R. B., 380
Wester, F., 263
Westinghouse, 218
White, H., 294, 325
White, J., 175, 178
White, R., 428
Whorfian hypothesis, 45
Wickens, C., 306
Wiebe, G. D., 314
Wiemann, J. M., 83, 84, 130, 160, 372, 405, 411
Wiener, N., 369
Wiener, S. E., 20
Wildawsky, A., 18
Wildman, S. S., 201
Willey, M. M., 362
Williams, R., 57, 220, 232, 243, 333, 352, 360, 364, 376
Willis, P., 219, 278–79, 280, 282
Winograd, T., 301
Wittgenstein, L., 45–46
Wolf, M., 115
Wollacott, J., 219

Women in development (WID), 226, 227–28
Women's perspective, as basis for development communication, 229–30
Wood, E. M., 342
Woodbury, J. R., 198
Woodson, W., 306
Worth, T. L., 32, 127
Worthington, N., 233n
Wright, D. K., 177
Wright, J. D., 153
Wrong, D., 323
Wundt, W., 371
Wurtzel, A., 183
Wyatt, N., 408

Y

Yamamoto, S., 316n

Z

Zaller, J. R., 193, 294, 297, 298, 299, 386
Zavala Rodriguez, C. J., 424
Zelizer, B., 324, 325, 326